Practical Chemoinformatics

Muthukumarasamy Karthikeyan • Renu Vyas

Practical Chemoinformatics

 Springer

Muthukumarasamy Karthikeyan
Digital Information Resource Centre
National Chemical Laboratory
Pune
India

Renu Vyas
Scientist (DST)
Division of Chemical Engineering and
Process Development
National Chemical Laboratory
Pune
India

ISBN 978-81-322-3491-3   ISBN 978-81-322-1780-0 (eBook)
DOI 10.1007/978-81-322-1780-0
Springer New Delhi Dordrecht Heidelberg London New York

Printed on acid-free paper

Springer is part of Springer Science+Business Media (www.springer.com)

Dedicated to our respected parents and loving children

Foreword

The term "cheminformatics" was only coined in 1998; nevertheless, in the last 15+ years this field has experienced a burgeoning growth with respect to the numbers of publications, conferences, specialized journals, and the diversity of research. The editorial published in the inaugural issue of the journal Cheminformatics in January of 2009 outlined major challenging problems facing cheminfomatics such as "overcoming stalled drug discovery ... advancing green chemistry ... understanding life from chemical prospective, and ... enabling the network of the world's chemical and biological information to be accessible and interpretable". This visionary editorial emphasized that despite their breadth and complexity cheminformatics embodies thenecessary concepts and tools to effectively tackle these vital problems.

Addressing challenges facing cheminformatics is exciting but it requires deep understanding of the cheminformatics theory as well as practical knowledge of the many important cheminformatics tools created by specialists working in the field. *Practical Chemoinformatics* by Karthikeyan and Vyas serves a critical purpose of bringing cheminformatics education and tools to researchers at all levels, from undergraduate students to specialists. The book incorporates ten excellently written chapters that cover cheminformatics methods and applications from A to Z. Not only do the authors provide critical summary of major cheminformatics concepts but most importantly they incorporate many case studies illustrating how typical research problems can be addressed and solved using proprietary as well as open source databases and computational tools.

I am confident that the book will be of interest to all scientists working in chemical biology and drug discovery but it will be particularly valuable for beginners and undergraduate, graduate or post-graduate students specializing in chemistry, biology and allied sciences.

Alexander Tropsha, PhD
UNC Eshelman School of Pharmacy
University of North Carolina at ChapelHill, USA

Preface

Chemoinformatics is a key technology for today's synthetic/medicinal chemist. People with extensive knowledge of chemistry and computer skills are immensely required by the industry. Database producers, chemical software developers, and chemical publishers offer attractive opportunities to the chemoinformaticians. The present book is intended to be a useful practical guide on chemoinformatics for the students at graduate, postgraduate, and Ph.D. levels. There are a couple of books on the theory of chemoinformatics and plenty of scattered information is available on the web but a well structured *Do it yourself* book is urgently required. The idea is that the reader of any background should be enthused to follow the book and start using the computer or a computer enthusiast can start learning the basics of computational chemistry. With this objective in mind, numerous step by step practice tutorials, source code snippets, and *Do it yourself* exercise have been given for quick grasp of the subject. The book intends to put the students in the driver's seat to test drive the software, code snippets, and practice tutorials. Rules of thumb have been provided at the end of every chapter for specific practical guidance. The language has been intentionally kept simple, technical jargon wherever used has been thoroughly explained. Adequate bibliography has been provided for readers seeking advanced knowledge on any of the given topics. The chapters in the book are linked to each other and at the same time are independent of each other.

The book begins with an elementary chapter on how to read and write molecules into a computer and basic file format conversions. The second chapter teaches how to compute properties of molecules and store them in a database. The third chapter delves into the use of computed property data to build models employing machine learning methods. The fourth and fifth chapters deal with protein active site prediction and docking studies, both of which are essential for any successful drug design experiment. The sixth and seventh chapter focus on use of reaction and NMR chemical shift based fingerprints respectively, and their use of virtual screening—an important component in chemoinformatics. The eighth chapter deals with text mining and its role in chemoinformatics methods to discover a lead molecule. The ninth and tenth are technology focused chapters that demonstrate ways to handle big data using today's state of art workflows, portals deployed in distributed, cloud computing platforms, and Android-based app development. To sum up, the purpose behind bringing out this book is to demystify and master chemoinformatics through a practical approach and make students aware of the latest developments in this field. After comprehending the entire book the reader will be able to appreciate the power of chemoinformatics tools and apply them for practical use.

Acknowledgments

The authors express their deep sense of gratitude and heart-felt thanks to all the contributors of this book without whose help the book would not have seen the light of the day. First and foremost thanks are due to the young enthusiastic team—Deepak Pandit, Chinmai P., Monalisa M., Soumya, Surojit Sadhu, Yogesh Pandit, Apurva for their tireless efforts in compiling data, checking code and proof reading the chapters. We wish to thank senior scientists and mentors Dr. B.D. Kulkarni and Dr. S.S. Tambe for being an inspiration for writing the chapter on machine learning and special guidance regarding the section on genetic programming. The help from academicians, Dr. Sankar and Dr. Agila for the reaction ontology discussion in the chapter on reaction fingerprint and modelling, is greatly acknowledged. The support from industry came from Mr. Sameer Choudhary and Ms. Sapna, CEO of Rasa Life Science Informatics for workflow related topics in chapters 5 and 9. We wish to thank Dr. S. Krishnan for nurturing and guiding the growth of chemoinformatics at NCL. Sincere thanks are due to former NCL directors Dr. R.A. Mashalkar, Dr Paul Ratanasamy, Dr. S. Shivram, and present director Dr. Sourav Pal for being the source of inspiration and constant encouragement. We also wish to express our gratitude towards all our chemoinformatics mentors, collaborators and colleagues whose valuable interactions have helped in career development- Dr J Gasteiger, Prof Alex Tropsha, Dr. Janest Ash, Dr. Wendy Warr, Dr. Peter Murray Rust, Dr. Peter Ertl, Dr Andreas Bender, Dr. Robert Glen, Dr Christopher Steinbeck, Prof Igor Tetko, Dr. Jonathan Goodman to name a few. Finally, we thank the publisher, Springer, for bringing out the book on time.

Contents

1 Open-Source Tools, Techniques, and Data in Chemoinformatics 1

  1.1 Chemoinformatics 2

     1.1.1 Open-Source Tools 2

     1.1.2 Introduction to Programming Languages 3

  1.2 Chemical Structure Representation 8

  1.3 Code for Including the Editor Applet in JChemPaint 9

  1.4 Definition of Templates (Polygons, Benzene, Bond, Atom, etc.) 9

  1.5 Free Tools 10

  1.6 Academic Programs 11

     1.6.1 Marvin Sketch 11

     1.6.2 ACD Labs 12

  1.7 Commercial Tools 12

     1.7.1 ChemDraw 12

     1.7.2 Schrodinger 14

     1.7.3 MOE (CCG) 14

     1.7.4 Accelrys 14

  1.8 A Practice Tutorial 15

     1.8.1 Interconversion of Name/SMILES to Structure
and Vice Versa 15

  1.9 Introduction to Chemical Structure Formats 20

     1.9.1 Linear Format 20

     1.9.2 Graph-based Representation (2D and 3D formats) 21

     1.9.3 Connection Tables 22

     1.9.4 FILE FORMATS 22

  1.10 2D and 3D Representation 30

     1.10.1 Code for 3D Structure Generation in ChemAxon 31

     1.10.2 A Practice Tutorial 31

  1.11 Abstract Representation of Molecules 32

  1.12 File Format Exchange 35

     1.12.1 A Practice Tutorial 36

     1.12.2 Code for Reading a Molecule, checking the Num-
ber of Atoms, and Writing a SMILES String 38

1.12.3   Code for Reading a SMILES String in Python 39
1.13   Similarity and Fingerprint Analysis 39
    1.13.1   Simple Fingerprints (Structural Keys) 41
    1.13.2   Hashed Fingerprints 42
    1.13.3   A Practice Tutorial 44
1.14   Molecular Similarity 45
    1.14.1   Exact Structure Search 46
    1.14.2   Substructure Search 47
    1.14.3   Similarity Search 48
    1.14.4   Subsimilarity Search 50
1.15   Search for Relationship 51
1.16   Similarity Measures 52
1.17   Molecular Diversity 55
1.18   Advanced Structure-handling Tools 56
    1.18.1   CCML 56
1.19   ChemXtreme 56
    1.19.1   Barcoding SMILES 57
    1.19.2   Chem Robot 57
    1.19.3   Image to Structure Tools 58
    1.19.4   CLide 59
    1.19.5   Advanced Structure Computation Platforms 59
1.20   Virtual Library Enumeration 59
1.21   Clustering 60
1.22   Databases 60
    1.22.1   Database Server My SQL 62
    1.22.2   Code for Connecting to a MySQL Database 63
    1.22.3   A Practice Tutorial 64
    1.22.4   Creating and Hosting Database 67
    1.22.5   A Practice Tutorial 67
    1.22.6   Hosting the Database 71
    1.22.7   Chemical Databases 74
    1.22.8   Do It Yourself (DIY) 85
    1.22.9   Questions 89
References 89

2   Chemoinformatics Approach for the Design and Screening
    of Focused Virtual Libraries 93
2.1   Introduction to Structure–Property Correlations 93
    2.1.1   Descriptors 94
    2.1.2   Online Property Prediction Tools 108
    2.1.3   Virtual Library Generation (Enumeration) 111
    2.1.4   Virtual Screening 121
    2.1.5   Thumb Rules for Computing Molecular Properties 128
    2.1.6   Do it Yourself 128
    2.1.7   Questions 129
References 129

3   Machine Learning Methods in Chemoinformatics for
     Drug Discovery   133
     3.1   Introduction   133
     3.2   Machine Learning Models for Predictive Studies   134
     3.3   Machine Learning Methods   136
     3.4   Open-Source Tools for Building Models for Drug Design   139
            3.4.1   Library for Support Vector Machines (LibSVM)   139
            3.4.2   Waikato Environment for Knowledge Analysis (WeKa) ...   141
            3.4.3   R Program   151
     3.5   Free Tools for Machine Learning   152
            3.5.1   An Example of SVR-based Machine Learning   152
            3.5.2   Rapid Miner   160
     3.6   Commercial Tools for Building ML Models   164
            3.6.1   Molecular Operating Environment (MOE)   164
            3.6.2   IBM SPSS   176
            3.6.3   Matrix Laboratory (MATLAB)   178
     3.7   Genetic Programming-Based ML Models   179
            3.7.1   A Practical Demonstration of GP-Based Software   185
     3.8   Thumb Rules for Machine Learning-Based Modelling   189
     3.9   Do it Yourself (DIY)   191
     3.10 Questions   191
     References   192

4   Docking and Pharmacophore Modelling for Virtual Screening   195
     4.1   Introduction   195
     4.2   A Practice Tutorial: Docking Using a Commercial Tool   196
     4.3   Docking Using Open Source Software   211
            4.3.1   Autodock Steps   212
            4.3.2   Docking Using AutoDock Vina   220
     4.4   Other Docking Algorithms   223
            4.4.1   Induced Fit Docking   224
            4.4.2   Flexible Protein Docking   225
            4.4.3   Blind Docking   226
            4.4.4   Cross Docking   226
            4.4.5   Docking and Site-Directed Mutagenesis   229
     4.5   Protein–Protein Docking   231
     4.6   Pharmacophore   234
            4.6.1   Pharmacophore Modelling in SCHRÖDINGER   235
            4.6.2   Finding Pharmacophore Features Using MOE   248
     4.7   Open Source Tools for Pharmacophore Generation   253
     4.8   Rules of Thumb for Structure-Based Drug Design   254
     4.9   Do it Yourself Exercises   260
     4.10 Questions   261
     References   267

5   **Active Site-Directed Pose Prediction Programs for Efficient**
     Filtering of Molecules   271
     5.1   Introduction   271
     5.2   A Practice Tutorial for Predicting Active Site Using SiteMap   272
     5.3   A Practice Tutorial for Active Site Prediction Using MOE   276
     5.4   Free Online Tools for Active Site Prediction   279
     5.5   Homology Modelling   282
     5.6   A Practice Tutorial for Homology Modelling   285
     5.7   Model Validation Using Online Servers   295
     5.8   Receptor-Based Pharmacophore   296
     5.9   Studies on Active Site Structural Features   298
            5.9.1   Application of Active Site Features in Chemoinformatics ...   300
     5.10 Thumb Rules for Active Site Identification and Homology
            Modelling   312
     5.11 Do it Yourself Exercises   313
     5.12 Questions   313
     References   313

6   **Representation, Fingerprinting, and Modelling of**
     Chemical Reactions   317
     6.1   Introduction   318
     6.2   Reaction Representation in Computers   318
     6.3   Computational Methods in Reaction Modelling   318
            6.3.1   Empirical and Semiempirical Methods   319
            6.3.2   Molecular Mechanics Methods   320
            6.3.3   Molecular Dynamics Methods   321
            6.3.4   Statistical Mechanics and Thermodynamics   321
            6.3.5   The Quantum Mechanical/molecular Mechanical
                     Approach   322
            6.3.6   Modelling the Transition State of Reactions   322
     6.4   TS Modelling of Organic Transformations   324
            6.4.1   Name Reactions   324
            6.4.2   A Practice Tutorial for Transition State and Intrinsic
                     Reaction Coordinate Modelling   326
            6.4.3   A Practice Tutorial Using Maestro–Jaguar   338
            6.4.4   A Practice Tutorial Using Spartan   344
     6.5   Reaction-Searching Approaches and Tools   347
            6.5.1   Chemical Ontologies Approach for Reaction Searching   351
            6.5.2   Reaction Searching Using Fingerprints-Based Approach ...   354
            6.5.3   Tools for Reaction Searching   359
     6.6   Reaction Databases   363
            6.6.1   Tools for Reaction Library Enumeration   364
            6.6.2   A Practice Tutorial   365
     6.7   Artificial Intelligence in Chemical Synthesis   366
     6.8   Modelling Enzymatic Reactions   369

6.9   Thumb Rules for Performing Reaction Representation,
       Fingerprints, and Modelling   369
6.10   Do it Yourself   371
6.11   Questions   371
References   371

7   **Predictive Methods for Organic Spectral Data Simulation**   375
7.1   Introduction   376
7.2   Fragment-Based Drug Discovery   378
7.3   Spectra Prediction Methods   384
7.4   Spectra Prediction Tools   384
7.5   Open-Source Tools   385
       7.5.1   GAMESS   385
7.6   Proprietary Tools   385
       7.6.1   ACD/NMR Predictors   385
       7.6.2   Cambridgesoft Chem3D   385
       7.6.3   Jaguar   385
       7.6.4   Gaussian   390
       7.6.5   ADF   391
       7.6.6   MestreNova   392
       7.6.7   Spartan   396
       7.6.8   Spectral Databases   399
7.7   Spectra Viewer Programs   404
7.8   In-House Tools for Spectra Prediction   404
7.9   Code to Generate Proton and Carbon NMR Spectrum   406
7.10   Thumb Rules for Spectral Data Handling and Prediction   409
7.11   Do it Yourself   410
7.12   Questions   411
References   412

8   **Chemical Text Mining for Lead Discovery**   415
8.1   What is Text Mining?   416
       8.1.1   Text Mining vis-a-vis Data Mining   416
       8.1.2   A Snippet of Java Code Using the Above URL   418
8.2   What are the Components of Text Mining?   419
8.3   Text-mining Methods   421
       8.3.1   Statistics/ML-based Approach   422
       8.3.2   Rule-based Approach   423
8.4   Why Text Mining   424
8.5   General Text-mining Tools   424
       8.5.1   A Practice Tutorial with an Open-source Tool   425
       8.5.2   R Program for Text Mining   430
8.6   Free Tools for Text Mining   434
8.7   Biomedical Text Mining   434
8.8   Chemically Intelligent Text-mining Tools   435

8.9   In-house Tools for Text-mining Applications for
       Chemoinformatics 437
       8.9.1   Java Code Snippet for Data Distribution 441
8.10   Thumb Rules While Performing and Using Text-mining Results ... 445
8.11   Do it Yourself 445
8.12   Questions 445
References 445

9 **Integration of Automated Workflow in Chemoinformatics
   for Drug Discovery** 451
   9.1   What is a Workflow? 451
   9.2   Need for Workflows 452
   9.3   General Workflows in Bioinformatics 453
   9.4   General Workflows in Chemistry Domain 453
          9.4.1   Accelrys Pipeline Pilot 453
          9.4.2   IDBS Chemsense (Inforsense Suite) 454
          9.4.3   CDK Taverna 455
          9.4.4   KNIME 455
          9.4.5   Workflow Examples 467
          9.4.6   Workflow for QSAR (Anti-cancer) 469
   9.5   Schrodinger KNIME Extensions 470
          9.5.1   A Practice Tutorial 473
   9.6   Other KNIME Extensions 481
          9.6.1   MOE(CCG) 481
          9.6.2   ChemAxon 483
   9.7   Protein–Ligand Analysis-Based Workflows for Drug Discovery 483
          9.7.1   A Practice Tutorial for Protein–Ligand Fingerprint
                   Generation 486
   9.8   Prolix 489
   9.9   J-ProLINE: An In-house-developed Chem-Bioinformatics
          Workflow Application 489
   9.10   Targetlikeness Score 496
   9.11   Databases and Tools 496
   9.12   Thumb Rules for Generating and Applying Workflows 496
   9.13   Do it Yourself 497
   9.14   Questions 497
   References 497

10 **Cloud Computing Infrastructure Development for
   Chemoinformatics** 501
   10.1   What is a Portal? 501
   10.2   Need for Development of Scientific Portals 502
   10.3   Components of a Portal 502
   10.4   Examples of Portal Systems 503

10.5   A Practice Tutorial for Portal Creation   504
        10.5.1   Custom Database connection and Display Table
                  with Paginator via portlet in Liferay Portal   509
10.6   A Practice Tutorial for Development of Portlets for
        Chemoinformatics   512
        10.6.1   Marvin Sketch Portlet   512
        10.6.2   JME Portlet   515
        10.6.3   Jchempaint Portlet   515
10.7   Mobile Computing   516
        10.7.1   Android Applications for Chemoinformatics   517
10.8   Need of High-Performance Computing in Chemoinformatics   526
10.9   Thumb Rules for Developing and Using Scientific
        Portals and Mobile Devices for Computing   526
10.10   Do it Yourself Exercises   526
10.11   Questions   527
References   527

Index   529

About the Authors

Muthukumarasamy Karthikeyan obtained his Bachelors and Masters Degree in Chemistry from Pondicherry University and Ph.D. (Chemistry) from National Chemical Laboratory (University of Pune) in the area of Organic Synthesis. He began his career as a scientist in Armament Research Development Establishment (Ministry of Defence, DRDO) Pune, and then joined CSIR-National Chemical Laboratory, Pune as a senior scientist; since then he is pursuing his research career in Chemoinformatics, especially in the area of high performance computing for molecular informatics, and its application in lead identification and lead discovery. In 2007 he organized the first International Conference on Chemoinformatics (http://moltable.ncl.res.in/). He has published several key papers in chemoinformatics handling large scale molecular data including entire PubChem repository (ChemStar) which currently holds more than 70 million entries and harvesting chemical information from Google (ChemXtreme) with more than 10 billion web pages. He is also the recipient of BOYSCAST Fellowship from Department of Science and Technology and Long term Overseas Associateship from Department of Biotechnology. He is a visiting scientist/professor at the University of North Carolina at Chapel Hill, USA. His current interest includes development of open source tools in visual computing for molecular informatics (ChemRobot), hybrid computing (distributed, parallel, cloud) using multicore CPU-GPU processors as a web-based problem solving environment in chemical informatics. He is a member on the executive advisory board of journal of Molecular Informatics from Wiley. Currently he is serving as a guest editor for a special issue on chemoinformatics for virtual screening.

Dr. Renu Vyas is currently a DST women scientist at National Chemical Laboratory Pune, India. She pursued her Ph.D. in synthetic organic chemistry at National Chemical Laboratory and postdoctoral studies at the University of Tennessee, USA. She is the recipient of several university and national level fellowships. She has a number of research publications in internationally renowned journals, reviews, and book chapters to her credit. She held high positions and possesses varied experience in research, teaching, administration, and software industry. Her research interests include molecular modelling in the twin domains of chemoinformatics and bioinformatics.

Chapter 1
Open-Source Tools, Techniques, and Data in Chemoinformatics

Abstract Chemicals are everywhere and they are essentially composed of atoms and bonds that support life and provide comfort. The numerous combinations of these entities lead to the complexity and diversity in the universe. Chemistry is a subject which analyzes and tries to explain this complexity at the atomic level. Advancement in this subject led to more data generation and information explosion. Over a period of time, the observations were recorded in chemical documents that include journals, patents, and research reports. The vast amount of chemical literature covering more than two centuries demands the extensive use of information technology to manage it. Today, the chemoinformatics tools and methods have grown powerful enough to handle and discover unexplored knowledge from this huge resource of chemical information. The role of chemoinformatics is to add value to every bit of chemical data. The underlying theme of this domain is how to develop efficient chemical with predicted physico-chemical and biological properties for economic, social, health, safety, and environment. In this chapter, we begin with a brief definition and role of open-source tools in chemoinformatics and extend the discussion on the need for basic computer knowledge required to understand this specialized and interdisciplinary subject. This is followed by an in-depth analysis of traditional and advanced methods for handling chemical structures in computers which is an elementary but essential precursor for performing any chemoinformatics task. Practical guidance on step-by-step use of open-source, free, academic, and commercial structure representation tools is also provided. To gain a better understanding, it is highly recommended that the reader attempts the practice tutorials, *Do it yourself* exercises, and questions given in each chapter. The scope of this chapter is designed for experimental chemists, biologists, mathematicians, physicists, computer scientists, etc. to understand the subject in a practical way with relevant and easy-to-understand examples and also to encourage the readers to proceed further with advanced topics in the subsequent chapters.

Keywords Chemical structure · Molecular modelling · Chemical databases · Open-source software · Drug discovery

M. Karthikeyan, R. Vyas, *Practical Chemoinformatics*,
DOI 10.1007/978-81-322-1780-0_1, © Springer India 2014

1.1   Chemoinformatics

Chemoinformatics has been defined in various ways [1], and the most popular one is by Greg Paris which states that "Chemoinformatics is a generic term which encompasses the design, creation, organization, management, retrieval, analysis, dissemination, visualization, and use of chemical information." The basic core operations of a chemical information system include storing, retrieving, and searching information/data and their relationships [2, 3]. Chemoinformatics helps to harvest large-scale chemical data from publicly available sources and design materials with desired characteristics through prediction methods that include physical, chemical, or biological properties of compounds, spectra simulation, structure elucidation, reaction modelling, synthesis planning, and frequently used drug design and lead optimization process. It is applied "mostly" to a large number of "small" molecules where $\#N \sim (10...100...1,000...10,000...10^6...10^{60}...)$.

The main applications of this subject are in the fields of medical science for developing novel and effective drugs and in material science to develop new and superior materials [4]. The other allied fields that benefit from the pursuit of chemoinformatics are agrochemicals and biotechnology [5]. These operations differ from the classical storage of data in a computer because the data associated with the chemical information system are mostly structural that require special algorithms and methods to handle unlike textual data.

With the availability and access to modern high-performance computing infrastructure, it is now possible to add value to the diverse field of chemoinformatics in terms of speed and efficiency. Open-source tools are now playing a pivotal role in revolutionizing the way chemoinformatics data can be handled in a high-throughput manner, and experiments requiring intensive computational power can be performed in an *in-silico* environment.

1.1.1   Open-Source Tools

Free Open-Source Software (FOSS) tools are defined as those programs which anyone can download and change the source code, provided that they make the changes publicly available again, according to the GNU Lesser General Public License (LGPL) [6]. Anyone is freely licensed to use, copy, study, and change the software in any way, and the source code is openly shared so that people are encouraged to voluntarily improve the design of the software. Some of the most popular open-source tools are Linux and OpenBSD [7] and are widely utilized today, powering millions of servers, desktops, smart phones (e.g., Google Android), and other devices. This is in contrast to proprietary software, where the software is under restrictive copyright and the source code is hidden from the users, so that the rights holders (the software publishers) can sell binary executables (Table 1.1).

Table 1.1 Open-source tools, languages, and resources available for performing chemoinformatics experiments

S. No.	Tools and platforms	Programming languages
1	Open Babel	C++
2	CDK	JAVA
3	RDKit	Python
4	Joelib	Perl
5	BlueDesc	Ruby
6	ISIDA	CUDA[1]
7	TEST	
8	MOLD2	
9	Bioeclipse	

[1] Compute Unified Device Architecture

Why to use open-source tools in chemoinformatics?
- Use them to "Handle" large-scale data through integration (linking multiple free tools, databases, etc.)
- Use through "Internet Access" to Web services (servers at various institutes)

1.1.2 Introduction to Programming Languages

It will be pertinent here to provide a brief discussion on the background computer knowledge required to master the subject of chemoinformatics. Though a number of software with graphical user interface (GUI) options are available, it is recommended that the users train themselves in some of the programming languages and be aware of the ongoing developments so as to become proficient in harnessing the computing power of the existing software applications for specific individual needs. The computer is one of the most important tools for the new generation of chemoinformaticians and bioinformaticians. Along with the evolution of computer hardware, operating systems and computer programming languages also evolved with time. One of the earliest scientific programming languages was FORTRAN developed in 1953 [8].

Writing a software code is not difficult at all, and what is required to learn programming is a bit of patience and perseverance. The choice of language for programming is left to the user. Here, we highlight few lines of codes in different programming environments to demonstrate simple input–output tasks related to chemical information. This would encourage the readers to go ahead with the selection of programming language and identification of tasks to be accomplished in chemoinformatics.

Choice of Operating Systems It is also important to be familiar with operating systems like Windows, Linux, and Mac OS. Students and faculties of chemoinformatics and bioinformatics should be able to execute commands in Linux/UNIX systems for computationally intensive tasks. Some of the most frequently used

UNIX/Linux commands are: [9] cat (displays the contents of the file), cd (changes directory), cp (copies file to a specified directory or copies to another file), grep (searches the mentioned files for lines containing regular expression), head (displays specified number of lines from a file), ls (lists the content of a directory), man (displays the manual page for the given command), mkdir (makes directory), more (displays the contents of a file(s) page by page on the screen), mv (moves files or directories or renames files/directories, etc.), pwd (presents working directory/current directory), rm (removes or deletes one or more files), rmdir (removes or deletes directories), tail (displays the last N lines of one or several files), telnet (establishes a connection to another computer via telnet protocol), and wc (counts the number of words/characters/lines in the file).

Internet and WWW Today, it is not necessary to introduce the Internet or World Wide Web (WWW) to a student of chemoinformatics or bioinformatics as they are already familiar with these resources for their day-to-day research or education. The Internet was originally designed by the US military in order to avoid total failure of the network. With the establishment of Transmission Control Protocol (TCP) and Internet Protocol (IP) also known as TCP/IP, the definition of the term Internet was born [10]. The WWW was developed by Tim Berners-Lee of CERN. File Transfer Protocol (FTP) is largely used in chemoinformatics and bioinformatics to get the scientific data (small molecules and sequences, structures, properties, activities, toxicity, and literature) from the Internet. Microsoft's Internet Explorer, open-source-based Mozilla, Opera, Google's Chrome, Safari, etc. are usually used to access Internet web pages using Hyper Text Transmission Protocol (HTTP) and FTP. FileZilla (filezilla-project.org), an open-source FTP client available under many platforms including Windows, Linux, and Mac OS, is also worth mentioning.

Some of the most frequently used FTP commands are as follows: ascii (changes mode to ASCII), bin (changes mode to Binary transport), bye (terminates the FTP session), get (gets the file), put (uploads the file to the FTP server), pwd (shows the current directory), and quit (terminates the FTP session).

Some of the most popular Internet browsers are: Internet Explorer, Mozilla Firefox, Google Chrome, Safari, Opera, etc.

In addition to learning about operating systems, commands to handle files, use of Internet browsers to search the right information from a volume of data from public resources, there is a need to learn a bit of programming to accomplish simple, routine tasks required in chemoinformatics.

Fig. 1.1 Chemical structure of Caffeine

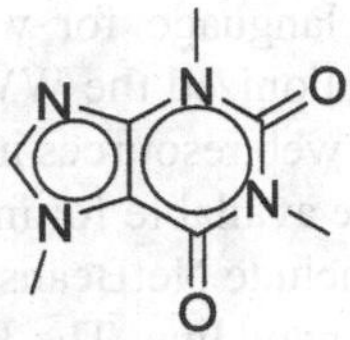

Introduction to basics of programming Here, we will print the name of a small
molecule of caffeine using the simplest code snippet in Fortran on a UNIX system.
Caffeine is a stimulant and drug molecule from the alkaloid family (Fig. 1.1).

```
program hello Caffeine
print *, 'Hello Caffeine!'
end program hello
Caffeine
```

Without doubt, the program which changed the world of computing and compiling
was "C" [11]. The GUI compilers like Turbo C or Borland were used in early days
of programming. A simple C program is written as

```
/* Hello Caffeine program
*/
#include<stdio.h>
main()
{
    printf("Hello
Caffeine");

}
```

Later, the concept of object-oriented programming evolved with C++ for better reusability of the codes [12].

```
#include <iostream.h>
main()
{
    cout << "Hello
Caffeine!";
    return 0;
}
```

Recently, another object-oriented language for web compatibility, namely Java [13], has been created and it revolutionized the WWW of the Internet age. Several specialized books and free Internet web resources in the area of computer programming, languages, compilers, etc. are available for interested readers. Integrated Development Environments (IDEs) include NetBeans and Eclipse. The java program is compiled using javac in the command line. The JDK needs to be installed.

```
public class HelloCaffeine {

    public static void
main(String[] args) {
        System.out.println("Hello,
Caffeine");
    }

}
```

1.1.2.1 Other Important Programming Languages

Practical Extraction Report Language Practical Extraction Report Language (Perl) is a free interpreted language mainly developed for text handling [14]. A collection of Perl code is available at the Comprehensive Perl Archive Network (CPAN; www.cpan.org). Bioperl provides many modules for sequences, data parsing, and databases very often used in bioinformatics. A perl code snippet is as follows:

```
#!/usr/bin/perl
print "Hello Caffeine\n";
```

Python Python is a free object-oriented, easy-to-learn programming language and is useful in application development [15]. It overcomes some of the drawbacks of Perl. It contains scalable, extendable scripting and can be embedded:

```
$ vim hellocaffeine.py
#!/usr/bin/python
# Hello caffeine python program
print "Hello Caffeine!";
```

R R is an open-source based powerful language that is very good for performing statistical operations on large datasets and runs on a wide variety of platforms [16]. It includes a subset of C language. It allows branching and looping as well as modular programming using functions. The bin/linux directory of the Comprehensive R Archive Network (CRAN) contains all the packages.

```
state start:
pstr("hello
caffeine\n");
halt;
```

Introduction to compilers Compilers are required to write a computer program and to create the executable codes. The purpose of traditional and modern compilers is to translate man-made computer programs into machine-readable codes. The sequence of operations involved in writing a source code, compiling them, and generating executable programs is depicted below:

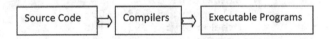

Hybrid computing Today, high-performance computing (HPC) platforms are reaching the home through cloud computing infrastructure. Like we access electricity at home, now with the help of the Internet, one can access tremendous computing power on demand, based on need and available resources. Supercomputers and virtual computers that are powered by both central processing units (CPUs) and graphics processing units (GPUs) are accessible through the Internet. Several academic institutions are providing access to high-performance computing to researchers through the Internet, and students with their mobile devices are able to harness the computing power through authentications. Therefore, it is necessary to learn more about emerging computing platforms and special programming skills to achieve the tasks in the shortest period of time. It is worth mentioning emerging modern programming languages like Cuda. GPUs usually used for high-end gaming are now being used for scientific computing including drug design, quantum chemistry, and weather forecasting. Today, simple GPU-based accessories with thousands of cores with high processing power are now accessible at moderate cost. There is a need to learn Cuda programming which is a parallel computing platform and is a boon for software developers and scientists [17]. Using Cuda, one gains access to specialized GPU processors-based computing to handle large data at extreme speed (teraflops) and carry out computer-intensive tasks. It supports programs written in languages like Java, C++ and Fortran, and there is no need for assembly language. Recent scientific applications include the development of high-throughput sequence alignment tools [18].

```
>> parallel.gpu.GPUDevice.current()

ans =

  parallel.gpu.CUDADevice handle
  Package: parallel.gpu

  Properties:
                    Name: 'Tesla K20c'
                   Index: 1
       ComputeCapability: '3.5'
           SupportsDouble: 1
            DriverVersion: 5.5000
        MaxThreadsPerBlock: 1024
         MaxShmemPerBlock: 49152
        MaxThreadBlockSize: [1024 1024 64]
             MaxGridSize: [2.1475e+09 65535]
                SIMDWidth: 32
              TotalMemory: 5.0330e+09
               FreeMemory: 4.9250e+09
       MultiprocessorCount: 13
             ClockRateKHz: 705500
              ComputeMode: 'Default'
      GPUOverlapsTransfers: 1
   KernelExecutionTimeout: 0
          CanMapHostMemory: 1
           DeviceSupported: 1
            DeviceSelected: 1

  Methods, Events, Superclasses
```

1.2   Chemical Structure Representation

Chemical structures are the international language of chemistry and their representation, interpretation, automatic generation, storage, searching them efficiently using mathematical approaches and analyzing them with chemical context are the most critical steps in solving chemical problems [19]. The basic requirement for building a chemical information system is the representation of molecules in a specific and generic way for fast processing by computers and easy understanding by chemists [20, 21]. The most widely known open-source and free tool for drawing chemical structures is JChemPaint (JCP) [22]. Currently, it is developed as a GitHub project which is the largest code hub in the world [23]. JCP can be used for educational purposes due to its capability of handling chemical structures in standard file formats (Simplified Molecular-Input Line-Entry System, SMILES; MOL; structure data file, SDF; Chemical Markup Language, CML, etc.) for easy exchange between the programs and also for managing chemical information [24]. JCP is the editor and viewer for two-dimensional (2D) chemical structures developed using Chemistry Development Kit (CDK) [25]. It is implemented in several forms including a Java application and two varieties of a Java applet. To use the JCP applet in web pages, one has to download the corresponding jar file and edit the HyperText Markup Language (html) page with the applet code including the dimension of applet, source of molecule file in the html document as shown below.

1.3   Code for Including the Editor Applet in JChemPaint

```
<applet
    code="org.openscience.cdk.applications.jchempaint.applet.JChemPaintEditorApplet"
    archive="jchempaint-applet-core.jar"
    name="Editor"
    width="600" height="500">
    <param name="load" value="a-pinene.mol">
</applet>
```

1.4   Definition of Templates (Polygons, Benzene, Bond, Atom, etc.)

The GUI helps to draw the chemical structure rapidly. The most frequently used molecular fragments are defined in the program as templates and are shown as icons in the user interface. It is easy for the user to select the icon, and clicking on an empty drawing area or workspace would place them appropriately. Once a template or a fragment is drawn, it can be modified by adding additional bonds, changing the bond types such as double, triple, or stereo-chemical (wedged or broken), fusing the additional rings, etc. These tools facilitate easy and rapid drawing of chemical structures and store them for reusability and inventory management. Still, without the aid of these tools, one can generate chemical structures by creating plain text files containing atoms (coordinate tables) and bond (connection table) information with some experience and expertise. However, drawing chemical structures using professionally designed software tools is encouraged to avoid inadvertent errors in the chemical structures. The graphical user programs help to draw chemical structures rapidly and facilitate the storage and interconversion in the standard file formats. The advanced programs are smart enough to monitor the progress of drawing or input by the user and alert them when they make mistakes (with wrong connectivity, exceeding atomic valency, etc.) and also auto-correct the structures dynamically. Now, with advancement in chemoinformatics tools, one can generate chemical names from the structures and vice versa. Some of these tools also help

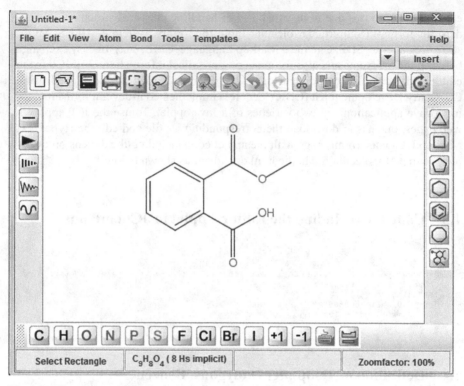

Fig. 1.2 JChemPaint graphical user interface displaying aspirin structure

to compute some of the primitive molecular descriptors like LogP (octanol–water co-efficient), total polar surface area (TPSA), chemical composition (percentage of atomic composition) and also to change the chemical structures into 3D formats as and when required (Fig. 1.2).

MCDL available at http://mcdl.sourceforge.net/ is another free open-source small Java molecular viewer/editor for chemical structures, stored in Modular Chemical Descriptor Language linear notation only [26].

1.5　Free Tools

Unlike the JCP program discussed above, where the source code for the program is available, there are other chemical structure drawing tools that are freely distributed as executable without the source code. A suitable example is JME Molecular Editor—a lightweight Java applet for web browsers which allows users to draw/edit molecules and reactions (including the generation of substructure queries) and to depict molecules directly within an HTML page [27]. The editor can generate Daylight SMILES or Molecular Design Limited (MDL) molfile of created structures [28]. The applet is widely known due to its ease of use in the input of molecules in the web servers to search the chemical structures or to predict the physicochemical

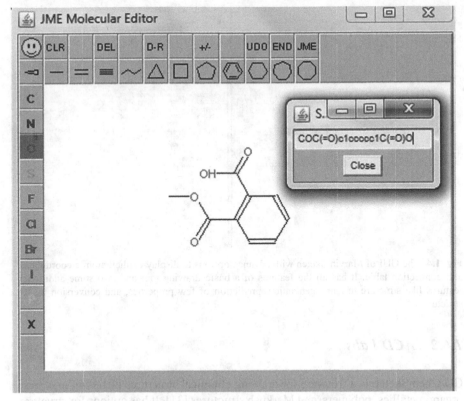

Fig. 1.3 A structure drawn using JME editor

properties. For example, molinspiration site provides space for the JME Home and helps with the installation and deployment of the JME [29]. The JME can be incorporated as an applet into an HTML page with the following code (Fig. 1.3):

```
<applet code="JME.class" name="JME" archive="JME.jar" width="360"
height="335"> <param name="options"value="listofkeywords"></applet>
```

1.6   Academic Programs

As these programs have different licensing options, there is a free version for academic users but a license fee is charged for corporate use.

1.6.1   Marvin Sketch

Marvin Sketch is a structure-editing tool, a component of java-based Marvin Tools provided via an academic license from the ChemAxon company [30] (Fig. 1.4).

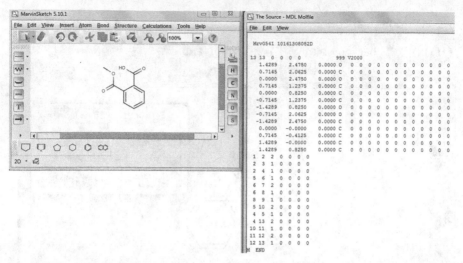

Fig. 1.4 The GUI of Marvin Sketch with advance options to display explicit atomic coordinates and connection table. It has all the features of a basic drawing tool and also some additional features like structure to name generation, prediction of few properties, and conversion to 3D structure

1.6.2   ACD Labs

CD/ChemSketch is a freeware for drawing chemical structures including organics, organometallics, polymers, and Markush structures [31]. It has options for structure cleaning, viewing and naming, inch conversion, stereo descriptors etc. For freeware, no technical support is provided and the functionalities are less compared to the commercial version which has structure search capabilities (Fig. 1.5).

1.7   Commercial Tools

A number of proprietary software programs are available for 2D structure creation and manipulation. In fact, all commercial software programs in the field of chemoinformatics and/or bioinformatics supply a drawing tool to the users.

1.7.1   ChemDraw

It is marketed by Cambridge soft as part of a suite of integrated tools called ChemOffice [32] (Fig. 1.6).

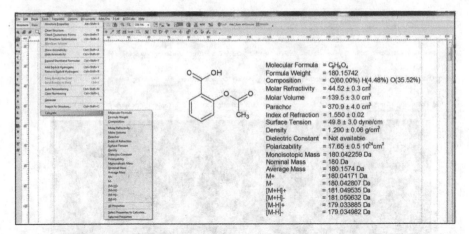

Fig. 1.5 A molecule drawn in ACD ChemSketch

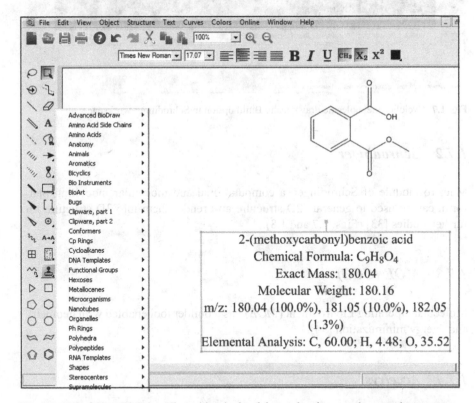

Fig. 1.6 GUI of ChemBioDraw Ultra with calculated data and options to select template structures

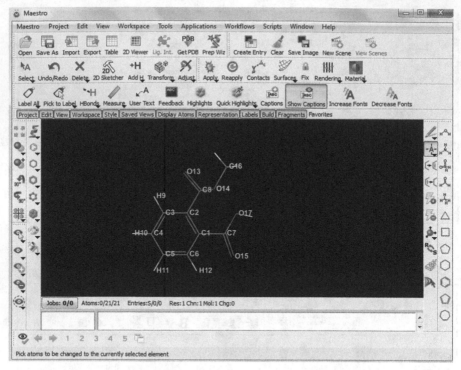

Fig. 1.7 Cyclohexane molecule drawn using Build option in Schrodinger workspace

1.7.2 Schrodinger

Maestro module of Schrodinger, a computational and molecular modelling platform, can be used to generate 2D structure and render them into 3D structure for further studies [33] (Figs. 1.7 and 1.8).

1.7.3 MOE (CCG)

Molecular Operating Environment (*MOE*) has a builder tool enabled with geometry and energy minimization [34].

1.7.4 Accelrys

Accelrys Draw 4.1 enables scientists to draw and edit complex molecules, chemical reactions, and biological sequences with ease, facilitating the collaborative

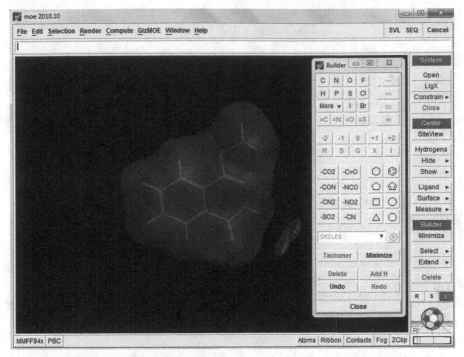

Fig. 1.8 A structure drawn using Builder option on the right-hand side (*RHS*) of the *MOE* GUI

searching, viewing, communicating, and archiving of scientific information [35] (Fig. 1.9).

Other chemical information service providers like Scifinder [36], ChemSpider [37], NIH [38], Beilstein [39], etc. provide their own drawing tools to the users.

1.8   A Practice Tutorial

1.8.1   Interconversion of Name/SMILES to Structure and Vice Versa

Chemical names are usually used for documentation and communication purposes. A molecule can have several valid chemical names including computer-generated International Union of Pure and Applied Chemistry (IUPAC) names, traditional name, common name, commercial name, company assigned identifiers, Chemical Abstracts Service (CAS) Registry number, and many other synonyms. It is challenging to generate chemical structures from the chemical names. In order to communicate effectively, line notations were developed for representing chemical struc-

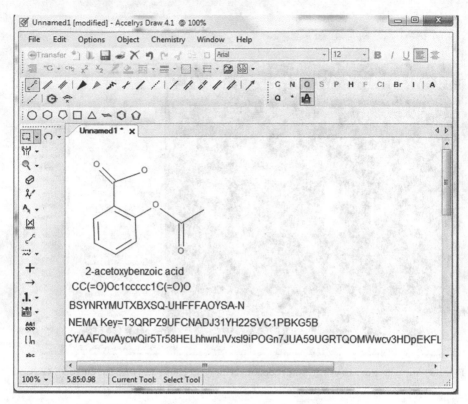

Fig. 1.9 Aspirin and its various line notations depicted in Accelrys 4.0

tures. SMILES are line notations used frequently in chemoinformatics especially for database operations. The SMILES format contains the details of connection table in a linear format. The details are described in the appropriate section of this chapter.

In this tutorial, the reader will learn how to get the SMILES/IUPAC name from the chemical structure and vice versa. Here, we selected two different software programs for demonstrating simple operations to handle chemical structures. ChemDraw from Perkinelmer informatics has been traditionally used by chemists for the past two decades especially for chemical documentation in particular for writing manuscript with chemical significance for the journals, patents, and PhD theses. ChemDraw is equipped with several templates to support these activities, for example, selection of templates suitable for organic chemistry journals, where the user will draw the reaction schemes and the dimensions would be automatically fixed according to the journal selected. In addition to this, ChemDraw programs were frequently used by organic chemists to generate IUPAC names, 1H and 13C predicted nuclear magnetic resonance (NMR) to assign particular peaks corresponding to the atomic environment in the molecule as a guideline and also to calculate primitive descriptors like atomic composition, molecular mass, logP, etc. In recent times, ChemAxon tools are becoming the most popular among the academic communities due to their

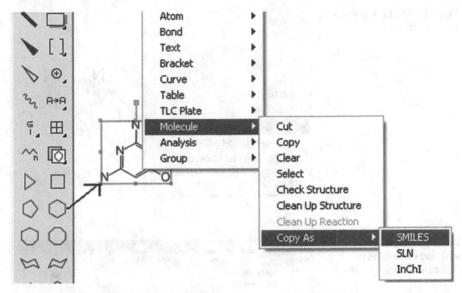

Fig. 1.10 ChemDraw GUI for copying chemical structures in SMILES format

flexibility in licensing policy and more features comparable and better than other commercial softwares. ChemAxon is the only software available today to handle millions of chemical structures in a database and enables to search them using exact structure-, substructure-, and similar structure-based queries in a relational database management system (RDBMS) environment. The number of chemical structures in the database is limited only by the hardware resources and database constraints. The 512 bits fragments-based binary fingerprinting algorithm implemented in ChemAxon tools is powerful enough to facilitate rapid searching in a large-scale database of chemical structures. ChemAxon also provides Java application programming interfaces (APIs) to extend and enhance the functionality of the program as per the user's needs. The details of advanced functionalities of open source and academic packages related to chemoinformatics are described in detail with practical do it yourself sections (Figs. 1.10, 1.11 and 1.12).

Do-it-yourself (*Requirement: ChemDraw software)
Structure to SMILES

- Start ChemDraw
- Open the chemdraw tool panel and draw the structure with the following tool bar
- After drawing the structure, right click on the structure, then select Molecule → copy as → SMILE → paste it where you want

Getting Structure from SMILES using ChemDraw

- Copy the SMILES from the source file
- Right click on the ChemDraw editor window
- Click on Paste SMILES

Fig. 1.11 SMILES to structure conversion

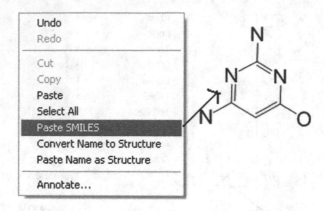

Fig. 1.12 Name to structure conversion in ChemDrawUltra

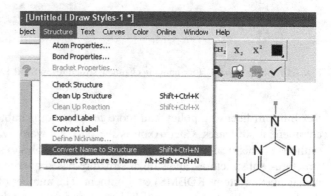

SMILES to Structure

- Open ChemDraw → edit → paste special → paste the desired format (SMILES) to retrieve the structure

Getting structure from IUPAC name using ChemDraw

- Copy the IUPAC name from the source file
- Open ChemDraw → structure → convert name to structure
- The output is the structure with the IUPAC name
- Conversely, one can convert name to structure in ChemDraw

Do-it-yourself (*Requirement: ChemAxon software)
Getting SMILES string from structure by MarvinSketch

- Draw the structure using the MarvinSketch window as shown in the figure
- After drawing the structure, select the structure and go to edit → Copy as SMILES as shown in Fig. 1.13

Similarly, one can insert the IUPAC name of the structure using the insert → IUPAC Name (Fig. 1.14)

Fig. 1.13 GUI for copying a structure in SMILES format using Marvin Sketch

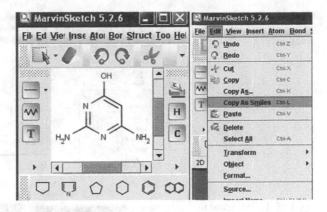

Getting SMILES/IUPAC name from structure by MarvinView Copy a valid SMILES string and paste into MarvinSketch or MarvinView panel to display the structure (Fig. 1.15).

- Start MarvinView
- Select Edit → Paste (Ctrl + V)
- To generate SMILES from the already drawn structure: Select Table → Select option Show SMILES

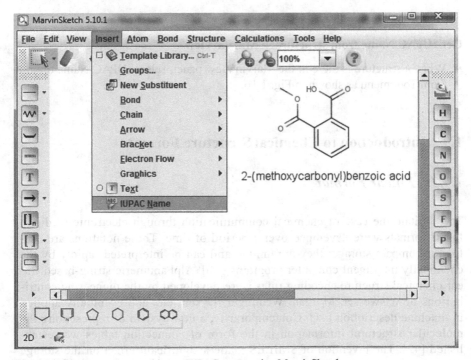

Fig. 1.14 Step to generate and insert IUPAC name using MarvinSketch

Fig. 1.15 MarvinView menu for generating SMILES to structure

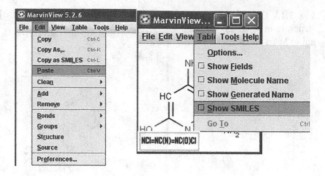

Fig. 1.16 IUPAC name to structure conversion in Marvin View

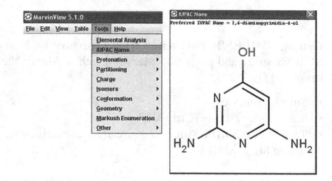

Get IUPAC Name from structure in MarvinView

- When a structure is there in the MarvinView panel, select IUPAC Name option from Tool menu as shown in Fig. 1.16.

1.9   Introduction to Chemical Structure Formats

1.9.1   Linear Format

To facilitate the ease of chemical communication through electronic medium, linear formats were developed over a period of time. These notations are useful for compact storage; they are unique and can be interpreted rapidly by the chemically intelligent computer programs [40]. Alphanumeric string-based linear chemical structure encoding rules were developed by the pioneering contributions of Wiswesser, Morgan, Weininger, Dyson, etc. and eventually applied in machine description [41]. Contemporarily, a new system of representing the molecular structural information in the form of connection tables was established [42]. The invention of SMILES made a significant effect on the storage methodology in chemical information systems and it has led to the development of the modern form of representing chemical structures [43]. This line notation system has several advantages over the older systems in terms of its compactness, simplicity, uniqueness, and human readability. A detailed description of many advanced versions of SMILES such as USMILES, SMILES Arbitrary Target Specification (SMARTS), STRAPS, and CHUCKLES can be found on the website www.daylightsmiles.com. SMARTS is basically an extension of SMILES used for describing molecular patterns and properties as well as for substructure searching [44]. In the early days of chemical structure representation, Sybyl Line Notation (SLN) was used extensively in American Standard Code for Information Interchange (ASCII) format which is almost similar to SMILES, the difference being mainly in the representation of explicit hydrogen atoms [45]. It can be used for substructure searching, Markush representation, database storage, and network communication, but the drawback is that it does not support reactions. An International Chemical Identifier (InChI) notation is a string of characters capable of uniquely representing a chemical substance. It is derived from a structural representation of that substance in a way designed to be independent of the way that the structure is drawn so that a single compound will generate the same identifier [46]. It provides a precise, robust, IUPAC-approved tag for representing a chemical substance. InChI is the latest and most modern of the line notations. It resolves many of the chemical ambiguities not addressed by SMILES, particularly with respect to stereo centers, tautomers, and other valence model problem. In modern-day chemical structure-based inventory management, canonical SMILES format is the most preferred due its uniqueness and compactness.

Sample line notations for Aspirin Molecule •
[#6]OC(=O)Cl=CC=CC=ClC(O)=O (SMARTS)
InChIKey = FNJSWIPFHMKRAT-UHFFFAOYSA-N (InChI Key)
COC(=O)clccccclC(O)=O (SMILES)

1.9.2 Graph-based Representation (2D and 3D formats)

According to graph theory, a chemical structure is a undirectional, unweighted, and labeled graph with atoms as nodes and bonds as edges [47]. Molecular graphs can be augmented with rings and functional groups by inserting additional vertices with corresponding edges [48]. Matrix representation of graph was also used to denote chemical structure with n atoms as an array of $n \times n$ entries [49]. There are several types of matrix representation, such as adjacency matrix, distance matrix, atom connectivity matrix, incidence matrix, bond matrix, and bond electron matrix, each with its own set of merits and demerits [50].

1.9.2.1   Code for obtaining the distance between pairs of points in a matrix

```
public double[][] getDistanceWithConnectPoints(double[][] matrix) {
    double[][] val = new double[50000][3];
    int cnt = 0;
    for (int i = 0; i < matrix.length; i++) {
        for (int j = i + 1; j < matrix.length; j++) {
            Point3d p1 = new Point3d(matrix[i]);
            Point3d p2 = new Point3d(matrix[j]);
            val[cnt][0] = p1.distance(p2);
            val[cnt][1] = (double) i;
            val[cnt][2] = (double) j;
            //     System.out.println(cnt + "\t" + p1.distance(p2));
            cnt++;
        }//j
    }//i
    System.out.println(cnt);

    double[][] val1 = new double[cnt][3];
    for (int i = 0; i < cnt; i++) {
        for (int j = 0; j < 3; j++) {
            val1[i][j] = val[i][j];
        }
    }
    return val1;
}
```

1.9.3   Connection Tables

A connection table is a list of atoms and bonds in a molecule which tells us the indices of the atoms connected to the reference atom i [51]. The bond table indexes between atom i and atom j. It enumerates the atoms and the bonds connecting specific atoms. The table provides the 3D (x, y, z) coordinates and the information about the bonds connecting the atoms along with the type of bonds (1 = single; 2 = double, etc.). Despite the size and format constraints, the connection tables are easily handled by the computers. However, the drawback is a lack of human interpretability of the structural information. Owing to the constraints, the connection tables have been widely adopted by the storage media. The present day's most important Chemical Abstract Service structure databases like Registry [52] contain the molecular information in connection table format only (MDL Mol).

1.9.4   FILE FORMATS

Chemical information can be downloaded, uploaded, and viewed as files or streams in multiple file formats with varying documentation difference. File formats are usually distinguished on the basis of three criteria [53]:

1. File extensions: They usually end in three letters, for example, .mol, .sdf, .xyz.
2. Self-describing file: The details of the file format are present in the file itself, for example, CML.
3. Chemical/MIME: They are provided by the server, "chemically-aware."

1.9.4.1 MOLFILE

Molfile was created by MDL (now Symyx). The Accelrys –Symyx merger has given its ownership to Accelrys. Molfile includes information on atoms, atomic bonds, connectivity, and the coordinates of the molecule. There are two versions of this file: V2000 and V3000, the former being the most accepted version. Most chemoinformatics softwares like Marvin, ACD ChemSketch, even Mathematica [54] support this format.

The following are the contents of Molfile for the given structure of aspirin (acetylsalicylic acid) (Fig. 1.17).

1.9.4.2 SDF FILE

SDF created by MDL is a chemical data file format and displays information on chemical structure [55]. SDF is an extension (additional information) of MDL Molfile. The first portion of the SDF file is the same as the MDL Molfile, and the second half contains additional information related to some molecular property. Delimiter is a set of specific characters used to segregate multiple compounds (Fig. 1.18).

Code for reading an sdf file

```
public String[] ReadSDF(String fname) {
        System.out.println(fname);
        int cnt = 0;
        String t = "";
        int mcnt = 1;
        double[][] dmatx = new double[mcnt][36];
        try {
            BufferedReader    br    =    new    BufferedReader(new    FileReader(new
File(fname)));
                String s1 = "";
                int lcnt = 0;
                int acnt = 0;
                int bcnt = 0;
                int[] v1 = new int[2];
                double[][] lcoord = new double[1000][3];
                double[][] bcon = new double[1000][3];
                int ac = 0;
                int bc = 0;
                while ((s1 = br.readLine()) != null && cnt < mcnt) {
                    lcnt++;
                    t += s1 + "\n";
                    try {
                        if (lcnt == 4) {
                            String[] t1 = stringToArray(s1);

                            acnt = Integer.valueOf(t1[0].trim());
                    bcnt = Integer.valueOf(t1[1].trim());
                        }
                        v1[0] = 4 + acnt;
                        if (lcnt > 4 && lcnt < v1[0]) {
                    String[] t2 = stringToArray(s1);
                    lcoord[ac][0] = Double.valueOf(t2[0]);
                            lcoord[ac][1] = Double.valueOf(t2[1]);
                            lcoord[ac][2] = Double.valueOf(t2[2]);
                            ac++;
```

```
            }
            v1[1] = v1[0] + bcnt;
            if (lcnt > v1[0] && lcnt < v1[1]) {
                String[] t3 = stringToArray(s1);
                bcon[bc][0] = Double.valueOf(t3[0]);
                bcon[bc][1] = Double.valueOf(t3[1]);
                bcon[bc][2] = Double.valueOf(t3[1]);
                bc++;
            }
            if (s1.contains("$$$$")) {
                double[] maxv = getMaxValue3(lcoord);
                double[] minv = getMinValue3(lcoord);
                double[][] gbx = BuildGridBox(minv, maxv);
                dmatx[cnt] = getDistance(gbx);
                cnt++;
                t = "";
                lcnt = 0;
                ac = 0;
                bc = 0;
            }
        } catch (Exception e) {
            t = "";
            lcnt = 0;
            ac = 0;
            bc = 0;
        }
    }
    br.close();
} catch (Exception e) {
    System.out.println(e);
}

for (int i = 0; i < dmatx.length; i++) {
    for (int j = 0; j < dmatx[0].length; j++) {
        System.out.print(df.format(dmatx[i][j]) + " ");
    }
    System.out.println();
}

String[] out = t.split("$$$$");
return out;
}
```

1.9.4.3   XYZ File

XYZ is a chemical file format that describes the geometry of the molecule [56]. This format is utilized in importing and exporting coordinates for chemical structures computationally. The units used in XYZ format are usually "angstroms."

File name extension: .XYZ The following are contents of the XYZ file for the given structure (acetylsalicylic acid) (Fig. 1.19).

1.9.4.4   PDB File Format

A PDB file is a topology file which describes the geometry of a protein or chemical structure [57]. It gives the coordinates for every atom or residue in the structure. Almost all the letters, numbers, and special characters are allowed in this format. There are certain mandatory fields based on the structure.

Mandatory fields in PDB format: HEADER, TITLE, COMPND, SOURCE, KEYWDS, EXPDTA, AUTHOR, REVDAT, REMARK 2, REMARK 3, SEQRES, CRYST1, ORIGX1 ORIGX2 ORIGX3, SCALE1 SCALE2 SCALE3, MASTER, END.

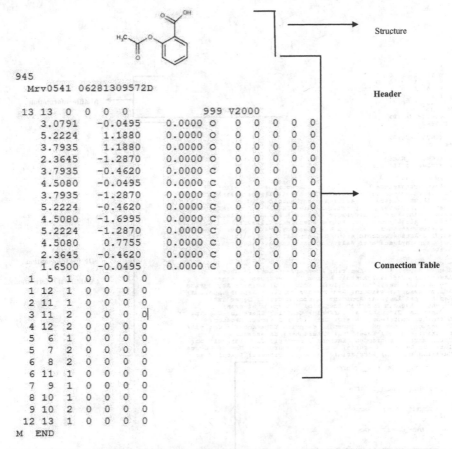

Lines	Section	Description
1-3	Header	
1	Molecule ID/ Molecule name	945-Drugbank ID (file was downloaded from drugbank)
2	File information	Mrv0541- Marvin ID(marvin view was used for visualisation) 06281309572D- 28-June-2013 09:57(time) 2-dimensional
3		Blank space
4-31	Connection Table	
4		Couns line: 13 atoms, 13 bonds V2000- version
5-17		Atom block (1 line for each atom): x, y, z coordinates
18-30		Bond block (1 line for each bond): 1st atom, 2nd atom, type,
31		M End-Properties block (empty)

Fig. 1.17 Depiction of a Molfile format

The following is a typical PDB text file (protein Lyase) (Fig. 1.20).

HEADER The HEADER record uniquely identifies a PDB entry through the idCode field.

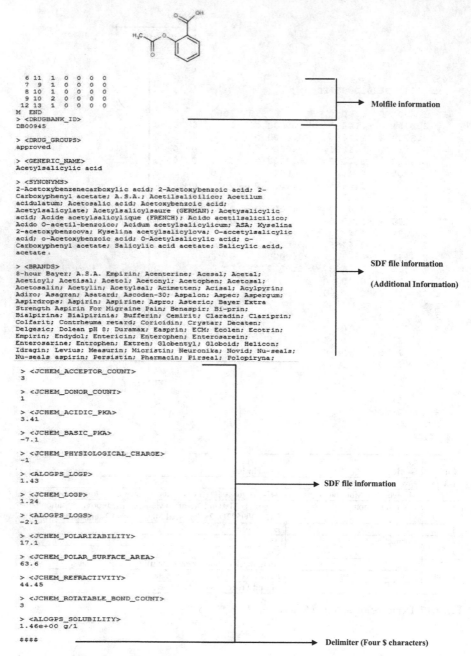

```
    6 11  1  0  0  0  0
    7  9  1  0  0  0  0
    8 10  1  0  0  0  0
    9 10  2  0  0  0  0
   12 13  1  0  0  0  0
M  END
> <DRUGBANK_ID>
DB00945

> <DRUG_GROUPS>
approved

> <GENERIC_NAME>
Acetylsalicylic acid

> <SYNONYMS>
2-Acetoxybenzenecarboxylic acid; 2-Acetoxybenzoic acid; 2-
Carboxyphenyl acetate; A.S.A.; Acetilsalicilico; Acetilum
acidulatum; Acetosalic acid; Acetoxybenzoic acid;
Acetylsalicylate; Acetylsalicylsaure (GERMAN); Acetysalicylic
acid; Acide acetylsalicylique (FRENCH); Acido acetilsalicilico;
Acido O-acetil-benzoico; Acidum acetylsalicylicum; ASA; Kyselina
2-acetoxybenzoova; Kyselina acetylsalicylova; O-accetylsalicylic
acid; o-Acetoxybenzoic acid; O-Acetylsalicylic acid; o-
Carboxyphenyl acetate; Salicylic acid acetate; Salicylic acid,
acetate .

> <BRANDS>
8-hour Bayer; A.S.A. Empirin; Acenterine; Acesal; Acetal;
Aceticyl; Acetisal; Acetol; Acetonyl; Acetophen; Acetosal;
Acetosalin; Acetylin; Acetylsal; Acimetten; Acisal; Acylpyrin;
Adiro; Asagran; Asatard; Ascoden-30; Aspalon; Aspec; Aspergum;
Aspirdrops; Aspirin; Aspirine; Aspro; Asteric; Bayer Extra
Strength Aspirin For Migraine Pain; Benaspir; Bi-prin;
Bialpirina; Bialpirinia; Bufferin; Cemirit; Claradin; Clariprin;
Colfarit; Contrheuma retard; Coricidin; Crystar; Decaten;
Delgesic; Dolean pH 8; Duramax; Easprin; ECM; Ecolen; Ecotrin;
Empirin; Endydol; Entericin; Enterophen; Enterosarein;
Enterosarine; Entrophen; Extren; Globentyl; Globoid; Helicon;
Idragin; Levius; Measurin; Micristin; Neuronika; Novid; Nu-seals;
Nu-seals aspirin; Persistin; Pharmacin; Pirseal; Polopiryna;

 > <JCHEM_ACCEPTOR_COUNT>
 3

 > <JCHEM_DONOR_COUNT>
 1

 > <JCHEM_ACIDIC_PKA>
 3.41

 > <JCHEM_BASIC_PKA>
 -7.1

 > <JCHEM_PHYSIOLOGICAL_CHARGE>
 -1

 > <ALOGPS_LOGP>
 1.43

 > <JCHEM_LOGP>
 1.24

 > <ALOGPS_LOGS>
 -2.1

 > <JCHEM_POLARIZABILITY>
 17.1

 > <JCHEM_POLAR_SURFACE_AREA>
 63.6

 > <JCHEM_REFRACTIVITY>
 44.45

 > <JCHEM_ROTATABLE_BOND_COUNT>
 3

 > <ALOGPS_SOLUBILITY>
 1.46e+00 g/l

$$$$
```

Molfile information

SDF file information

(Additional Information)

SDF file information

Delimiter (Four $ characters)

Fig. 1.18 Depiction of an sdf file format

	Description
1	Number of atoms
2	Molecule name/molecule ID
3-15	Atomic coordinates

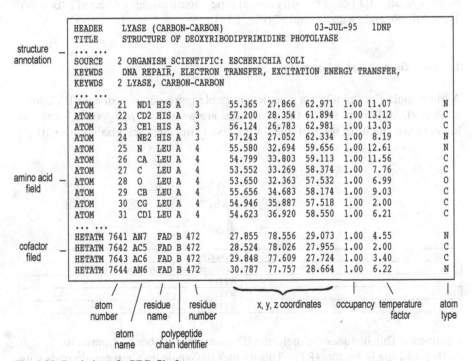

```
13
945
        O       5.74765      -0.09240      0.00000
        O       9.74848       2.21760      0.00000
        O       7.08120       2.21760      0.00000
        O       4.41373      -2.40240      0.00000
        C       7.08120      -0.86240      0.00000
        C       8.41493      -0.09240      0.00000
        C       7.08120      -2.40240      0.00000
        C       9.74848      -0.86240      0.00000
        C       8.41493      -3.17240      0.00000
        C       9.74848      -2.40240      0.00000
        C       8.41493       1.44760      0.00000
        C       4.41373      -0.86240      0.00000
        C       3.08000      -0.09240      0.00000
```

structure (Acetylsalicylic acid)   XYZ file

Fig. 1.19 XYZ format

```
          HEADER    LYASE (CARBON-CARBON)                    03-JUL-95   1DNP
          TITLE     STRUCTURE OF DEOXYRIBODIPYRIMIDINE PHOTOLYASE
structure ... ...
annotation SOURCE    2 ORGANISM_SCIENTIFIC: ESCHERICHIA COLI
          KEYWDS    DNA REPAIR, ELECTRON TRANSFER, EXCITATION ENERGY TRANSFER,
          KEYWDS    2 LYASE, CARBON-CARBON
          ... ...
          ATOM     21  ND1 HIS A   3      55.365  27.866  62.971  1.00 11.07      N
          ATOM     22  CD2 HIS A   3      57.200  28.354  61.894  1.00 13.12      C
          ATOM     23  CE1 HIS A   3      56.124  26.783  62.981  1.00 13.03      C
          ATOM     24  NE2 HIS A   3      57.243  27.052  62.334  1.00  8.19      N
          ATOM     25  N   LEU A   4      55.580  32.694  59.656  1.00 12.61      N
          ATOM     26  CA  LEU A   4      54.799  33.803  59.113  1.00 11.56      C
amino acid ATOM     27  C   LEU A   4      53.552  33.269  58.374  1.00  7.76      C
field     ATOM     28  O   LEU A   4      53.650  32.363  57.532  1.00  6.99      O
          ATOM     29  CB  LEU A   4      55.656  34.683  58.174  1.00  9.03      C
          ATOM     30  CG  LEU A   4      54.946  35.887  57.518  1.00  2.00      C
          ATOM     31  CD1 LEU A   4      54.623  36.920  58.550  1.00  6.21      C
          ... ...
          HETATM 7641 AN7  FAD B 472      27.855  78.556  29.073  1.00  4.55      N
cofactor  HETATM 7642 AC5  FAD B 472      28.524  78.026  27.955  1.00  2.00      C
filed     HETATM 7643 AC6  FAD B 472      29.848  77.609  27.724  1.00  3.40      C
          HETATM 7644 AN6  FAD B 472      30.787  77.757  28.664  1.00  6.22      N
```

atom number / residue name / residue number — x, y, z coordinates — occupancy temperature factor — atom type

atom name   polypeptide chain identifier

Fig. 1.20 Depiction of a PDB file format

HET HET records are used to describe nonstandard residues, such as prosthetic groups, inhibitors, solvent molecules, and ions, for which coordinates are supplied.

HETNAM This record gives the chemical name of the compound with the given hetID.

HETSYN This record provides synonyms, if any, for the compound in the corresponding (i.e., the same hetID) HETNAM record.

FORMUL The FORMUL record presents the chemical formula and charge of a nonstandard group.

The END record marks the end of the PDB file.

1.9.4.5　CML File Format

CML is an Extensible Markup Language (XML) format for chemical information [58]. CML reads multiple information elements from the structure file: molecule, atom, bond, name, formula, and the attribute: hydrogenCount, formalCharge, isotope, isotopeNumber, spinMultiplicity, radical (from Marvin), atomRefs4 (for atomParity), atomID (<atom>: id), elementType, atomRefs, atomic bond (<Bond>). The CML file ends with "</cml>" (Fig. 1.21).

1.9.4.6　Topos MOL2 Format

A tripos mol2 file (.mol2) is a complete portable representation of a SYBYL molecule [59]. Mol2 is an ASCII file. Mol2 files are written in "*free format.*" The following are contents of the SDF file for the given structure (acetylsalicylic acid).

Line	Description
1-5	Comments on the structure
6-19	RTI1
20-33	RTI2
34-35	RTI3

Comments: This includes the molecule ID/name, the number of atoms, etc.

Record Type Indicators (RTI): This divides the whole text into certain parts with relevant information about the structure (Fig. 1.22).

```xml
<?xml version="1.0"?>
<cml xmlns="http://www.xml-cml.org/schema" xmlns:convention="http://www.xml-
cml.org/convention" convention="convention:molecular"
xmlns:marvin="http://www.chemaxon.com/marvin/marvinDictRef" version="ChemAxon file
format v5.9.0, generated by v5.10.1">
<molecule id="m1">
  <atomArray>
    <atom id="a1" elementType="O" x2="11.302358174514687" y2="6.324999655485152"/>
    <atom id="a2" elementType="C" x2="9.968679052686651" y2="5.554999655485155"/>
    <atom id="a3" elementType="O" x2="8.634999930858617" y2="6.324999655485158"/>
    <atom id="a4" elementType="C" x2="9.968679052686650" y2="4.0149996554851555"/>
    <atom id="a5" elementType="C" x2="8.634999930858614" y2="3.244999655485155"/>
    <atom id="a6" elementType="C" x2="7.301320809030578" y2="4.0149996554851555"/>
    <atom id="a7" elementType="O" x2="5.967641687202542" y2="3.244999655485156"/>
    <atom id="a8" elementType="O" x2="7.301320809030580" y2="5.5549996554851555"/>
    <atom id="a9" elementType="C" x2="5.967641687202544" y2="6.324999655485156"/>
    <atom id="a10" elementType="C" x2="8.634999930858614" y2="1.704965774367074"/>
    <atom id="a11" elementType="C" x2="9.968669271906872" y2="0.9349488338080336"/>
    <atom id="a12" elementType="C" x2="11.302338612955129" y2="1.704965774367074"/>
    <atom id="a13" elementType="C" x2="11.302338612955129" y2="3.244999655485155"/>
  </atomArray>
  <bondArray>
    <bond atomRefs2="a1 a2" order="2"/>
    <bond atomRefs2="a2 a3" order="1"/>
    <bond atomRefs2="a2 a4" order="1"/>
    <bond atomRefs2="a5 a6" order="1"/>
    <bond atomRefs2="a6 a7" order="2"/>
    <bond atomRefs2="a6 a8" order="1"/>
    <bond atomRefs2="a8 a9" order="1"/>
    <bond atomRefs2="a5 a10" order="2"/>
    <bond atomRefs2="a4 a5" order="1"/>
    <bond atomRefs2="a4 a13" order="2"/>
    <bond atomRefs2="a10 a11" order="1"/>
    <bond atomRefs2="a11 a12" order="2"/>
    <bond atomRefs2="a12 a13" order="1"/>
  </bondArray>
</molecule>
</cml>
```

Fig. 1.21 Depiction of CML file format of Aspirin

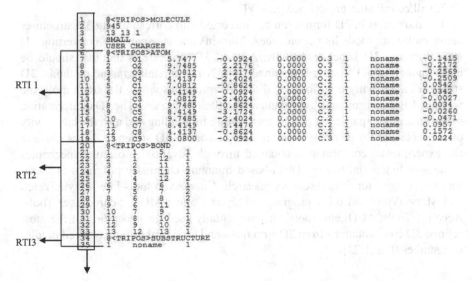

Fig. 1.22 Depiction of a mol2 format (lines are numbered for convenience and are not part of Mol2)

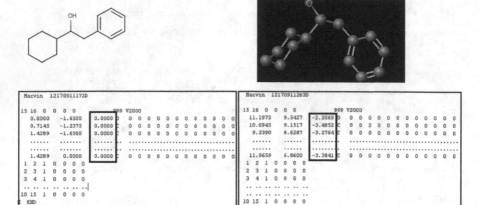

Fig. 1.23 2D and 3D conversion using MarvinView

1.10   2D and 3D Representation

Connections between the atoms specify the topology, but the relative spatial arrangement of atom in the configuration should also be defined. There are molecules with the same connectivity patterns but different spatial arrangement termed as stereoisomers which need to be distinguished. The spatial dimension of the building atoms defines the dimension of the molecule as:

0D all atoms are in [0, 0, 0]

2D z coordinates is 0, [x, y, 0]

3D all coordinates are defined [x, y, z]

The molecules in 2D format can be converted into corresponding 3D structures using molecular modelling approaches. MarvinView is capable of converting 2D structures into 3D structures rapidly. A good 3D structure is one that should be close enough to 3D structures obtained from X-ray crystallographic methods. 3D structures are usually used for drug discovery programs where the small molecule is docked against protein targets of interest in their active site. The 3D conformation of the structure in particular pose is responsible for binding and also the bioactivity of the molecule for that target. Generation of correct 3D structure that is close to the experimental conformation obtained through advanced molecular mechanics or density functional theory (DFT)-based quantum chemistry programs is therefore encouraged for drug discovery research. ChemAxon tools like MarvinSketch and MarvinView and other programs, such as Corina, MOE, Schrodinger Tools, Accelrys Tools, ACDLabs Tools, etc., are usually used for generation of the most refined 3D conformations from 2D structures and used further for advanced prediction studies (Fig. 1.23).

. In the 2D representation of a molecule, the values of the z coordinates of all the atoms are all set to "0," whereas in the case of a 3D structure, the z coordinates are generated based on the lowest energy conformation generated by the program. Molsoft has an interactive 2D to 3D molecule converter which can also be viewed using mobile apps [60].

1.10.1 Code for 3D Structure Generation in ChemAxon

```
// read input molecule
MolImporter mi = new MolImporter("test.mol");
Molecule mol = mi.read(); mi.close();
// create plugin
ConformerPlugin plugin = new ConformerPlugin();
// set target molecule
plugin.setInputMolecule(mol);
// set parameters and run calculation
plugin.setMaxNumberOfConformers(400);
plugin.setTimelimit(900);
plugin.run();
// get and process results
Molecule[] conformers = plugin.getConformers();
for (int i = 0; i < plugin.getConformerCount(); ++i) {
Molecule m = conformers[i];
// do something with the conformer ...
}

cxcalc conformers -m 250 -s true test.sdf

molconvert sdf -3:"S{fine}E" 0D.smi > 3D.sdf
```

The Corina program can generate 3D coordinates for 2D structures rapidly [61] With the help of a 3D structure, it is possible to calculate energy of the molecule, volume, interatomic charge distribution, and other 3D descriptors required for quantitative structure–activity relationship (QSAR)-based predictive studies (Fig. 1.24).

1.10.2 A Practice Tutorial

Interconversion of 2D to 3D optimization techniques
   Using MarvinView:

- Create and open the molecule in MarvinView as discussed previously
- Then go to edit → clean → 3D → clean in 3D
- The output will be the 3D structure of the molecule as shown in Fig. 1.25

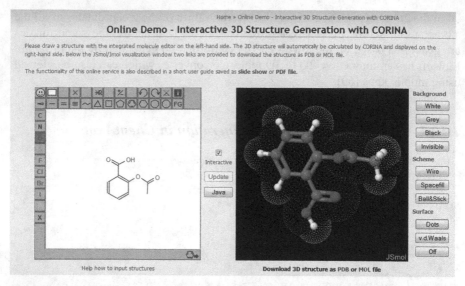

Fig. 1.24 Interactive 3D structure generation with CORINA (molecular networks)

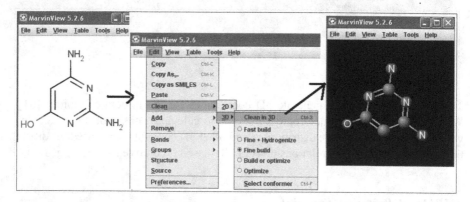

Fig. 1.25 2D to 3D structure conversion in MarvinView

Using ChemDraw:

- Create and open the molecule in ChemDraw as described above
- Then use edit → get 3D model
- The output will be the 3D structure of the molecule as shown in Fig. 1.26

1.11   Abstract Representation of Molecules

Sometimes, molecules are represented as Markush structures in a generic context to cover a family of molecular structures which can go beyond millions [62]. Markush structures are generic structures used in patent databases such as MARPAT main-

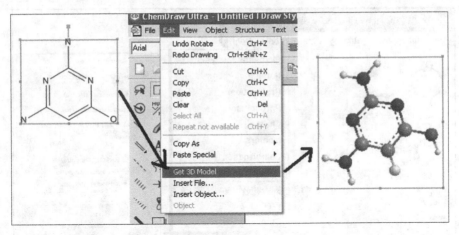

Fig. 1.26 2D to 3D conversion in ChemDrawUltra

Fig. 1.27 A Markush structure showing R groups

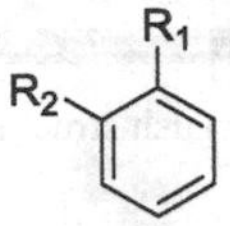

$$R_1 = OH, \ NH2....$$

$$R_2 = CH_3, \ CH_3HC_2, \ ....$$

tained by CAS for protecting intellectual chemical information in patents. An R group is a collection of possible substituent fragments that can be part of a molecule at a specific location. The complexity of such chemical structure representations cannot be captured by one single molecule object (Fig. 1.27).

Markush structures are used in patents, combinatorial library generation, depiction of polymers, etc.

ChemAxon provides plugins for generating Markush structures from a given library of molecules (Fig. 1.28).

The Markush viewer is another module to view R group definitions of a molecule in a hierarchical graphical form. It classifies scaffolds and R groups in a given molecule file. The markush structure of aspirin molecules is shown here with R1 and R2 group definitions (Fig. 1.29).

Markush structures can be enumerated efficiently using command line options. The command line syntax is >cxcalc randommarkushenumerations -f sdf -C 2:t5000 filename.mol and the output can also be piped to MarvinView. In the current version of the program, Instant JChem can be used to determine the Markush space density of a patent molecule.

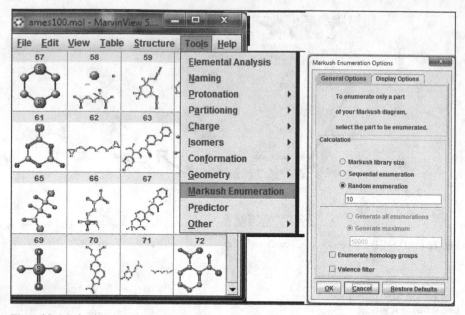

Fig. 1.28 Markush structures generation using ChemAxon

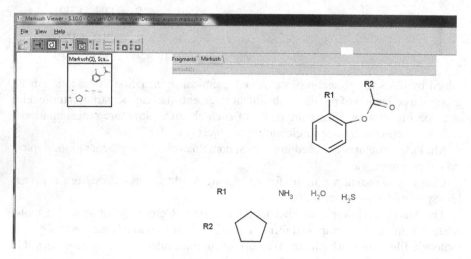

Fig. 1.29 Markush Viewer program in ChemAxon

1.12   File Format Exchange

A tool for the interoperability of native format for particular software to standard file formats is essential for reusability in chemoinformatics programs for property prediction, docking, QSAR model building, etc. Software programs like OpenBabel can interconvert molecules over 50 standard file formats required by several computational chemistry- and chemoinformatics-oriented programs [63]. MolConverter is a command line program in Marvin Beans and JChem that converts between various file types [64].

molconvert [options] outformat[:exportoptions] [files…]

The outformat argument must be one of the following strings:

mrv	(document formats)
mol, rgf, sdf, rdf, csmol, csrgf, cssdf, csrdf,	(molecule file formats)
cml, smiles, cxsmiles, abbrevgroup, peptide,	
sybyl, mol2, pdb, xyz, inchi, name, cdx, cdxml, skc	
jpeg, msbmp, png, pov, ppm, svg, emf	(graphics formats)
gzip, base64	(compression and encoding)

molconvert [options] query-encoding [files…]

to query the automatically detected encodings of the specified molecule files. From files having doc, docx, ppt, pptx, xls, xls, odt, pdf, xml, html or txt format, molconvert is able to recognize the name of compounds and convert it to any of the above-mentioned output formats. Some common commands for molconvert are given below:

1. molconvert smiles caffeine.mol (printing the SMILES string of a molecule in a molfile)
2. molconvert smiles:-*a* -s "clccccl" (dearomatizing an aromatic molecule)
3. molconvert smiles:*a* -s "Cl=CC=CC=Cl" (aromatizing a molecule)
4. molconvert smiles:*a_bas* -s "CN1C=NC2=ClC(=O)N(C)C(=O)N2C" (aromatizing a molecule using the basic algorithm)
5. molconvert mol caffeine.smiles -o caffeine.mol (converting a SMILES file to MDL Molfile)
6. molconvert sdf *.mol -o molecules.sdf (making an SDF from molfiles)
7. molconvert query-encoding *.sdf (printing the encodings of SDfiles in the working directory)
8. molconvert -2:2e mol caffeine.smiles -o caffeine.mol (SMILES to Molfile with optimized 2D coordinate calculation, converting double bonds with unspecified cis/trans to "either")
9. 2D coordinate calculation with optimization and fixed atom coordinates for atoms 1, 5, 6:
10. molconvert -2:2:F1,5,6 mol caffeine.mol (import a file as XYZ; do not try to recognize the file format: molconvert smiles "foo.xyz{xyz:}")

11. molconvert smiles "foo.xyz{fl.4C4}" (import a file as XYZ, with bond length cut-off=1.4, and maximum number of carbon connections=4, export to SMILES)
12. molconvert smiles "foo.xyz.gz{gzip:xyz:fl.4C4}" (import a file as Gzipped XYZ)
13. molconvert smiles -c "ID<=1000&logP>=-2&logP<=4" -T ID:logP foo.sdf (import an SDF and export a table containing selected molecules with columns: SMILES, ID, and logP)
14. molconvert mrv in.mrv -R2:1 rdef.mrv (fuse R2 definition from file; filter fragments with 1 attachment point)
15. molconvert mrv in.mrv -R frags.mrv (fuse fragments from file; note, that the input molecule, which the fragments are fused to, should also be specified)
16. molconvert "name:common, all" -s tylenol (generate all common names for a structure)
17. molconvert "name:common, all" -s tylenol (generate the most popular common name for a structure)
18. molconvert smiles foo.html (generate SMILES from those molecules whose names are mentioned in a file foo.html)

1.12.1 A Practice Tutorial

This tutorial deals with interconversion between various file formats using command prompt in ChemAxon tool molconvert and OpenBabel file conversion programs (Fig. 1.30).

In ChemAxon: Create a test file (testsmiles1.smi) containing SMILES (using text editor or MarvinSketch)
ClCCCCCl cyclohexane
ClCCCCCl benzene
Cl(Cl)C(C)CCCCl 1-chloro-2methylbenzene
Use molconvert to generate 2D coordinates for the SMILES.

```
C:\Program Files\ChemAxon\JChem\bin>molconvert
Molecule File Converter, version 5.2.6, (C) 1999-2008 ChemAxon Ltd
Usage: molconvert [options] outformat[:export-opts] [files...]
```

Molconvert is a utility for molecule file conversion from ChemAxon Ltd; it provides several other options which are listed once you type "*molconvert*" in command prompt.

```
C:\PROGRA~1\ChemAxon\JChem\bin>type testsmiles1.smi
C1CCCCC1           cyclohexane
c1ccccc1           benzene
c1(Cl)c(C)cccc1 1-chloro-2methylbenzene
```

Verify the smiles using Marvinview

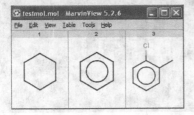

Fig. 1.30   Validation of SMILES in MarvinView

Usage: molconvert [options] outformat[:export-opts] [files…]

SMILES to Molfile   Syntax- **molconvert −2:e mol foo.smiles -o foo.mol**

So to convert our testsmiles1.smi into MOLfile, we need to type in the following in the command prompt.

molconvert −2:e mol testsmiles1.smi -o testmol.mol

Input SMILES: ClCCCCCl

Output MOL format:

```
C:\Program Files\ChemAxon\JChem\bin>type testmol.mol
cyclohexane
  Marvin  01151012172D

  6  6  0  0  0  0            99 V2000
    0.7145    1.2375    0.0000 C   0  0  0  0  0  0  0  0  0  0  0  0
    1.4289    0.8250    0.0000 C   0  0  0  0  0  0  0  0  0  0  0  0
    1.4289   -0.0000    0.0000 C   0  0  0  0  0  0  0  0  0  0  0  0
    0.7145   -0.4125    0.0000 C   0  0  0  0  0  0  0  0  0  0  0  0
    0.0000    0.0000    0.0000 C   0  0  0  0  0  0  0  0  0  0  0  0
    0.0000    0.8250    0.0000 C   0  0  0  0  0  0  0  0  0  0  0  0
  1  2  1  0  0  0  0
  2  3  1  0  0  0  0
  3  4  1  0  0  0  0
  4  5  1  0  0  0  0
  5  6  1  0  0  0  0
  1  6  1  0  0  0  0
M  END
```

The other two SMILES are also converted into MOLfile, which is not displayed here.

We can also create the 3D MOL file as shown in the following figure. We need to type the following to get the desired result.

molconvert −3:e mol testsmiles1.smi -o testmol.mol

```
C:\Program Files\ChemAxon\JChem\bin>molconvert -3:e mol testsmiles1.smi -o testm
ol1.mol
C:\Program Files\ChemAxon\JChem\bin>type testmol1.mol
cyclohexane
  Marvin  01151012313D            10.95000

  6  6  0  0  0  0                999 V2000
    0.0417   -0.0140    0.1063 C   0  0  0  0  0  0  0  0  0  0  0  0
    1.5764   -0.2397    0.1464 C   0  0  0  0  0  0  0  0  0  0  0  0
    2.3317    0.7716   -0.7562 C   0  0  0  0  0  0  0  0  0  0  0  0
    1.9557    2.2387   -0.4183 C   0  0  0  0  0  0  0  0  0  0  0  0
    0.4209    2.4644   -0.4585 C   0  0  0  0  0  0  0  0  0  0  0  0
   -0.3344    1.4531    0.4441 C   0  0  0  0  0  0  0  0  0  0  0  0
  1  2  1  0  0  0  0
  2  3  1  0  0  0  0
  3  4  1  0  0  0  0
  4  5  1  0  0  0  0
  5  6  1  0  0  0  0
  1  6  1  0  0  0  0
M  END
```

Create Image from molecules

molconvert "jpeg:w100,Q95,#ffffff" testmol1.mol -o nice.jpg

The above code creates a 100×100 Joint Photographic Expert Group (JPEG) image on a yellow background, with 95 % quality.

Open Babel is a chemical toolbox designed to speak the many languages of chemical data. It is an open, collaborative project allowing anyone to search, convert, analyze, or store data from molecular modelling, chemistry, solid-state materials, biochemistry, or related areas. It has ready-to-use programs and provides a complete programmer's toolkit. It can read, write, and covert over 110 chemical file formats, besides filtering and searching molecular files using SMARTS and other methods.

1.12.2   Code for Reading a Molecule, checking the Number of Atoms, and Writing a SMILES String

```
#include <iostream.h>

    // Include Open Babel classes for OBMol and OBConversion
    #include <openbabel/mol.h>
    #include <openbabel/obconversion.h>

    int main(int argc,char **argv)
    {
        // Read from STDIN (cin) and Write to STDOUT (cout)
        OBConversion conv(&cin,&cout);

        // Try to set input format to MDL SD file
        // and output to SMILES
        if(conv.SetInAndOutFormats("SDF","SMI"))
        {
            OBMol mol;
            if(conv.Read(&mol))
            {
                // ...manipulate molecule
                cerr << " Molecule has: " << mol.NumAtoms()
                    << " atoms." << endl;
            }

            // Write SMILES to the standard output
            conv->Write(&mol);
        }
        return 0; // exit with success
    }
```

All of the main classes, including OBMol and OBConversion, include example code designed to facilitate using the Open Babel code in real-world chemistry (Fig. 1.31).

1.12.3 Code for Reading a SMILES String in Python

```
import openbabel as ob
# Initialize the OBConversion object
conv = ob.OBConversion()
if not conv.SetInFormat('smi'):
  print 'could not find smiles format'
# Read the smiles string
mol = ob.OBMol()
if not conv.ReadString(mol, 'CCCC'):
  print 'could not read the smiles string'
# ... Use OBMol object ...
```

After understanding and practicing the practical approaches and techniques described in the above sections, the reader should be able to draw molecules on a computer and get the SMILES for them. One should also be able to view molecules in 3D for a better understanding of the molecules. In the next section, we describe some advanced techniques which allow us to draw molecular structures on a computer and store them in reusable formats for various chemoinformatics applications.

1.13 Similarity and Fingerprint Analysis

Molecule A	1000100011100010100001000000100101	a=11
Molecule B	0001100001000010010001000000100101	b=9
Similarity (A and B)	0000100001000010000001000000100101	c=7

It is a well-established fact that common sub-structural fragments often tend to share similar biological activity. Molecular similarity deals with finding molecules which have a comparable amount of structural similarity [65]. This is used to find structures that are similar to a molecule with less information. Molecular similarity is very handy in drug designing, because it reduces the amount of animal testing, as the recorded data can be extrapolated. In this chapter, we learn the basic concepts of molecular fingerprints, similarity measures, and the use of molecular fingerprints in similarity search.

Searching a molecule in a database involves matching it against all the molecules present in the database. It requires lots of time and highly expensive computational

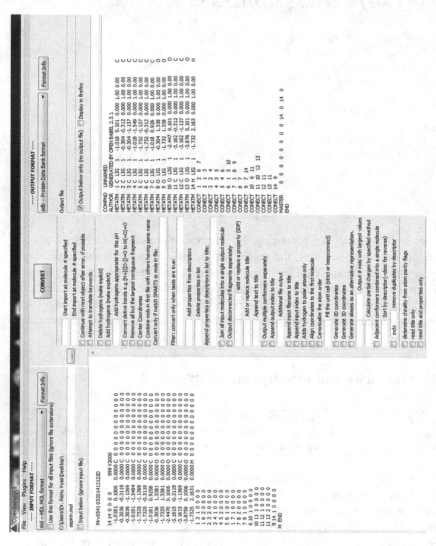

Fig. 1.31 Conversion of a molfile of curcumin molecule to pdb file format in OpenBabel GUI

Fig. 1.32 Molecule and its
constituent fragments

-COOH

-NH₂

-OH

-phenyl

-path of connected atoms

- rings

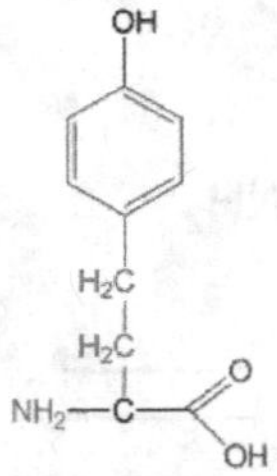

- atom and its nearest neighbors along with the bonds between them

facilities for its completion, making it impractical. In order to search a database or
to find similar structures, the molecule (graph) is fragmented into various logical
fragments (subgraphs), such as functional groups, rings, etc. From Fig. 1.32, we can
create several subgraphs of fragments.

Consider the case of a text search where we combine several keywords to form
a specific query to meet our requirements, and so is the case here; each fragment is
like a keyword, which can be combined to perform a specific structure search. When
we use a particular fragment as our query or as a part of our query, the retrieved
structure must contain that fragment. The list of retrieved structures will include all
those structures that contain the fragment in the specified manner in their structure.

1.13.1 Simple Fingerprints (Structural Keys)

Structural key is basically a string of values that describes the chemical composition and/or structural motifs that are present in the chosen substructure and each
molecule in the database [66]. A structural key is usually represented as a *boolean
array,* an array in which each element is TRUE or FALSE. A given bit is set to 1
(**True**), if a particular structural feature is present and a given bit is set to 0 (**False**),
if it is not as shown in the following figure. A *structural key* is a bitmap in which
each bit represents the presence (TRUE) or absence (FALSE) of a specific structural
feature (pattern). The I-th bit of this array, for example, can be used to represent any
structural feature of the molecule. This list can include:

- Any number of occurrences of a particular element or a particular atom type
- Presence of a particular functional group
- Presence of other structural elements, etc.

One important point to emphasize in the use of a structural key is that the definition
of a particular array element must be chosen initially. This has the disadvantage
that this key can become extremely long and is inflexible. Conversely, it is possible
to optimize this structural key for the class of compounds present in the database
(Fig. 1.33).

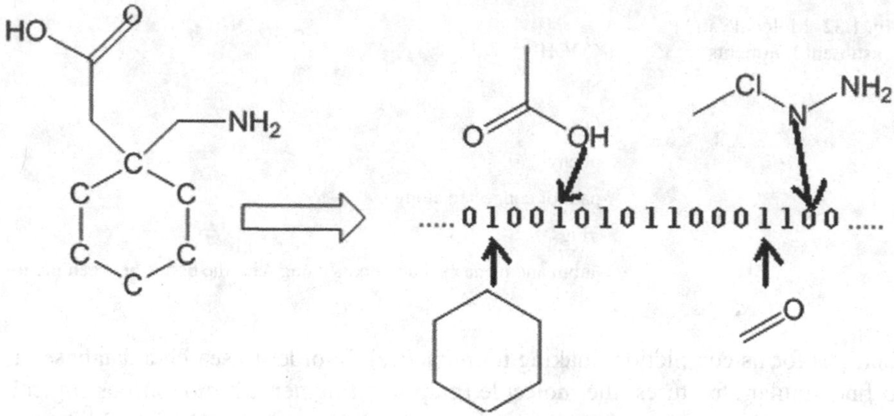

Fig. 1.33 Fingerprint generation from a molecule

1.13.2   Hashed Fingerprints

Molecular "fingerprints" are composed of bits of molecular information (fragments), such as types of rings, functional groups, and other types of molecular and atomic data. Comparing fingerprints will allow one to determine the similarity between two molecules, search databases, etc., but does not include full structural data (i.e., coordinates). A "fingerprint" is made up of a set of *descriptors* for a molecule. Each descriptor describes (usually the presence or absence of) a particular 2D structural feature in the concerned molecule. Most fingerprints are binary strings made up of zeros and ones. Each 0 or 1 can be represented as a single bit in the computer (a "bitstring"). The 0s represent the absence of the fragment in the molecule and the 1s represent the presence of the fragment. Fingerprints are generally 150–2,500 bits long. The fingerprint characterizes the molecule, but does not uniquely describe it. It is useful in many applications we will come to later, e.g., similarity, clustering, diversity.

For example, the fingerprint of methane (CH_4) is
..........0000000000100000000000.......

The patterns for a molecule's fingerprint are generated from the molecule itself. When we create fingerprints for a molecule, the fingerprinting algorithm generates the following after scrutinizing the concerned molecule:

- It creates a pattern for each and every molecule
- Each atom and its nearest neighbors, along with the bonds between them, are represented using specific patterns
- Each group of atoms and bonds connected by paths up to 2 bonds long are represented using a pattern
- Patterns are created for representing atoms and bonds connected by paths up to 3 bonds long
- ...continuing, with paths up to 4, 5, 6, and 7 bonds long.

For example, the molecule in the figure would generate the following patterns:

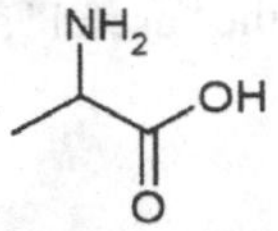

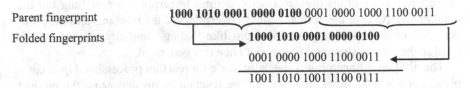

0-bond paths (i.e. atoms): C, N, O

1-bond paths: C-C, C-N, C=O, C-O

2-bond paths: C-C-C, C-C-N, C-C=O, C-C-O, O-C=O

3-bond paths: C-C-C-O, C-C-C=O, N-C-C-O, N-C-C=O

Etc.

Fig. 1.34 All possible fragments in a compound (all sequences of atoms from 2–7 atoms, augmented atoms, atom pairs)

Parent fingerprint   1000 1010 0001 0000 0100 0001 0000 1000 1100 0011

Folded fingerprints   1000 1010 0001 0000 0100

0001 0000 1000 1100 0011

1001 1010 1001 1100 0111

Fig. 1.35 Depiction of hashed fingerprints

For example, the molecule in Fig. 1.34 would generate the following patterns:

The number of fragments represented can be huge (100,000 for just the 2–7-length sequences for C, N, S, O, P, not considering bond types or generalizations). These are hashed onto a fixed number of bits (e.g., 1,024). Bits and fragment are not directly related and unlike structural keys, no predefined dictionary is required.

The amount of information conveyed by fingerprint is directly proportional to its information density; information density indicates the ratio of the "on" bits, i.e., 1s to the total number of bits, i.e., all 1s and 0s. Fingerprints have a fixed size; this makes the representation of information of large molecules a difficulty, because if the fingerprint length is small, there will be maximum "on" bits, whereas if the fingerprints are large, they will contain mostly "off" bits and waste space. To avoid these problems, the concept of *folding* fingerprints/hashed fingerprints [67] was proposed, where the long fingerprints are folded to make them compact. The fingerprints are folded when the size of the fingerprint becomes quite large. Then, the fingerprint is folded into two equal parts as shown above and then they are combined using a logical OR operator. We can repeatedly fold the fingerprint until the desired information density (called the *minimum density*) is reached or exceeded (Fig. 1.35).

Advantages of hashed fingerprints:

- Hashed fingerprints do not need a preexisting dictionary or library—every fragment/group present will be encoded in the fingerprint
- Novel substructures are not missed
- Easily calculated—their calculation does not require a substructure matching step

Disadvantages of hashed fingerprints:

- Mapping every substructure present has the potential to swamp the "useful" substructures
- In reality, mapping of fragments overlaps and so
  - Some information may be lost
  - Interpretation of the fingerprint is not straightforward
- It is impossible to recover the structure from the fingerprint
  - Also, multiple counts of the same path are not accounted for

Fingerprints are also used for reaction processing. Daylight provides two distinct types of fingerprints for this purpose, namely "normal" structural fingerprints and "difference" fingerprints [69]. Normal structural fingerprints are nothing but the combination (OR) of the normal hashed fingerprints of the reactants and the products. All the normal fingerprint operations like folding, similarity, etc. can be applied to the normal structural fingerprint once it is generated.

The difference fingerprint is specially made for reaction processing. Upon completion of a stoichiometric reaction, all the reactant atoms appear on the product side but the bonds between the atoms change during this process. The changes in bonds can be detected by a change in the fingerprint of the reactant molecules and the product molecules. Similar to the "normal structural fingerprints," once the different fingerprints are created, all the fingerprint operations are applicable on it.

1.13.3 A Practice Tutorial

Creating molecular fingerprint using ChemAxon tools

- Using command prompt, enter the bin directory of JChem.
- Type "generfp—h," this will display the options available as described below.

Usage: generfp [options]<inputfile>outputfile

```
Options:
 -h            display this help and exit
 -fl <length>  fingerprintlength length in bytes (default: 64)
 ..........
  -f<format>    format of the output
   -fb          binary
   -fl          ones and zeros (001011011...) (default)
 ..........
   -stat        generate statistics
 -s <separator> separator between numbers in case of text output
    Separators:    'n'o separator
                'c'omma (default),  't'ab, 's'pace
```

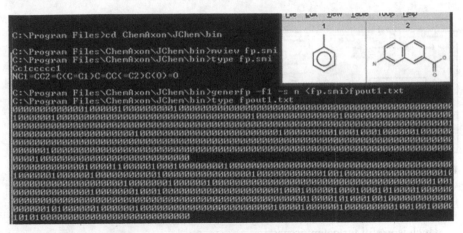

Fig. 1.36 Molecular fingerprints generation in JChem

- To generate the fingerprint of the molecules stored in some file (e.g., fp.smi), type the following:

"generfp -f1 -s n<fp.smi>fpout.txt"

These fingerprints can be used to calculate similarity measures using various formulas as described in the previous sections (Fig. 1.36).

1.14 Molecular Similarity

At present, a large number of chemical databases are available that provide molecular structure. These databases are very important in modern chemical research, most importantly in drug discovery studies. The aim of using computational tools in drug discovery is to find compounds that possess drug-like properties as early as possible so that further studies, synthetic and biological, can be carried out. Similarity search methods and other computational methods have proved to be very useful in this respect [68]. A query can be formed and the required database can be searched for the target structure. It is a proven fact that structurally similar molecules are expected to exhibit similar properties or biological activities; other than that, there are several other reasons for using similarity methods which include:

- Formulation of a query requires very little information; initially it is immaterial which part of the query molecule confers activity.
- Searching large databases can be easily performed because many implementations of similarity methods are computationally inexpensive.
- These methods help us find a particular molecule rank a set of molecules in the database based on our requirements.

Query structure　　　　　　　　　　　　　　　　　　　**Retrieved Hits**

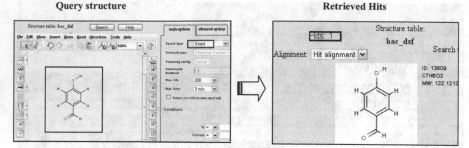

Fig. 1.37 An exact structure search in JChem

- These methods also help us to know whether a new structure is unique or not, which is useful for patent issues.
- These methods make screening and clustering of a database easier.

Similarity is very subjective, as it depends on what are we looking for and from what point of view we are looking. For example, from a mathematical point of view, we would denote two molecules as similar if they have common features in three dimensions, whereas if we take a chemical approach, we would denote two molecules as similar if they had similar physical properties. Similarity-based methods have gained popularity due to the rapid technological progress and increased number of entries in chemical databases. This has made the application of computational search methods a necessity.

Similarity measures are generally based on the presence and/or absence of features in two molecules. Similarity can be measured by numerical or distance measures. The former involves the expression of similarity by a numerical value in the range of 0–1, while the latter involves the expression of similarity in numerical value not less than 0. These measures are discussed in detail later in the chapter. In the next sections of this chapter, you will come to know about the various similarity-searching techniques and similarity matrices.

1.14.1　Exact Structure Search

Exact structure search involves the searching of exactly the same structure in a database [69]. The retrieved structure is exactly similar to the query molecule. In a database with unique structures, exact structure returns either one (exactly same molecule) or it does not return any hits indicating the absence of such structures in the database. Figure 1.37 shows an exact structure search using ChemAxon application JChem [70].

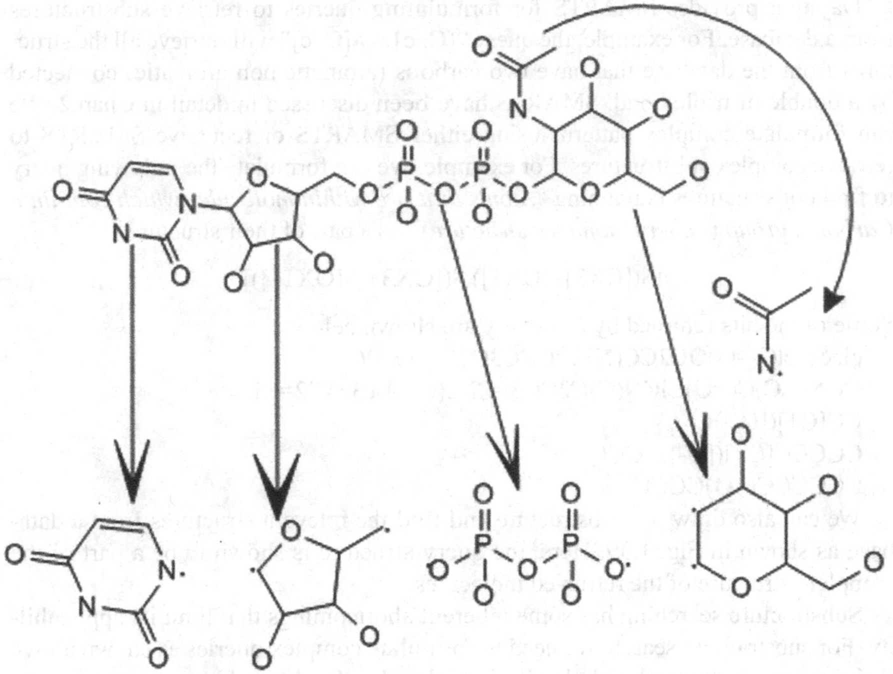

Fig. 1.38 Substructures of a big molecule

1.14.2 Substructure Search

Substructure searching is a method of retrieving chemical structure from a database based on the input query [71]. This approach retrieves all the structures from the database that contains the query structure as a part of their structure. The substructure is normally a functional group or core structure representing a class of molecules. This approach is very helpful when we want target structures that contain fragments or a functional group of interest. Indexing of chemical fragments decreases the search time drastically. Substructure indexing is a precomputing process in which the stored contents are indexed according to some specific criteria so that the answer for the expected question in a shorter duration of time can be obtained. For example, a famous search engine loads the screen with the search results within a minute for the given search term "formaldehyde." This is so because the documents are already indexed by the provider (Fig. 1.38).

But, the same engine may give the search results for the same query within 1 or 2 years in the absence of indexing. In chemoinformatics, we use the index of substructures instead of indexing words; they decompose the molecule into smaller bits and index them appropriately as shown in the figure alongside. Substructure search can also be performed in 3D. 3D substructure search allows the user to find atoms in correct spatial orientation relative to each other.

Daylight provides SMARTS for formulating queries to retrieve substructures from a database. For example, the query "[C, c] = ,#[C, c]" will retrieve all the structures from the database that have two carbons (aromatic/non-aromatic) connected by a double or triple bond. SMARTS have been discussed in detail in Chap.2. We can formulate complex patterns using either SMARTS or recursive SMARTS to retrieve complex substructures. For example, we can formulate the following query to find out structures containing "*Atoms that are within molecules which contain a Carbonyl group (either resonance structure)*" as a part of their structure.

$$[\$([CX3]=[OX1]),\$([CX3+]-[OX1-])]$$

Some of the hits returned by this query are shown below:

clccccclC(=O)OC2CC(N3C)CCC3C2C(=O)OC
CCN(CC)C(=O)ClCN(C)C2CC3=CNc(ccc4)c3c4C2=Cl
CC[C+]([O-])C
CCCCC[C+]([O-])CCCC
CCCCCC(=O)CCCC

We can also draw the substructure and find the relevant structures from a database as shown in Fig. 1.39. Here, the query structure is shown to be a part of the complete structure of the retrieved molecules.

Substructure searching has some inherent shortcomings that limit its applicability. For substructure search, we need to formulate complex queries as shown above and the results obtained include all the molecules that have the query structure as a part of their complete structure. Sometimes, huge numbers of hits are obtained which reduce the efficiency of the search, whereas a highly specific query does the opposite, that is, it retrieves very less number of hits, again decreasing the efficiency. Basically, the substructure search divides the database into two parts: one that contains the substructure query and the other that does not contain it. For example, if you want to search molecules that have similar properties to your query structure based on the presence of functional groups, you will retrieve a list of molecules that contain the specified functional groups, but there is no way to find out which molecule among the retrieved list is likely to have the closest resemblance with the query. In other terms, there is no mechanism to rank the retrieved hits in terms of similarity (Fig. 1.40).

Another problem associated with substructure search is that it will not enlist the structures (structure 2) with minor differences (presence of a single bond in place of a double bond) even if they are highly similar to the query structure and are expected to have properties similar to the query structure.

1.14.3   Similarity Search

Similarity search was developed as an effort to remove the limitations of substructure search [72]. A similarity search compares a set of characteristics describing the target structure with the corresponding structure with the set of characteristics of

Query Structure   Retrieved Hits

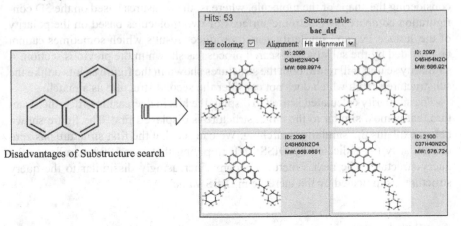

Disadvantages of Substructure search

Fig. 1.39 Substructure searching using JChem

Fig. 1.40 Figure depicting
the limitation of substructure
searching

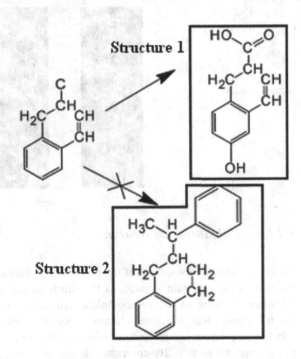

the structures available in the database. Query of a similarity search is usually the
full structure that the user wants to retrieve. However, the retrieved structures may
be a substructure of another larger molecule. The measure of similarity between
the target and the database structures is calculated based on the degree of resemblance of the two sets of characteristics. Measures based on 2D topology compare

the 2D topology considering only the atoms and bonds of the molecule without considering the shape of the molecule, whereas, the measures based on the 3D configuration compare the electronic surfaces of two molecules based on the polarity of the surface. A good similarity search produces results which sometimes cannot be provided by the substructure search process as shown in the previous section. A similarity search will return both the structures shown in the figure as hits unlike the substructure search which does not consider the second structure as a match.

The similarity calculated is used to display the hits in decreasing order; the structure that is most similar to the target structure is displayed first. The figure shown below explains the similarity searching. We can see that the first structure has zero dissimilarity (as indicated by DISS: 0.0) implying that is it 100 % similar to the query structure. The next structures become increasingly dissimilar to the query structure (as indicated by the increase in DISS value).

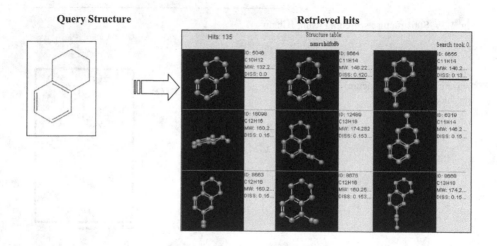

1.14.4   Subsimilarity Search

Similarity search solves most of the problems associated with substructure search, but it does have certain drawbacks that limit its usage. Similarity search is useful when we want to retrieve complete structures similar to the query structure, but it becomes less effective when we need to retrieve structures that contain a substructure which is similar to a target structure or target substructure as shown in the previous section. By contrast, substructure search helps us get a list of molecules that have the query structure as a part of its structure, but it does not rank the molecules, so there is no way to know which molecule is most similar to the query structure. As discussed in the previous section, substructure search does not enlist molecules with minor differences but highly similar. To attend to these kinds of search problems, a new searching approach was devised, termed as, subsimilarity searching. Subsimilarity searching or substructure similarity searching is a kind of local similarity search. It basically combines substructure search and similar structure search into a single search discipline. It involves a detailed similarity calculation and takes into consideration the parts of the molecules that are being compared. The similarity measure utilized is based on the number of bonds or atoms in the maximal common substructure (MCS) between the target structure and each database structure [67]. The largest substructure present in both the structures is the MCS for that particular pair of structures. Similar compounds are likely to share large MCS. Subsimilarity search uses a simple fragment-based similarity search to calculate the maximum size of the MCS and then uses it to rank the database structures. Using the same query structure as used in the similarity search, the hits retrieved using subsimilarity search are shown below. The query structure is present as a substructure in most of the hits and the hits are ranked in descending order based on DISS values.

Query Structure   Retrieved Hits

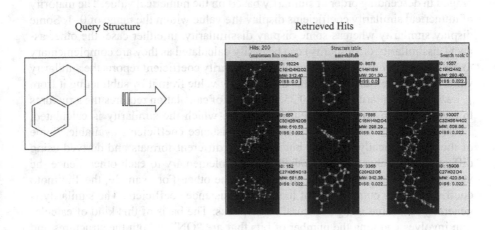

1.15   Search for Relationship

This involves the retrieval of physico-chemical and pharmacological properties with respect to a specific structure such as melting point, boiling point, log P, pka, QSAR, QSPR etc. The logP (o/w) of the following structure is retrieved as (Fig. 1.41):

Fig. 1.41 The log p value for a molecule possessing phenanthrene-3-one ring system

logP(o/w) = 2.62

1.16   Similarity Measures

Similarity or dissimilarity can be measured either in terms of numerical value or in terms of distance measures [73]. In other terms, we can measure similarity as similarity coefficient or distance coefficient. If we can measure similarity or dissimilarity, then it will help us to

- Group structures
- Characteristics of each group can be easily analyzed once they are grouped
- Efficiently organize and retrieve information
- Classify new structures into a specific group
- Property of the new structure can be predicted based on the group it belongs to

Numerical similarity methods calculate the numerical similarity between the query molecule and the molecules in the database and return a list. The molecules are arranged in descending order of similarity based on the numerical value. The majority of numerical similarity coefficients display the value within the range of 0–1. Some display similarity whereas some display dissimilarity, in either case, the other aspect (dissimilarity or similarity) can be easily calculated as they are complementary to each other. For example, if a particular similarity coefficient reports the similarity value as 0.65, we can calculate the dissimilarity value from it by subtracting it from 1, i.e., the dissimilarity value is 0.35. This kind of calculation requires the structures to have common structural features based on which the similarity is calculated. There are a large number of similarity and distance coefficients available. Some of them are basically the same but written in different formats and derived using different approaches, whereas others are complementary to each other, hence the value calculated by one can be predicted by the other. For example, the Tanimoto coefficient is the complement of the Soergel distance coefficient. The similarity is generally measured using structural fingerprints. The basis of this kind of calculation involves counting the number of bits that are "ON" in both the structures and then calculating the similarity using a distance metrics or similarity coefficient. Similarity coefficients are often referred to as association coefficients. *Monotonic coefficients* are those coefficients that rank the objects identically based on their similarity to a specified target. Distance coefficients correspond to the distances in multidimensional space, but they are not necessarily the same. A distance coefficient is described as a metric, if it satisfies the following four criteria:

1. Distance values must be zero or positive, and the distance from an object to itself must be zero

$$\text{Distance}_{A, B} \geq 0 \text{ or } \text{Distance}_{A, A} = \text{Distance}_{B, B} = 0$$

2. Distance values must be symmetric

$$\text{Distance}_{A,B} = \text{Distance}_{B, A}$$

3. Distance values must obey the *triangular inequality*

$$\text{Distance}_{A, B} \leq \text{Distance}_{A,C} + \text{Distance}_{C, B}$$

4. The distance between nonidentical objects must be greater than zero.

$$A \neq B \leftrightarrow \text{Distance}_{A, B} > 0$$

The following table enlists the symbols used in the similarity coefficient and distance matrices in the following sections.

i, j	attributes
A, B	objects (or molecules)
n	total number of attributes of an object (e.g., bits in a fingerprint)
X_A	attribute vector describing object A
xj_A	value of jth attribute in object A
a	number of bits "on" in molecule A
b	number of bits "on" in molecule B
c	number of bits "on" in both molecules A and B
d	number of bits "off" in both molecules A and B
χ_A	set of "on" bits in binary vector XA
$S_{A, B}$	similarity between objects A and B
$D_{A, B}$	distance between objects A and B

As mentioned earlier, there are a number of similarity coefficients and distance matrices. Most of the coefficients can be calculated by two different formulas; one is used for continuous variables, whereas the other one is used for binary variables or dichotomous variables. Similarity can be better defined when continuous variables are used as descriptors rather that the "ON" "OFF" bits of fingerprints. The descriptors, on the other hand, are basically molecular properties, which have a wide range of values. So, they are normalized in the range of zero to one.

The Tanimoto coefficient or Jaccard similarity coefficient is a statistic used for comparing the similarity and dissimilarity of structures [74]. It is one of the most commonly used similarity coefficient used in chemoinformatics, because it allows rapid calculation due to its simple nature and absence of complex mathematical operators. In general, the complement of the Tanimoto coefficient does not follow the triangular inequality. The Tanimoto coefficient is calculated as follows:

For dichotomous variables:

$$S_{A, B} = c/[a + b - c]$$

$$\text{Range} = 0 \text{ to } +1$$

For continuous variables:

$$S_{A,B} = \left[\sum_{j=1}^{j=n} x_{jA} x_{jB} \right] / \left[\sum_{j=1}^{j=n} (x_{jA})^2 + \sum_{j=1}^{j=n} (x_{jB})^2 - \sum_{j=1}^{j=n} x_{jA} x_{jB} \right]$$

Range $= -0.333$ to $+1$

As can be seen from the formula given above, Tanimoto coefficient takes into account only those bits that are "ON." Note that the OFF bits do not determine the similarity. In other words, if some molecular features are absent in both molecules, then that is not taken as an indication of similarity between the two. If two molecules have Tanimoto coefficient equal to 1, it indicates that the molecules have identical fingerprint patterns, it however does not indicate the presence of identical molecules, because identical fingerprints do not always designate identical molecules. On the contrary, if the value is zero for dichotomous variables, it indicates complete dissimilarity. The following example will make it clearer:

Molecule A 100010001110001010000100000100101 a $= 11$
Molecule B 000110000100001001000100000100101 b $= 9$
A and B   000010000100001000000100000100101 c $= 7$

So tanimoto coefficient, $\mathbf{S_{A,B}} = 7 / [11 + 9 - 7]$
$$= \mathbf{0.53}$$

Hence, we can say that the structures A and B are 53 % similar. It should be noted that the complement of the Tanimoto coefficient is identical to the Soergel distance.

There are other coefficients like the Dice coefficient, Cosine coefficient, simple matching coefficient, and Tversky similarity coefficient.

Distance is complementary to similarity. A few lines have been discussed on distance coefficients in the previous sections. The complementary relationship between the similarity and distance coefficients allows the calculation of one from the value provided for the other by subtracting it form one, that is,

Distance $= 1 -$ Similarity.

However, care should be taken that this expression is true for only those similarity coefficients that have their value within the range of zero–one. For example,

Soergel Distance $= 1 -$ Tanimoto Coefficient

Distance coefficients are also called as distance matrices when they obey the criteria discussed previously. Hamming distance and Soergel Distance are examples of metric distance coefficients.

Euclidean distance The Euclidean distance or Euclidean metric is the "ordinary" distance between two points that one would measure with a ruler and is given by the Pythagorean formula. It can be calculated using the following formula:

For dichotomous variables:
$$D_{A,B} = [a + b - 2c]^{1/2}$$
$$\text{Range} = n \text{ to } 0.$$

For continuous variables:
$$D_{A,B} = \left[\sum_{j=1}^{j=n} (x_{jA} - x_{jB})^2 \right]^{1/2}$$
$$\text{Range} = \infty \text{ to } 0.$$

It follows all the four metric properties and is monotonic with Hamming distance. For dichotomous variables, (Euclidean distance)^2 = Hamming Distance.

1.17 Molecular Diversity

We have seen in the previous sections that molecular similarity plays a major role in clustering sets of molecules together based on their degree of similarity. The same measures that are used to find the similarity can also be used to find the molecular dissimilarity. As already discussed, many similarity coefficients provide the dissimilarity value when their complement is considered. Molecular dissimilarity provides an important means to study molecular diversity [75]. Consider a case where we study only similar molecules; in that case, the chemistry space will be very limited. By contrast, if we used molecular diversity to study dissimilar molecules, then we can span the entire chemistry space rather than limiting us to a cluster of molecules. Molecular diversity comes in very handy when dealing with a selection of new compounds. It also proves to be a great tool for designing combinatorial libraries. Compound selection using molecular diversity involves the selection or identification of structurally dissimilar compounds or sets of compounds that can be tested for their bioactivity. Using a diverse set of compounds generates a greater amount of information related to the structure–activity relationship. Molecular diversity also helps find out the molecules of interest from a database on which a similarity search has been performed. These molecules are essentially dissimilar to the query structure, but, as mentioned earlier, are very useful in drug designing. A diverse subset can be generated from a library of molecules using MOE program. After importing the dataset in database viewer of MOE, one can proceed to compute the diverse subset. There are three methods available by which diversity between two database entries can be assessed viz. descriptors, fingerprint data or conformation data.

1.18   Advanced Structure-handling Tools

Due to the sophistication in techniques of combinatorial chemistry, availability of high-throughput screening data, and computational power, there is a need to develop advanced structure-handling methods for fast processing of data [76]. Some of the efforts in this direction are highlighted below.

1.18.1   CCML

One of the major breakthroughs due to the progress of the WWW system was the evolution of content-based markup language based on XML syntax, the CML developed by Peter Murray-Rust. Currently, CML has become a valuable tool with the functionalities to describe atomic, molecular, and crystallographic information. CML captures the structural information through a concise set of tags with the associated semantics. CCML is a methodology for encoding chemical structures as compressed CML generated by popular chemical structure-generating programs like JME [77]. The CCML format consists of both SMILES and/or equivalent data along with coordinate information about the atom for generating chemical structures in plain text format. Each structure generated by JME in standalone or generated by virtual means can be stored in this format for efficient retrieval, as it requires about one-tenth or below of actual CML file format, since the SMILES describes the interconnectivity of the molecule. The CCML format is compatible for automated inventory application and is a commonly used technique in security and inventory management [78].

1.19   ChemXtreme

ChemXtreme is a java-based computer program to harvest chemical information from Internet web pages using Google search engine and applying distributed computing environment [79]. ChemXtreme employs the "search the search engine" strategy, where the uniform resource locators (URLs) returned from the search engines are analyzed further via textual pattern analysis. This process resembles the manual analysis of the hit list, where relevant data are captured and, by means of human intervention, are mined into a format suitable for further analysis. ChemXtreme, transforms chemical information automatically into a structured format suitable for storage in databases and further analysis and also provides links to the original information source. The query data retrieved from the search engine by the server are encoded, encrypted, and compressed and then sent to all the participating, active clients in the network for parsing. Relevant information identified by the

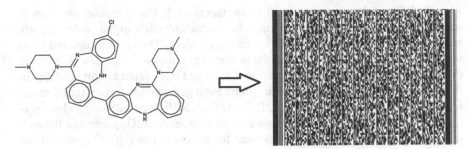

Fig. 1.42 A barcode representation of a molecule

clients on the retrieved websites is sent back to the server, verified, and added to the database for data mining and further analysis. The chemical names including global identifiers like InChI or corporate identifiers like CAS registry numbers, Beilstein registry number, etc. could be mapped to corresponding structural information in relational database systems.

1.19.1   Barcoding SMILES

Chemical structures can be encoded and read as 2D barcodes (PDF417 format) in a fully automated fashion [80]. A typical linear barcode consists of a set of black bars of varying width separated by white spaces, encoding alphanumeric characters. To reduce the amount of data that has to be encoded on the barcode, a template-based chemical structure-encoding method was developed, the Automatic Chemical Structure (ACS) file format. This method is based on the Computer Generated Automatic Chemical Structure Database (CG-ACS-DB) originally developed to create a virtual library of molecules through enumeration from a selected set of scaffolds and functional groups. Scaffolds and groups are stored in ACS format as a plain text file. In this ACS format, the most commonly used chemical substructures are represented as templates (scaffolds or functional groups) through reduced graph algorithm along with their interconnectivity rather than atom-by-atom connectivity information. The barcoded chemical structures can be used for error-free chemical inventory management. One of the molecules containing over thousands of atoms can be easily represented as barcoded and can be decoded automatically and accurately in seconds without manual intervention (Fig. 1.42).

1.19.2   Chem Robot

An open source-based computer program called Chem Robot is developed which can use digital video devices to capture and analyze rapidly hand-drawn or computer-generated molecular structures from plain papers [81]. The computer program is capable of extracting molecular images from live streaming digital video signals and prerecorded chemistry-oriented educational videos. The images captured from these sources are further transformed into vector graphics for edge detection, node detection, Optical Character Recognition (OCR) and interpreted as bonds, atoms in the molecular context. The molecular information generated is further transformed into reusable data formats (MOL, SMILES, InCHI, SDF) for modelling and simulation studies. The connection table and atomic coordinates (2D) generated through this automatic process can be further used for generation of IUPAC names of the molecules and also for searching the chemical data from public and commercial chemical databases. Applying this software, the digital webcams and camcorders can be used for recognition of molecular structure from hand-drawn or computer-generated chemical images. The method and algorithms can be further used to harvest chemical structures from other digital documents or images, such as PDF and JPEG formats. Effective implementation of this program can be further used for automatic translation of chemical images into common names or IUPAC names for chemical education and research. The performance and efficiency of this workflow can be extended to mobile devices (smart phones) with Wi-Fi and camera.

1.19.3 Image to Structure Tools

Yet another upcoming technology based on Optical Character Recognition (OCR) can recognize molecular structures from scanned images of printed text that can recognize structures, reactions, and text from scanned images of printed chemistry literature. This can save users valuable time of redrawing structures from printed material, as it directly transforms the "images" into "real structures" that can then be saved into chemical databases. Programs such as CLiDE [82], OSRA [83], and ChemOCR [84] are the known relevant softwares that recognize structures, reactions, and text from scanned images of printed chemistry literature. OSRA is a utility designed to convert graphical representations of chemical structures, as they appear in journal articles, patent documents, textbooks, trade magazines, etc. into SMILES (see http://en.wikipedia.org/wiki/SMILES) or SD files—a computer recognizable molecular structure format. OSRA can read a document in any of the over 90 graphical formats parseable by ImageMagick—including GIF, JPEG, PNG, TIFF, PDF, PS, etc. and can generate the SMILES or SDF representation of the molecular structure images encountered within that document (http://cactus.nci.nih.gov/cgi-bin/osra/index.cgi).

A Practice tutorial

- Select the file one wants to process or enter a URL (http://…) pointing to an image and click the "Submit" button. Any of the over 90 image formats recognized by ImageMagick including GIF, JPEG, PNG, PDF, PS, and TIFF can be processed.

- Correct recognized structures using the JME Molecular Editor.
- Preview the 3D structure if necessary. Note—the generated 3D image is for demonstration purposes only, e.g., to help disambiguate bridge bonds, etc. OSRA only generates the connection table, not the 3D coordinates.
- Click on the "Get SMILES" button to obtain the SMILES of the structure. One can then use the provided live links to convert SMILES to other chemical formats or to locate the structure in Chemical Structure Lookup Service. "Get SD File" button will be active only after checking all the structures recognized in the document. Download the SD file containing all the recognized structures.

1.19.4 CLide

CLiDE is a chemistry intelligent equivalent of OCR software. Just as an OCR can recognize characters from scanned images of printed text, CLiDE can recognize structures, reactions, and text from scanned images of printed chemistry literature. The software saves users hours of redrawing structures from printed material, as it transforms the "images" into "real structures" that can then be input into databases. It is available at http://www.simbiosys.com/clide/.

1.19.5 Advanced Structure Computation Platforms

HPC/Cloud computing tools, which can handle millions of structure, are discussed in detail in the last chapter. An HPC script generator has been developed that can perform 100,000 per hour large-scale docking in an automated fashion. JAVA RMI-based open-source methods have been employed to compute structural properties on a large scale [85].

1.20 Virtual Library Enumeration

In order to design a better lead molecule, one has to perform a sequence of several steps starting from collecting molecular data with known bioactivity, analysis of those chemical structures to extract significant features related to activity of interest, and rebuild new molecules with promising and favorable bioactivity profiles. Virtual library of diverse molecules which are not yet synthesized can be enumerated from a set of scaffolds and functional groups by combinatorial means [86] Here, the scaffold represents a molecule containing at least one ring or several rings which are connected by linker atoms. Scaffolds can be generated from complex molecular structures by a systematic disconnection of functional groups connected by single bonds. The scaffolds and functional groups generated could be further enumerated to build virtual library of diverse organic molecules. An alternate approach namely "lead hopping" is also available to replace common scaffold by chemically and spatially equivalent core fragments.

1.21   Clustering

Clustering is a process of finding the common features from a diverse class of compounds that requires multivariate analysis methods [87]. It is a type of important unsupervised learning approach used in machine learning. One of the most suitable methods for this study is clustering where the consensus score and distance between set of compounds can be easily measured through mean/Euclidean distance measures. This score reflects the similarity or dissimilarity between classes of compounds and helps identify potential active or toxic substances through predictive studies. Cluster 3.0 is an open-source program that was developed to analyze gene expression data that employs routines for hierarchical (pairwise simple, complete, average, and centroid linkage) clustering, k-means and k-medians clustering, and 2D self-organizing maps [88]. The routines are available in the form of a C clustering library, an extension module to Python, a module to Perl, as well as an enhanced version of Cluster, which was originally developed by Michael Eisen of Berkeley Lab. The Jarvis Patrick algorithm is useful for clustering chemical structures on the basis of 2D fragment descriptors. The Lipinski rule of five is one such example where the similar characteristics of drug molecules can be derived by clustering a large number of drugs and lead molecules. Javatreeview is an open-source, cross-platform rewrite that handles very large datasets well and supports extensions to the file format that allow the results of additional analysis to be visualized and compared [89]. An applet version is also available that can be used on any website with no special server-side setup. ChemAxon provides clustering tools to analyze hundreds and thousands of molecules (Library MCS) via maximum common substructures [90]. JKlustor provides many methods for clustering molecules. Molecule datasets can be clustered on the basis of similarity, descriptors, structure, diversity, scaffolds, etc. Using the command line option *Compr[<options>]*, we can compare large databases with millions of entries to obtain their diversity and similarity statistics in batch mode (Fig. 1.43).

1.22   Databases

Database is a collection of information, usually, kept in a list or table(s) on a particular subject. It helps organize the data for easy retrieval through simple querying. Using a database storage, one can reduce the number of files in a computer by storing the information in database tables. Databases usually contain many tables. All the tables can be linked by a common identifier such as a primary key within the database or through foreign key association [91].

Fig. 1.43 Clustering of
molecules related to malaria
(five clusters are visible) in
JAVAtreeview

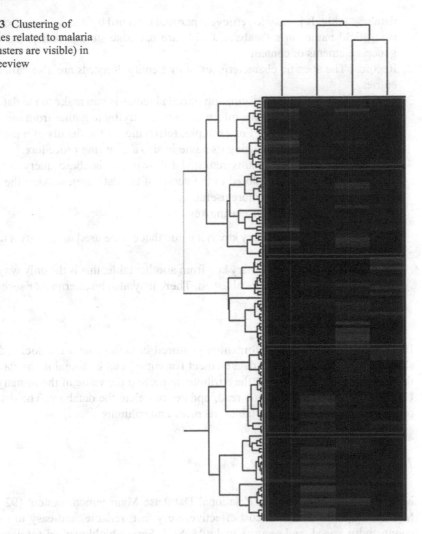

Some of the most familiar terms used in databases are:

Entity: object, concept, or event (subject)

Attribute: a characteristic of an entity

Row or Record: the specific characteristics of one entity

Table: a collection of records

Database: a collection of tables

Parts of a database:

database contains fields, records, queries, and reports.

1. Fields: In the design of database table, information is stored under a particular
   field (for example, column names in a table). Field names should be unique in the

database table. It is easy to retrieve a particular record by accessing information using field names in a database. Fields are database storage units, also called generic elements of content.

2. Records: The specific characteristics of one entity. Records are also called data entries.

3. Queries: Queries are the information retrieval requests you make to the database. Your queries are all about the information one is trying to gather from the stored information in a database. For example, retrieving all the details of a molecule from a corporate database using its name is also a querying procedure.

4. Reports: The retrieved results returned following a database query is called reports. Reports can be tailored to the needs of the data user, making the information they extract much more useful.

   a. Linking data in a database using keys

      – Primary key: A primary key is a value that can be used to identify a unique row in a table.
      – Foreign key: The primary key from another table, this is the only way joint relationships can be established. There may also be alternate or secondary keys within a table.

   b. Relational database

In relational database, the information is stored in tables that are associated with shared attributes (keys). Any data element (or entity) can be found in the database through the name of the table, the attribute name, and the value of the primary key. Using database, one can create, read, update, or delete the database. The database operations occur at all levels: tables, records, and columns.

1.22.1   Database Server   MySQL

MySQL is a freely available Relational Database Management system [92]. The MySQL Database Server is cost effective, very fast, reliable, and easy to use. Its connectivity, speed, and security make MySQL Server highly suited for accessing databases on the Internet. The MySQL Database Software is a client/server system that consists of a multi-threaded SQL server that supports different backends, several different client programs and libraries, administrative tools, and a wide range of application programming interfaces (APIs). A password system for MySQL is very flexible and secure and allows host-based verification. The WWW Links are MySQL, Oracle, Postgre SQL.

1.22.2 Code for Connecting to a MySQL Database

```java
public String[] ReadSDF(String fname) {
        System.out.println(fname);
        int cnt = 0;
        String t = "";
        int mcnt = 1;
        double[][] dmatx = new double[mcnt][36];
        try {
                BufferedReader    br    =    new    BufferedReader(new    FileReader(new
File(fname)));
                String s1 = "";
                int lcnt = 0;
                int acnt = 0;
                int bcnt = 0;
                int[] v1 = new int[2];
                double[][] lcoord = new double[1000][3];
                double[][] bcon = new double[1000][3];
                int ac = 0;
                int bc = 0;
                while ((s1 = br.readLine()) != null && cnt < mcnt) {
                    lcnt++;
                    t += s1 + "\n";
                    try {
                        if (lcnt == 4) {
                            String[] t1 = stringToArray(s1);

                            acnt = Integer.valueOf(t1[0].trim());
                bcnt = Integer.valueOf(t1[1].trim());
                        }
                        v1[0] = 4 + acnt;
                        if (lcnt > 4 && lcnt < v1[0]) {
                String[] t2 = stringToArray(s1);
                lcoord[ac][0] = Double.valueOf(t2[0]);
                            lcoord[ac][1] = Double.valueOf(t2[1]);
                            lcoord[ac][2] = Double.valueOf(t2[2]);
                            ac++;
                        }
                        v1[1] = v1[0] + bcnt;
                        if (lcnt > v1[0] && lcnt < v1[1]) {
                            String[] t3 = stringToArray(s1);
                            bcon[bc][0] = Double.valueOf(t3[0]);
                            bcon[bc][1] = Double.valueOf(t3[1]);
                            bcon[bc][2] = Double.valueOf(t3[1]);
                            bc++;
                        }
                        if (s1.contains("$$$$")) {
                            double[] maxv = getMaxValue3(lcoord);
                            double[] minv = getMinValue3(lcoord);
                            double[][] gbx = BuildGridBox(minv, maxv);
                            dmatx[cnt] = getDistance(gbx);
                            cnt++;
                            t = "";
                            lcnt = 0;
                            ac = 0;
                            bc = 0;
                        }
                    } catch (Exception e) {
                        t = "";
                        lcnt = 0;
                        ac = 0;
                        bc = 0;
                    }
                }
            br.close();
        } catch (Exception e) {
            System.out.println(e);
        }

        for (int i = 0; i < dmatx.length; i++) {
            for (int j = 0; j < dmatx[0].length; j++) {
                System.out.print(df.format(dmatx[i][j]) + " ");
            }
            System.out.println();
        }

        String[] out = t.split("$$$$");
        return out;
    }
```

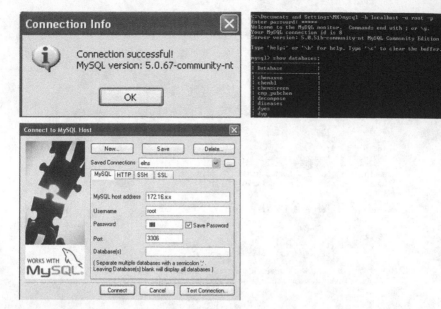

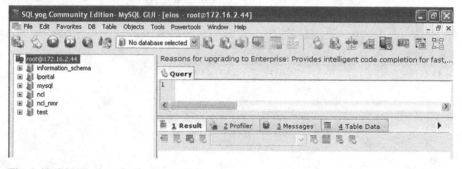

Fig. 1.44 Connecting to the MySQL server

Fig. 1.45 SQLYog interface

1.22.3   A Practice Tutorial

1. Install MySQL locally in the computer (skip this step if already installed).
2. Create user with privileges (Admin/ User/ Guest).
3. Check the status of MySQL (if not active start the MySQL server).
4. Learn to use SQLYog as GUI for MySQL server.
5. One can explore existing databases, tables, data after authentication.
6. Simple GUI of SQLYog.
7. Next, click the databases to expand.
8. Sample query to Create Table in MySQL (Figs. 1.44, 1.45, 1.46, and 1.47).

To view the contents of table: click Tables>>Right mouse button>>select View Data

Fig. 1.46 Opening a
database

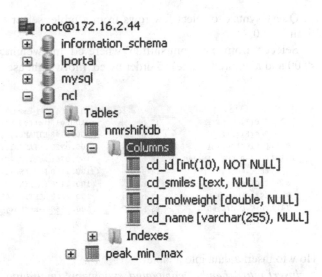

root@172.16.2.44
- information_schema
- lportal
- mysql
- ncl
  - Tables
    - nmrshiftdb
      - Columns
        - cd_id [int(10), NOT NULL]
        - cd_smiles [text, NULL]
        - cd_molweight [double, NULL]
        - cd_name [varchar(255), NULL]
      - Indexes
    - peak_min_max

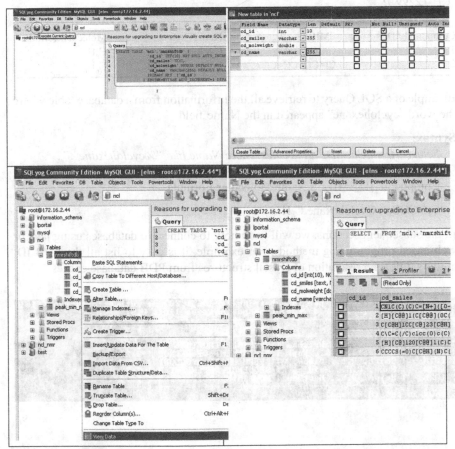

Fig. 1.47 Steps for creating and viewing a database table

Query syntax to select few rows from a table: select * from 'ncl'.'nmrshiftdb' limit 0, 500;

Select * from 'ncl'.'nmrshiftdb' where cd_molweight>100 and cd_molweight <500 and length(cd_name)>5 order by cd_molweight asc

cd_id	cd_smiles		cd_molweight	cd_name
33305	F\C(F)=C(\F)F	13 b	100.015	tetrafluoroethene
2570	OCC(F)(F)F	10 b	100.0398	ETHANOL,2,2,2-TRIFLUORO
3870	O=C1CCC(=O)O1	13 b	100.0728	BUTANEDIOIC ACID,ANHYDRIDE SUCCINIC
16522	COC(=O)[CH-][N+]#N	18 b	100.0761	ACETIC ACID,DIAZO,METHYL ESTER
3682	CCOC(=O)C=C	11 b	100.1158	PROPENOIC ACID,ETHYL ESTER (ETHYLACF
3698	COC(=O)C(C)=C	13 b	100.1158	PROPENOIC ACID,2-METHYL,METHYLESTER
3820	CC(=C)OC(C)=O	13 b	100.1158	ACETIC ACID,ISOPROPENYL ESTER
7418	CO\C=C/C(C)=O	13 b	100.1158	4-methoxy-3-buten-2-one

How to insert a data into a table?

Insert into 'chembl'.'compound_synonyms' (molregno, synonyms)values ('97', 'CP-12299');

Compound_ID	SMILES	Name	Molecular formula
1	ClCCCCC1	Cyclohexane	C_6H_{12}
2	Cl=CC=CC=Cl	Benzene	C_6H_6

Example of a SQL Query to retrieve all the information from a database table where the word "cyclohexane" appeared in the Name field.

Syntax:

*Select * from ChemDB.Molecules where Name like "%cyclohexane%";*
Output of Query:
Compound_ID SMILES Name Molecular Formula

1. ClCCCCC1 Cyclohexane C_6H_{12}

In the subsequent sections, we will learn how to connect to databases using java or web-based programming methods. For example, it is easy to list all the PDB ID, authors, title, and resolution of crystal structures from PDB database entries.

```
mysql> select field1 as pdb_id,substring(field6,1,20) as authors,substring(field
4,1,20) as title, field7 as res  from pdb_entries limit 0,5;
+--------+----------------------+----------------------+------+
| pdb_id | authors              | title                | res  |
+--------+----------------------+----------------------+------+
| 100D   | Ban, C., Ramakrishna | CRYSTAL STRUCTURE OF | 1.9  |
| 101D   | Goodsell, D.S., Kopk | REFINEMENT OF NETROP | 2.25 |
| 101M   | Smith, R.D., Olson,  | SPERM WHALE MYOGLOBI | 2.07 |
| 102D   | Nunn, C.M., Neidle,  | SEQUENCE-DEPENDENT D | 2.2  |
| 102L   | Heinz, D.W., Matthew | HOW AMINO-ACID INSER | 1.74 |
+--------+----------------------+----------------------+------+
5 rows in set (0.00 sec)
```

Table 1.2 Example format

Field1	Field2	Field3	Field4
Rec1	Entry1	Entry2	Entry3
Rec2	Entry4	Entry5	Entry6

Table 1.3 Example of ChemDB molecules

Compound_ID	SMILES	Name	Molecular formula
1	ClCCCCCl	Cyclohexane	C_6H_{12}
2	Cl=CC=CC=Cl	Benzene	C_6H_6

Please follow the instructions from http://moltable.ncl.res.in/ to install MySQL and connect to database, define, and build user query (create table, insert/delete/update data, query tables, etc.) for chemoinformatics data.

1.22.4 Creating and Hosting Database

In this section, we will learn to create a database and host it over the Internet. We create huge amounts of data, but if they are not stored properly, they might be lost. We have learnt some of the basic computing skills in the previous section here, we will use them and some other tools to create databases. In this section, we will learn to create a database using SQL commands and SQLyog (a MySQl GUI).

Steps for creating database and tables using SQL are as follows:

Step 1: Determine the entities involved and create a separate table for each type of entity (thing, concept, event, and theme) and name it.
Step 2: Determine the Primary Key for each table.
Step 3: Determine the properties for each entity (the non-key attributes).
Step 4: Determine the relationships among the entities.

1.22.5 A Practice Tutorial

Creating database using SQL command prompt
In this tutorial, we will use MySQL database; some example syntax for creating tables in a database are given below (Table 1.2 and 1.3):

Syntax *CREATE TABLE TableName(columnname1 datatype (size),……., columnname4 datatype (size));*
    Rows (Rec1, Rec2, etc.)
    Columns or Field Names (Field1–4)
    An example of an SQL Query to retrieve all the information from a database table where the word "cyclohexane" appeared in the Name field.

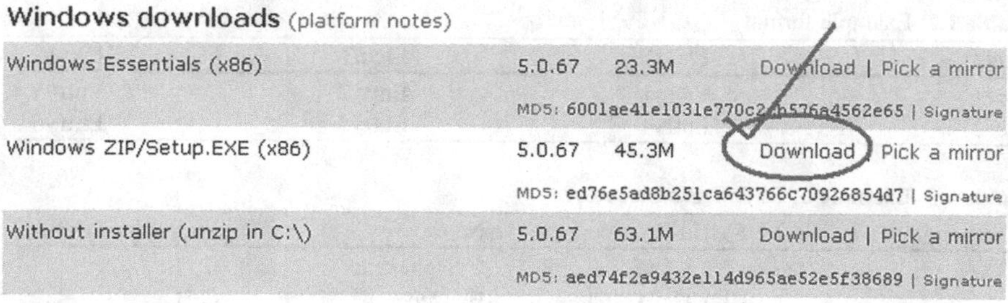

Windows downloads (platform notes)

Windows Essentials (x86)	5.0.67	23.3M	Download | Pick a mirror
	MD5: 6001ae41e1031e770c2ch536a4562e65 | Signature		
Windows ZIP/Setup.EXE (x86)	5.0.67	45.3M	Download | Pick a mirror
	MD5: ed76e5ad8b251ca643766c70926854d7 | Signature		
Without installer (unzip in C:\)	5.0.67	63.1M	Download | Pick a mirror
	MD5: aed74f2a9432e114d965ae52e5f38689 | Signature		

Fig. 1.48 Windows option for downloading the MySQL program

Syntax: *Select * from ChemDB.Molecules where Name like "%cyclohexane%";*
   Output of Query:

Compound_ID	SMILES	Name	Molecular formula
1	ClCCCCCl	Cyclohexane	C_6H_{12}

Example for Alter Table ChemDB.Molecules for change field name "Name" to "CompoundName."

Syntax: ALTER TABLE 'ChemDB.Molecules' CHANGE 'Name' 'Compound-Name' varchar(255) NOT NULL
   Output of Query:

Compound_ID	SMILES	Compound name	Molecular formula
1	ClCCCCCl	Cyclohexane	C_6H_{12}
2	Cl=CC=CC=Cl	Benzene	C_6H_6

Creating a database using MySQL, SQLyog, JChemManager

1. Download MySQL from the link provided below
   URL: http://dev.mysql.com/downloads/mysql/5.0.html#downloads

If you are using Windows, select the following link (or equivalent depending on the updates or your operating system) (Fig. 1.48)

2. Save the file and install it by following the instructions.
3. SQLyog:

Creating and managing databases using the SQL queries can be cumbersome sometimes; to avoid that and manage databases easily, one can use SQLyog. SQLyog is a MySQl GUI that helps us create and manipulate tables and databases using a user-friendly easy-to-use interface. The Community Edition is Free and Open Source under GPL license. It can be downloaded free of cost from the following link:
   URL: http://code.google.com/p/sqlyog/downloads/list
   Once downloaded, it can be installed easily following the instructions.

a. To create a database as shown in the following figure right-click on the

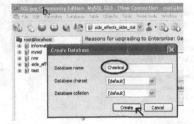

c. root@localhost → create database → type

the name of the database and click create.

Fig. 1.49 Creating a new database

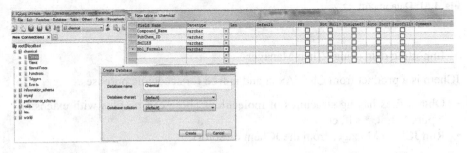

Fig. 1.50 Creating a new table

In the following sections, you will see the usefulness of this GUI tool.

- Creating the database "Chemical" in MySQL using SQLyog interface
  - To create a database as shown in Fig. 1.49, right-click on the
  - root@localhost → create database → type the name of the database and click create.
  - Creating table "chemicals" in the database "Chemical"
- *Right-click* on the *tables* icon within the newly created database
- Click on the *create-table* option.
- Fill in the required fields and other parameters and click on *create table, then enter the table name and click OK* (Fig. 1.50)
  - Importing or adding data to the created table:

Using SQLyog, we can easily add data to the table one has created. One can also import data from any .csv file by clicking some buttons as shown in the following figures.

- Select the table and right-click
- Select import option as shown in Fig. 1.51.
- Browse and select the required file and import it.

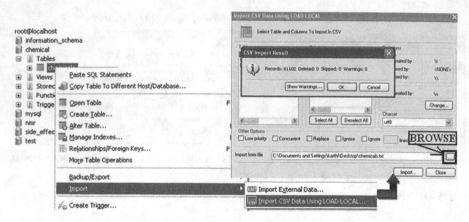

Fig. 1.51 Data import

4. Importing data using JChemManager

JChem is a product from ChemAxon and is free for academic purposes.

- Obtain files having structures of molecules. These can be files with extension *.pdb, *.mol, *.sdf, etc.
- Run JChem Manager from the JChem directory.

Following window pops up (Fig. 1.52):

- Enter the details as shown in Fig 1.52. Here, **chemical** is the database created in MySQL Server 5.0 using SQLyog.
- In JChem Manager, go to File→Create Table.
- Enter the name for your Table (**QSAR**). Leave rest of the columns as they are.

A table with the following attributes will be created:

CREATE TABLE qsar(
        cd_id INTEGER AUTO_INCREMENT NOT NULL PRIMARY KEY,
        cd_structure MEDIUMBLOB NOT NULL,
        cd_smiles TEXT,
        . . . . .
        cd_timestamp DATETIME NOT NULL,
        cd_fp1 INTEGER NOT NULL,
        . . . . .
        . . . . .
        cd_fp16 INTEGER NOT NULL
)

- Then click on the **Import** button. Select the database table you want to put the structures in and select the file containing structures.

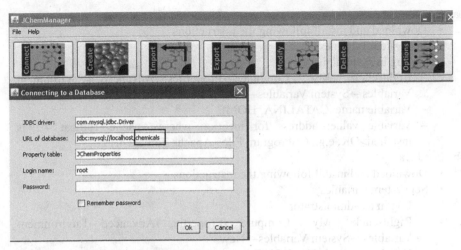

Fig. 1.52 JChemManager homepage

A table named **"qsar"** will be created in the **"chemical"** database containing all the structures.

1.22.6   Hosting the Database

For hosting the database created over the web, you need the following tools. All the tools used here are freely available. So firstly, they need to be downloaded from their respective sites. Once downloaded, install them on to the system. The system needs to be preloaded with the Java Runtime Environment.

1. JAVA—http://www.java.com/en/download/
2. MySQL Server 5.0—http://dev.mysql.com/downloads/mysql/5.0.html#downloads
3. SQLyog—http://code.google.com/p/sqlyog/downloads/list
4. JChem—http://www.chemaxon.com/jchem/download.html
5. Marvin Beans—http://www.chemaxon.com/marvin/download.html
6. Apache Tomcat 4.1—*http://tomcat.apache.org/*
7. MySQL JDBC Drivers 5.0—http://dev.mysql.com/downloads/connector/j/

MySQL Server 5.0, SQLyog, JChemManager, and marvin have already been dealt in the previous section of this chapter. The rest of the tools will be dealt in detail in subsequent chapters. In this chapter, we will mainly focus on their use for hosting a database.

Download and set system variables. One needs to download Java and Tomcat from the links provided in the preceding section and install them following the simple instructions that the respective installers display. Then, you need to set the system variables as shown in the following section.

a. For Tomcat
  • Download and install following the instructions
  • Set system variables
    − Log in as administrator
    − Right-click   My   Computer→Properties→Advanced→Environment Variables→System Variables→New
    − Variable name: CATALINA_HOME
    − Variable value:<address for location where Apache Tomcat 4.1 is installed>OK, e.g., C:\Program Files\Apache Tomcat 4.1
b. For Java
  • Download and install following the instructions
  • Set system variables
    − Log in as administrator
    − Right-click   My   Computer→Properties→Advanced→Environment Variables→System Variables→New
    − Variable name: JAVA_HOME
    − Variable value:<address for location where JDK directory is present>OK
    − e.g., C:\Program Files\Java\jdk1.6.0

Configuring JChem Manager and Creating a Database

Configure JChem Manager and Create a Database as shown in the previous section.

Hosting the Database, Configuring Tomcat 4.1:

• Go to\Apache Tomcat 4.1\bin and start the service by double clicking **startup**
• Open Internet Explorer. Type the following in the address bar:
• http://localhost:7070/, where 7070 is the port set while installing Apache Tomcat 4.1

If the following Tomcat homepage is seen, it means that the setup has been done successfully (Fig. 1.53)

• Click on Tomcat administration tool. It leads you to the Tomcat Web Server Administration Tool.
• Type in the User name and Password created while installing Apache Tomcat 4.1.
• Once logged in, go to Tomcat Server→Service (Tomcat Standlone) →Host (localhost) →Host Actions.
• Select Create New Context.
• Document Base:<address for location where JChem is installed>
  e.g, C:\Program Files\BioInformatics\JChem

Path:/jchem

• Save the changes and Commit changes.
• Then, in your Internet Explorer type http://localhost:7070/jchem/index.html

If the following JChem homepage is seen, it means the setup has been done successfully (Fig. 1.54).

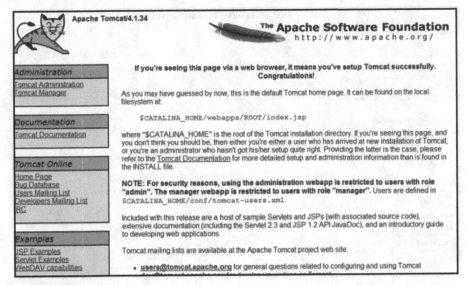

Fig. 1.53 Apache Tomcat

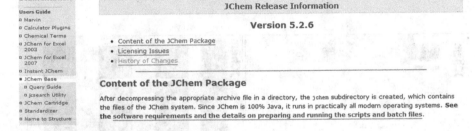

Fig. 1.54 Installing the JChem homepage

- Then, in the Internet Explorer type http://localhost:7070/jchem/examples/jsp1_x/setup.jsp

The following page will be displayed (Fig. 1.55)

- Enter the details as shown and done previously, and save the changes.
- Select the database Table (**qsar**), OK. The database has been hosted. The page displayed would look like this (Fig. 1.56):

An *.sdf file containing the structures was imported into the database.

All these structures can be viewed. More structures can be imported. The selected structures can be exported to any of the following viz. MOLFILE, SDFILE, SMILES, JTF, RDF, Marvin Document. The structures can be modified as well. A query of 2D structure can also be placed to be searched within the database. For querying, MarvinSketch application from the JChem package is used.

This database can be hosted by anyone to use through a website.

JSP Database Example Setup Page

JDBC driver class name:	com.mysql.jdbc.Driver
URL for JDBC connection:	jdbc:mysql://localhost/ chemical
Property table name:	JChemProperties
Database user login name:	root
Database user password:	••••••••
Read-only tables:	SCOTT.JSPEXAMPLE;SCOTT.TABLE
Chemical Terms filter file:	
Searches to remember:	None ▾

Additional properties:

```
jspexample.form.jchemform.cell=ID: <cd_id>;<$cd_structure>;<cd_formula>;MW:
<cd_molweight>
jspexample.form.jchemform.rows=6
jspexample.form.jchemform.layout=:3:2:L:0:0:1:2:w:n:0:10:M:1:0:2:1:c:n:1:10:L:1:1:1:1:w:n
:L:2:1:1:1:nw:n
jspexample.form.jchemform.param=:L:10:M:150:150:L:11b:L:10
SCOTT.JSPEXAMPLE.queryConditions=cd_id#id;name
jspexample.form.jchemform.celldim=240:180
jspexample.form.jchemform.cols=3
```

Fig. 1.55 JSP Database

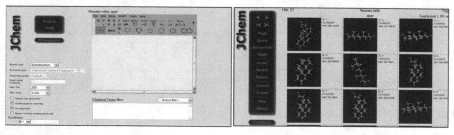

Fig. 1.56 The JChem interface

1.22.7 Chemical Databases

Chemistry is one of the first scientific disciplines that employed databases to store the chemical information. There are a wide variety of chemical databases available in chemistry. Here, we describe the list of available chemical databases which are very useful and frequently used for computational modelling and chemoinformatics activities. Recently, National Institute of Health (NIH) took initiatives to collect molecular structures from publicly available resources and organized them in a single database called PubChem Database containing over 30 millions of unique molecular entries and made it available for free to the public [93]. Due to the huge and continuously increasing amount of data related to chemical information, it is impossible to handle the data in file systems. Using database system and other additional chemoinformatics methods, we can manage the contents of this large resource for research and educational purposes.

1.22.7.1 Literature (textual) Databases

This type includes mainly bibliographic and also full text database containing the individual publication from the primary literature as objects using character strings. Some such databases are listed below:

CAS: CAS is a division of the American Chemical Society. CAS database provides literature information from more than 10,000 journals and 60 patent authorities related to chemistry, biomedical sciences, engineering, materials science, agricultural science, and many more. It is updated daily and made accessible through state-of-the-art information services. CAS is a commercial database and is not available for free.

(URL: www.cas.org/)

Medline: MEDLINE (Medical Literature Analysis and Retrieval System Online) is a bibliographic database. It contains more than 16 million references to journal articles starting from 1949 till present. All the records in MEDLINE are indexed with Medical Subject Headings (MeSH). MEDLINE is a part of PubMed and covers the subjects of biomedicine and health, chemical sciences, bioengineering, etc. MEDLINE is the primary component of PubMed (http://pubmed.gov); a link to PubMed is found on the National Library of Medicine (NLM) home page at http://www.nlm.nih.gov. The result of a MEDLINE/PubMed search is a list of citations (including authors, title, source, and often an abstract) to journal articles and an indication of free electronic full-text availability.

(URL: www.nlm.nih.gov/databases/databases_medline.html)

PubMed The US NLM at the NIH maintains PubMed as part of the Entrez information retrieval system. It is a free search engine for searching citations in MEDLINE. PubMed also provides access and links to the other Entrez molecular biology resources. PubMed also provides links to other sites providing full-text articles.

(URL: www.ncbi.nlm.nih.gov/pubmed/)

MeSH MeSH is a huge controlled vocabulary (or metadata system) for the purpose of indexing journal articles and books in the life sciences. Created and updated by the US NLM, it is used by the MEDLINE/PubMed article database and by NLM's catalog of book holdings. MeSH can be browsed and downloaded free of charge on the Internet. The yearly printed version was discontinued in 2007.

(URL: http://www.nlm.nih.gov/mesh/)

NIOSHTIC-2 NIOSHTIC-2 is a searchable bibliographic database of occupational safety and health publications, supported in whole or in part by the National Institute for Occupational Safety and Health (NIOSH). NIOSHTIC-2 is updated continuously. At a minimum, each citation contains the author's name or names, the title, and sufficient source information to facilitate retrieval, including the publication name, publication date, publication number(s), and pagination. Abstracts, key terms, and links to full text are also provided when available. Additional citation information may be available under the "Full View" option. NIOSHTIC-2 contains 44,568 occupational safety and health information resource citations. Each month, approximately 70 current citations are added with an annual yearly yield of more than 800 new current NIOSH-funded citations. Retrospective material is also added at about the same rate resulting in a total annual increase of approximately 1,600 citations. A significant portion of the citations (39,000) dates from 1971 to the present. An additional 13,800 resources in NIOSHTIC-2 are publications dating from the 1930s to the present from the NIOSH Mining Safety & Health Research Laboratories (formerly the US Bureau of Mines). There are several valuable search tools encoded into NIOSHTIC-2 records. They are intended to make searching easier and more productive. They include Standard Industrial Classification (SIC) codes, North American Industry Classification System (NAICS) codes, and CAS registry numbers.

Other databases in this category include NLM, ACS Journals (paid-service), Elsevier (paid-service), Science Direct (paid-service), etc.

(URL: http://www2a.cdc.gov/nioshtic-2/)

Factual (alphanumeric) Databases They provide the required textual or alphanumeric information such as physical properties, spectral data, description of research projects, legal information, etc. They also provide the literature references to the origin of the data represented so that the user need not go back to the primary literature as with bibliographic databases. Some such databases are listed below:

Cambridge Structural Database Cambridge Structural Database (CSD) is a repository for small organic and metal-organic molecule crystal structure. CSD contains structures that are mostly determined by X-ray diffraction or neutron diffraction and deposited directly to CDS or are present in publications in the open literature. It provides bibliographic, chemical, and crystallographic information of small molecules and excludes polypeptides and polysaccharides having more than 24 units, oligonucleotides and Metals and Alloys.

(URL: www.ccdc.cam.ac.uk/products/csd/)

Beilstein database The database covers the scientific literature from 1771 to the present and contains experimentally validated information on millions of chemical reactions and substances from original scientific publications. The electronic database was based on Beilstein's Handbook of Organic Chemistry. In this database, each compound is given a unique Beilstein Registry Number which helps in their easy identification. Each substance has up to 350 fields containing chemical and physical data. References to the literature in which the reaction or substance data appears are also given. The content is made available through the "CrossFire Beilstein" database.

(URL: http://info.crossfiredatabases.com/)

Some other examples of factual databases are Gmelin, SpecInfo, MDL, CHEMCATS, ChemSource, etc.

Structural (topological) Databases The structural databases play a central role in chemistry because they contain information on chemical structures. Examples of this type are CAS registry, National Cancer Institute (NCI) database, Crystallographic Structure Database (ICSD), CSD, Protein Data Bank (PDB), etc. The structure databases are usually designed to store chemical structural information representing the chemical bonds and atoms in such a way to use them for computational operations, such as structure search, data mining, etc.

There are two principal techniques for representing chemical structures in digital databases: as connection tables or adjacency matrices—MDL Molefile, PDB, CML—or as linear string notations—SMILES, SMARTS, WLN, InChI.

Some of the structural databases are listed below:

PubChem A chemical database is a database specifically designed to store chemical information. Chemical structures are traditionally represented using lines indicating chemical bonds between atoms and drawn on paper (2D structural formulae). Various chemical databases are available on the Internet which are free for all. Large chemical databases are expected to handle the storage and searching of information on millions of molecules. PubChem is one of the free chemical databases which is developed by the National Center for Biotechnology Information (NCBI). More than 24 millions of compound structures and descriptive datasets can be freely downloaded from PubChem. PubChem is a user-friendly database, we can search the compounds by compound name/keyword, and we can also search the compound by chemical properties. We can download the compounds in SDF format which is the standard one for various structural viewers. PubChem has three components, namely PubChem Compounds, PubChem Substances, and PubChem BioAssay described below.

(URL: http://pubchem.ncbi.nlm.nih.gov/)

PubChem Compounds The PubChem Compounds Database contains validated chemical depiction information provided to describe substances in PubChem Substance. Structures stored within PubChem Compounds are pre-clustered and cross-referenced by identity and similarity groups. We can search unique chemical structures using names, synonyms, or keywords. Links to available biological property information are also provided for each compound.

PubChem Substances The PubChem substance database contains chemical structures, synonyms, registration IDs, description, related urls, and database cross-reference links to PubMed, protein 3D structures, and biological screening results. We can search deposited chemical substance records using names, synonyms, or keywords. Links are also provided to biological property information and depositor websites.

PubChem BioAssay The PubChem BioAssay Database contains BioActivity screens of chemical substances described in PubChem Substance. It provides searchable descriptions of each BioAssay, including descriptions of the conditions and readouts. We can search bioassay records using terms from the bioassay description, for example "cancer cell line." Links are available to active compounds and bioassay results.

ChemIndustry ChemIndustry is a comprehensive directory and search engine for chemical and related industry professionals. It contains more than 45,000 chemical industry-related entities and contain the full text of millions of pages.

(URL: http://www.chemindustry.com)

ChemExper ChemExper is a company that joins together the areas of chemistry, computer science, and telecommunication. The ChemExper Chemical Directory is a free service that allows finding a chemical by its molecular formula, IUPAC name, common name, CAS number, catalog number, substructure or physical characteristics, as well as chemical suppliers. This database contains currently more than 500,000 chemicals, 16,000 material safety data sheet (MSDS), 10,000 infrared (IR) spectra, and more than 500 chemical suppliers.

(URL: http://www.chemexper.com/)

PDB The Protein Data Bank (PDB) is a repository for 3D structural data of proteins and nucleic acids. These data, typically obtained by X-ray crystallography or NMR spectroscopy and submitted by biologists and biochemists from around the world, are released into the public domain and can be accessed for free (see also protein structure). As of 24 June 2008, the database contained 51,491 released atomic coordinate entries (or "structures"), 47,526 of those entries were proteins, the rest being nucleic acids, nucleic acid–protein complexes, and a few other molecules. About 5,000 new structures are released each year.

(URL: http://www.rcsb.org/pdb/home/home.do)

The databases described below are classified here according to their use.

Databases dedicated for QSAR/QSPR WOMBAT (Drug Target)

WOMBAT (World of Molecular BioAcTivity) is a flagship product of Sunset Molecular Discovery. WOMBAT-PK is the reference Database for Clinical Pharmacokinetics and Drug Target Information. In this database, drugs are indexed from multiple literature sources. WOMBAT-PK 2009 contains over 13,000 clinical pharmacokinetic measurements. Each drug is represented in neutral species. WOMBAT can calculate physico-chemical properties like % oral bioavailability, % urinary excretion, % plasma protein binding, systemic clearance, Cl (mL/min*kg), nonrenal clearance (fractional), volume of distribution, VDss (L/kg), half-life, T1/2 (hrs), MRTD (mM/kg- bw/day), in vitro binding data (from WOMBAT), LogD7.4 (measured), LogPoct (measured), pKa (measured), water solubility (measured), blood brain barrier permeability, cardiac toxicity (Torsades des Pointes), LD50 (mammal data), BDDCS annotation, phase 1 metabolizing enzymes, drugs target annotation, and drugs annotated with anti-targets. These properties are very important in computational drug discovery.

(URL: http://www.sunsetmolecular.com)

ChemSpider ChemSpider is a chemistry search engine. ChemSpider is a free access service providing a structure-centric community for chemists. It provides access to millions of chemical structures and integrates a multitude of other online services. ChemSpider is the richest single source of structure-based chemistry information. It has been built with the intention of aggregating and indexing chemical structures and their associated information into a single searchable repository and makes it available to everybody, at no charge. ChemSpider is a value-added offering of publicly available chemical structures since many additional properties have been added to each of the chemical structures.

(URL: http://www.chemspider.com/)

DrugBank The DrugBank database is a unique bioinformatics and chemoinformatics resource that combines detailed drug (i.e., chemical, pharmacological, and pharmaceutical) data with comprehensive drug target (i.e., sequence, structure, and pathway) information. The database contains nearly 4,800 drug entries including >1,350 Food and Drug Administration (FDA)-approved small molecule drugs, 123 FDA-approved biotech (protein/peptide) drugs, 71 nutraceuticals, and around 3,243 experimental drugs. Additionally, more than 2,500 non-redundant protein (i.e., drug target) sequences are linked to these FDA-approved drug entries. Each DrugCard entry contains more than 100 data fields with half of the information devoted to drug/chemical data and the other half devoted to drug target or protein data.

(URL: http://www.drugbank.ca/)

ZINC ZINC is a free database of commercially available compounds for virtual screening. ZINC contains over 21 million purchasable compounds in ready-to-dock, 3D formats. ZINC is provided by the Shoichet Laboratory in the Department of Pharmaceutical Chemistry at the University of California, San Francisco (UCSF), CA, USA.

Databases dedicated for QSTR DSSTox

Distributed Structure-Searchable Toxicity (DSSTox) Database Network is a freely available chemical database developed by EPA. It can be used for the structure–activity and predictive toxicology studies. The DSSTox provides a public forum for publishing downloadable, structure-searchable, standardized chemical structure files associated with toxicity data. We can search the molecule and its similar chemicals using compound-by-compound name/keyword/smile string or directly we can draw the structure on the screen of JME editor in the result we will get the number of hits and detail information of the compound. This database is very helpful for structure–activity and predictive toxicology; hence, it is useful for people who deal in QSAR/QSTR.

(URL: http://www.epa.gov/ncct/dsstox/index.html)

Registry of Toxic Effects of Chemical Substances (Toxicity Data)

The Registry of Toxic Effects of Chemical Substances (RTECS) is a comprehensive database of basic toxicity information for over 150,000 chemical substances including prescription and nonprescription drugs, food additives, pesticides, fungicides, herbicides, solvents, diluents, chemical wastes, reaction products of chemical waste, and substances used in both industrial and household situations. Reports of the toxic effects of each compound are cited. In addition to toxic effects and general toxicology reviews, data on skin and/or eye irritation, mutation, reproductive consequences, and tumorigenicity are provided.

Material Safety Data Sheet An MSDS is a form containing data regarding the properties of a particular substance. An important component of product stewardship and workplace safety, it is intended to provide information such as physical data (melting point, boiling point, flash point, etc.), toxicity, health effects, first aid, reactivity, storage, disposal, protective equipment, and spill-handling procedures. MSDS is a widely used system for cataloging information on chemicals, chemical compounds, and chemical mixtures. MSDS information may include instructions for the safe use and potential hazards associated with a particular material or product. MSDS can be found anywhere chemicals are being used. There are several other databases that provide chemical structure and information useful for drug discovery. Some of them are mentioned below

UMLS The Unified Medical Language System (UMLS) is a compendium of many controlled vocabularies in the biomedical sciences. It provides a mapping structure among these vocabularies and thus allows one to translate among the various terminology systems; it may also be viewed as a comprehensive thesaurus and ontology of biomedical concepts. UMLS further provides facilities for natural language processing. It is intended to be used mainly by developers of systems in medical informatics. UMLS consists of Metathesaurus as the core database of the UMLS, a collection of concepts and terms from the various controlled vocabularies and their relationships; Semantic Network is a set of categories and relationships that are being used to classify and relate the entries in the Metathesaurus. SPECIALIST Lexicon is a database of lexicographic information for use in natural language processing.

(URL: http://www.nlm.nih.gov/research/umls/)

ChemBank ChemBank is a public, web-based informatics environment created by the Broad Institute's Chemical Biology Program and funded in large part by the National Cancer Institute's Initiative for Chemical Genetics (ICG). This knowledge environment includes freely available data derived from small molecules and small-molecule screens, and resources for studying the data so that biological and medical insights can be gained. ChemBank is intended to guide chemists synthesizing novel compounds or libraries, to assist biologists searching for small molecules that perturb specific biological pathways, and to catalyze the process by which drug hunters discover new and effective medicines. ChemBank stores an increasingly varied set of cell measurements derived from, among other biological objects, cell lines treated with small molecules. Analysis tools are available and are being developed that allow the relationships between cell states, cell measurements, and small molecules to be determined.

(URL: http://chembank.broadinstitute.org/welcome.htm)

eMolecules (http://www.emolecules.com/)

eMolecules is a search engine for chemical molecules. The system was first launched in November 2005. The standard search allows querying for names, substructures, and suppliers. The expert search allows interactive searching using a molecular weight range, CAS numbers, suppliers, etc. Search by File upload (SD or MOL file, i.e., MDL format)

eMolecules Search page for Substructure Search Hits for Aspirin structure and the results after clicking on the first hit (Fig. 1.57). The eMolecules result for Aspirin gives information on molecular weight, molecular formula, CAS number, Links to Focus synthesis and Activate Scientific, etc.

FDA The US FDA is an agency of the US Department of Health and Human Services and is responsible for the safety regulation of most types of foods, dietary supplements, drugs, vaccines, biological medical products, blood products, medical devices, radiation-emitting devices, veterinary products, and cosmetics. The FDA also enforces section 361 of the Public Health Service Act and the associated regulations, including sanitation requirements on interstate travel as well as specific rules for control of disease on products ranging from pet turtles to semen donations for assisted reproductive medicine techniques.

(URL: http://www.fda.gov/)

SPECS Specs, founded in 1987, provides chemistry and chemistry-related services that are required in drug discovery. Specs is one of the world's leading providers of compound management services besides being a main supplier of screening compounds and building blocks to the life science industry. They have a diverse in-house chemical collection, consisting of single synthesized, well-characterized, and drug-like small molecules; it has been built through global acquisition programs utilizing a network of more than 2,000 academic sources worldwide. In addition to providing compound-handling services and high-quality compounds, Specs offers a diverse and unique set of about 400 isolated or synthesized natural products and derivatives thereof from natural sources like plants, fungi, bacteria, sea organisms, etc. These compounds range from common to very complex and rare natural products. Specs' selection of natural products consists of purely isolated or synthesized and well-characterized compounds. This means that no extracts are offered. All natural products offered have been checked by 1H NMR and/or LC/MS to ensure the integrity of the structure and a purity >80%.

(URL: http://www.specs.net/snpage.php?snpageid=home)

MDDR MDDR is a database covering the patent literature, journals, meetings, and congresses. Produced by Symyx and Prous Science, the database contains over 180,000 biologically relevant compounds and well-defined derivatives, with updates adding about 10,000 a year to the database. The MDDR Finder allows you to search the database by structure or across relevant data fields. Symyx also offers MDDR-3D. It is basically a structural database for use with MDL Information Systems, Inc.'s MACCS-II and ISIS/Host software.

MOLTABLE Web Portal MOLTABLE has several databases of both chemical and pharmaceutical importance. MOLTABLE goals are now being redefined to extract and analyze molecular data from literature and patents to support chemical, pharma-

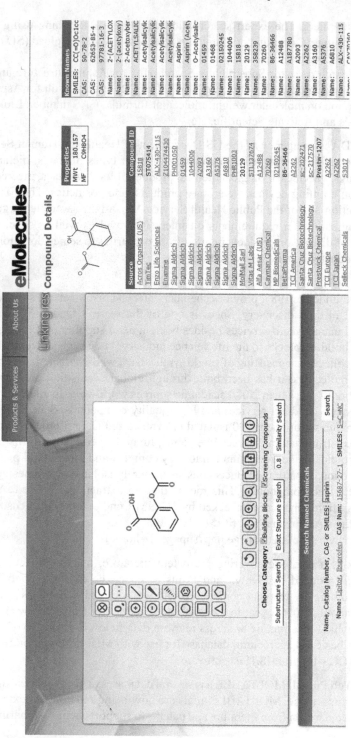

Fig. 1.57 Search and results page for file upload (SD or MOL file, i.e., MDL format)

ceutical, strategic, and other industrial research sectors. The MOLTABLE intends to discover drug candidates against potentially devastating infectious diseases through chemoinformatics research. Dynamic QSAR initiatives through "focused" virtual library design and the results will be made "open access" through MOLTABLE portal. MOLTABLE hosts information on ChemXtreme, a program to harvest chemical information such as properties, activities, and toxicity of molecules from Internet web pages. ChemStar highlights the use of distributed computing environment for calculating molecular properties for large collection of PubChem. Every molecule in the collection is generated with molecular fingerprints for substructure, exact structure, and similar structure analysis. All the molecules are computed for both 2D and 3D descriptors along with physico-chemical properties like solubility, molar refractive index, etc., which is essential for identifying drug-likeliness. The source code and data are freely accessible. MOLTABLE portal can be used for searching chemical information from published literature especially on drug design (8,000+ journals, 4 decades, 18 million articles) (Fig. 1.58)

(URL: http://moltable.ncl.res.in/)

Chemoinformatics.org This is a noncommercial website which compiles information on chemoinformatics web resources and provides links to chemoinformatics programs. It also provides datasets for QSAR, QSPR, BBB penetration, $CaCO_2$ permeability, etc. There are a total of 44 datasets, which are freely downloadable. It also provides links to molecular similarity search, online diversity assessment. The datasets are divided according to the use into binary (active/inactive) datasets, QSAR datasets, QSPR datasets, toxicity datasets, metabolism datasets, permeability datasets, docking datasets, mechanistic datasets, and mixed/other datasets.

(URL: http://www.cheminformatics.org/menu.shtml)

Biological databases Biological databases are libraries of life sciences information, collected from scientific experiments, published literature, high-throughput experiment technology, and computational analyses. They contain information from research areas including genomics, proteomics, metabolomics, microarray gene expression, and phylogenetics. PDB, DNA Data Bank of Japan (DDBJ), European Molecular Biology Laboratory (EMBL), and GenBank are some biological databases which are free on the Internet.

PROSITE PROSITE is a database of protein families and domains. It consists of entries describing the domains, families, and functional sites as well as amino acid patterns, signatures, and profiles in them. These are manually curated by a team of the Swiss Institute of Bioinformatics and tightly integrated into Swiss-Prot protein annotation. It provides additional information about functionally or structurally critical amino acids. The rules contain information about biologically meaningful residues, like active sites, substrate- or co-factor-binding sites, posttranslational modification sites, or disulfide bonds, to help function determination. These can automatically generate annotation based on PROSITE motifs.

(URL: http://www.expasy.ch/prosite/)

Fig. 1.58 Homepage of MOLTABLE portal

EMBL The EMBL is a molecular biology research institution supported by 20 European countries and Australia as an associate member state. It is Europe's primary nucleotide source. We can find out nucleotide sequences and much more data from it. It is the main source for DNA and RNA sequences. The database is a result of the collaboration between GenBank (USA) and the DDBJ.

(URL: http://www.ebi.ac.uk/embl/)

OMIM The Mendelian Inheritance in Man project is a database that catalogs all the known diseases with a genetic component, and, when possible, links them to the relevant genes in the human genome and provides references for further research and tools for genomic analysis of a catalogued gene. OMIM is a comprehensive, authoritative, and timely compendium of human genes and genetic phenotypes. OMIM contains information on all known Mendelian disorders and over 12,000 genes. OMIM focuses on the relationship between phenotype and genotype.

(URL: http://www.ncbi.nlm.nih.gov/omim/)

NCBI The NCBI is part of the USNLM, a branch of the NIH. The NCBI houses genome sequencing data in GenBank and an index of biomedical research articles in PubMed Central and PubMed, as well as other information relevant to biotechnology. All these databases are available online through the Entrez search engine. It contains more than 1,500,000 articles from more than 450 journals.

(URL: http://www.ncbi.nlm.nih.gov/)

1.22.8   Do It Yourself (DIY)

1. Determine the chemical structure using the Connection Tables given below:

```
SMI2MOL
2 1 0 0 0 0 0 0 0 0999 V2000
-0.5100 1.5300 0.0000 C 0 0 0 0 0 0 0 0 0 0 0 0
0.5100 1.5300 0.0000 C 0 0 0 0 0 0 0 0 0 0 0 0
1 2 1 0 0 0 0 0
M END
```

```
9 8
1.0303 0.8847 0.9763 C 0 0 0 0 0 0 0 0 0 0 0 0
1.8847 1.9889 1.5717 C 0 0 0 0 0 0 0 0 0 0 0 0
3.1883 1.4807 1.7425 O 0 0 0 0 0 0 0 0 0 0 0 0
...
1.4753 2.3225 2.5456 H 0 0 0 0 0 0 0 0 0 0 0 0
3.7056 2.1820 2.1139 H 0 0 0 0 0 0 0 0 0 0 0 0
1 2 1 0 0 0 0 0
1 4 1 0 0 0 0 0
1 5 1 0 0 0 0 0
```

```
1 6 1 0 0 0 0
2 3 1 0 0 0 0
2 7 1 0 0 0 0
2 8 1 0 0 0 0
3 9 1 0 0 0 0
M END
```

```
APtclserve04110610582D 0 0.00000 0.00000NCI NS
10 10 0 0 0 0 0 0 0 0 0999 V2000
3.732 2.250 0.000 C 0 0 0 0 0 0 0 0 0 0 0 0
3.732 1.250 0.000 C 0 0 0 0 0 0 0 0 0 0 0 0
2.866 0.750 0.000 N 0 0 0 0 0 0 0 0 0 0 0 0
2.866 -0.250 0.000 C 0 0 0 0 0 0 0 0 0 0 0 0
...
3.732 -0.750 0.000 C 0 0 0 0 0 0 0 0 0 0 0 0
2.329 1.060 0.000 H 0 0 0 0 0 0 0 0 0 0 0 0
1 2 1 0 0 0 0
2 3 1 0 0 0 0
3 4 1 0 0 0 0
4 5 2 0 0 0 0
5 6 1 0 0 0 0
6 7 2 0 0 0 0
7 8 1 0 0 0 0
8 9 2 0 0 0 0
4 9 1 0 0 0 0
3 10 1 0 0 0 0
M END
```

2. Draw the structures for the following SMILES strings:

1.	CCO
2.	CC(=O)O
3.	CC(=O)OCC.O
4.	C=CCBr
5.	C#N
6.	CCN(CC)CC
7.	C(C(C(=O)O)N)O
8.	OC(=O)C(Br)(Cl)N
9.	ClC(Br)(N)C(=O)O
10.	O=C(O)C(N)(Br)C

3. Write the SMILES strings for the following structures:

4. Draw the structures from the following CML code

```
<molecule title="?" id="m1">
       <atomArray>
 <atom id="c1" elementType="C" hydrogenCount="3" />
 <atom id="o1" elementType="O" hydrogenCount="1" />
       </atomArray>
       <bondArray>
 <bond id="b1"atomRefs2="c1 o1" order="S" />
       </bondArray>
</molecule>

<molecule title="?" id="m2">
       <atomArray>
  <atom id="n1" elementType="N" hydrogenCount="3" />
       </atomArray>
 </molecule>

<molecule title="?" id="m3">
       <atomArray>
  <atom id="b1" elementType="B" hydrogenCount="0" >
      <lectron id="e1" count="2"/>
  </atom>
  <atom id="f1" elementType="F" hydrogenCount="0" />
  <atom id="f2" elementType="F" hydrogenCount="0" />
   <atom id="f3" elementType="F" hydrogenCount="0" />
       </atomArray>
       <bondArray>
  <bond id="b1f1"atomRefs2="b1 f1" order="S" />
  <bond id="b1f2"atomRefs2="b1 f1" order="S" />
  <bond id="b1f3"atomRefs2="b1 f1" order="S" />
       </bondArray>
</molecule>

<molecule title="? " id="m4">
       <atomArray>
  <atom id="c1" elementType="C" hydrogenCount="3" />
  <atom id="c2" elementType="C" hydrogenCount="1" />
  <atom id="o1" elementType="O" hydrogenCount="0" />
  <atom id="o2" elementType="O" hydrogenCount="1" />
       </atomArray>
       <bondArray>
  <bond id="b1"atomRefs2="c1 o1" order="S" />
  <bond id="b2"atomRefs2="c2 o1" order="S" />
  <bond id="b3"atomRefs2="c2 o2" order="D" />
       </bondArray>
</molecule>
```

5. Input SMILES of the top ten drugs in the field of medicine and generate 3D structures using Corina and ChemAxon tools; also perform similarity searching in PubChem and Scifinder.

1.22.8.1   Thumb Rules for Structure Representation

- Please take care while converting a structure from one file format to another in a software to make sure all the information is retained like hydrogens, charges, ionic state, etc., before proceeding to the next step.

- Always save your chemical structures in the global formats like *.smi or *.sdf rather than the software-specific format for easy interoperability and compatibility.

1.22.9 Questions

1. What are the known structure representation methods in computer?
2. Write short notes on the databases useful in drug designing experiments.
3. What are the structure-searching methods that you are aware of? Elaborate on any one.
4. Give a brief note on the file conversion programs generally used in chemoinformatics.

References

1. Leach A (2007) An introduction to chemoinformatics. Springer
2. Gasteiger J, Engel T (eds) (2003) Chemoinformatics: a textbook. Wiley-VCH
3. Gasteiger J (ed) (2003) Handbook of chemoinformatics: from data to knowledge. Wiley-VCH
4. Umashankar V, Gurunathan S (2011) Chemoinformatics and its applications. General applied and systems toxicology. Wiley
5. Acton A (ed) (2011) Issues in biotechnology and medical technology research and application (Scholarly Editions)
6. Muffatto M (2006) Open source: a multidisciplinary approach. Imperial College Press
7. http://www.openbsdindia.org/
8. Ortega JM (1994) An introduction to fortran 90 for scientific computing. Oxford University Press
9. http://www.computerhope.com/unix.htm. Accessed on 22 Oct 2013
10. Douglas EC Internetworking with TCP/IP—Principles, Protocols and Architecture
11. Kernighan BW, Ritchie DM (1978) The C programming language, 1st ed. Prentice Hall, Englewood Cliffs
12. Stroustrup B (1997) "1". The C++ Programming Language, 3rd ed. Addison-Wesley
13. Fan Li (2006) Developing chemical information systems: an object oriented approach using enterprise Java. Wiley
14. http://www.perl.org/
15. http://www.python.org/
16. http://www.r-project.org/
17. http://www.nvidia.com/object/cuda_home_new.html
18. Schatz MC, Trapnell C, Delcher AL, Varshaney A (2007) High through put sequence alignment using graphics processing units. BMC Bioinformat 8:474
19. Ash JE, Warr WA, Willett P (1991) Chemical structure systems: computational techniques for representation, searching, and process of structural information. Ellis Horwood, New York
20. Gluck DJ (1964) A chemical structure storage and search systems developed at Du Pont. J Chem Informat Model 5:43–51
21. Warr WA (2011) Representation of chemical structures. WIREs Comput Mol Sci 1(4):557–579

22. Krause S, Willighagen E, Steinbeck C (2000) Using the collaborative forces of the internet to develop a free editor for 2D chemical structures. Mol 5:93–98
23. https://github.com/features/projects
24. http://www.xml-cml.org/
25. Steinbeck C, Han Y, Kuhn S, Horlacher O, Luttmann EE, Willighagen E (2003) The chemistry development kit(CDK): an open source JAVA library for Chemo-and Bioinformatics. J Chem Informat Model 43:493–500
26. http://mcdl.sourceforge.net/
27. Ertl P (2010) Molecular structure input on the web. J Cheminformatics 2:1
28. Bienfait B, Ertl, P (2013) JSME: a free molecule editor in JavaScript. J Cheminformat 5:24
29. http://www.molinspiration.com/. Accessed on 22 Oct 2013
30. http://www.chemaxon.com/. Accessed on 22 Oct 2013
31. http://www.acdlabs.com/resources/freeware/chemsketch/. Accessed on 22 Oct 2013
32. http://www.cambridgesoft.com/Ensemble_for_Chemistry/ChemOffice/. Accessed on 22 Oct 2013
33. http://www.schrodinger.com/. Accessed on 22 Oct 2013
34. http://www.chemcomp.com/. Accessed on 22 Oct 2013
35. http://accelrys.com/products/informatics/cheminformatics/draw/. Accessed on 22 Oct 2013
36. https://www.cas.org/products/scifinder. Accessed on 22 Oct 2013
37. http://www.chemspider.com/. Accessed on 22 Oct 2013
38. http://www.nih.gov/. Accessed on 22 Oct. 2013
39. http://www.beilstein-journals.org/bjoc/home/home.htm. Accessed on 22 Oct 2013
40. Sorter PF, Granito CE, Gilmer JC, Alan G, Metcalf EA (1963) Rapid structure searches via permutated chemical line notation. J Chem Doc 4(1):56–60
41. Fritts LE, Schwind MM (1982) Using the Wiswesser line Notation (WLN) for online, interactive searching of chemical structures. J Chem Inf Comput Sci 22:106–109
42. Dalby A, Nourse JG, Hounshell WD, Gushurst AKI, Grier DL, Leland B A, Laufer J (1992) Description of several chemical structure file formats used by computer programs developed at molecular design limited. J Chem Informat Model 32(3):244
43. Weininger D (1990) SMILES Graphical depiction of chemical structures J Chem Inf Comput Sci 30:237–243
44. www.daylight.com/dayhtml/doc/theory/theory.smarts.html
45. Cline AS, Homer MA, Hurst RW, Smith T, Gregory B (1997) SYBYL Line Notation (SLN): a versatile language for chemical structure representation. J Chem Inf Comput. Sci 37:71–79
46. Alan M (2006) The IUPAC international chemical identifier: In Chl. Chemistry International (IUPAC) 28 (6) http://www.iupac.org/publications/ci/2006/2806/4_tools.html.
47. King RB (ed) (1983) Chemical applications of topology and graph theory. Elsevier
48. Grave K D, Costa F (2010) Molecular graph augmentation with rings and functional groups. J Chem Inf Model 50:1660–1668
49. Santagata LN, Suvire FD, Enriz RD (2001) A matrix representation for the geometrical algorithm to search the chemical space. J Mol Struct Theochem 571:91–98
50. http://www.ccl.net/cca/documents/molecular-modeling/node3.html
51. www.lohninger.com/helpcsuite/connection_table.htmm
52. http://www.cas.org/content/chemical-substances
53. http://accelrys.com/products/informatics/cheminformatics/ctfile-formats/no-fee.php
54. http://www.wolfram.com/
55. http://cactus.nci.nih.gov/SDF_toolkit/
56. http://www.cgl.ucsf.edu/chimera/docs/UsersGuide/xyz.html
57. http://www.wwpdb.org/docs.html
58. Phadungsukanan W, Kraft M, Townsend JA, Murray-Rust P (2012) The semantics of chemical markup language(CML) for computational chemistry. J Cheminform 4(1):15
59. http://www.tripos.com/tripos_resources/fileroot/pdfs/mol2_format.pdf

60. http://www.molsoft.com/2dto3d.html

61. http:// www.molecular-networks.com

62. Barnard JM, Lynch MF, Welford S M (1981) Computer storage and retrieval of generic chemical structures in patents. GENSAL, a formal language for the description of generic chemical structures. J Chem Inf Comput Sci 21:151–161

63. O'Boyle NM, Banck M, James CA, Morley C, Vandermeersch T, Hutchison GR (2011) Open babel: an open chemical toolbox. J Cheminform 3:33

64. http://www.chemaxon.com/marvin/help/applications/molconvert.html

65. Bath, PAP, Andrew R, Willett P, Allen, FH (1994) Similarity searching in files of three-dimensional chemical structures: comparison of fragment-based measures of shape similarity. J Chem Inf Comput Sci 34:141–147

66. Wang Y, Bajorath J (2010) Advanced Fingerprint methods for similarity searching: balancing molecular complexity effects. Comb Chem High Throughput Screen 13:220–228

67. Wipke W T, Krishnan S, Ouchi G I (1978) Hash functions for rapid storage and retrieval of chemical structures. J Chem Inf Comput Sci 18:32–37

68. Takahashi Y, Sukekawa M, Sasaki S (1992) Automatic identification of molecular similarity using reduced-graph representation of chemical structure. J Chem Inf Comput Sci 32:639–43

69. http://www.cas.org/etrain/stn/exactfamilysearch.html

70. http://www.chemaxon.com/jchem/intro/index.html

71. http://www2.chemie.uni-erlangen.de/software/wodca/subsearch.html

72. Vogt M, Bajorath J (2013) Similarity searching for potent compounds using feature selection. J Chem Inf Model 53(7):1613–1619

73. Sayle RA, Batista JJ, Grant A (2013) An efficient maximum common subgraph(MCS) searching of large chemical databases. J Cheminformat 5(1):O15

74. Chen X, Reynolds CH (2002) Performance of similarity measures in 2D fragment-based similarity searching: comparison of structural descriptors and similarity coefficients. J Chem Inf Comput Sci 42:1407–1414

75. Holliday JD, Salim N, Whittle M, Willett P (2003) Analysis and display of the size dependence of chemical similarity coefficients. J Chem Inf Comput Sci 43:819–828

76. Weiss G (2007) Exploring the milky way of molecular diversity combinatorial chemistry and molecular diversity. Curr Opin Chem Biolo 11:241–243

77. Karthikeyan M, Vyas R (2012) Chemical structure representation and applications in computational toxicology. In: Reisfield B, Mayeno AN (ed) Computational toxicology. Springer, pp 167–192

78. Karthikeyan M, Uzagare D, Krishnan S (2003) Compressed chemical markup language for compact storage and inventory applications. 225th ACS Meeting New Orleans. CG ACS, pp 23–27

79. Karthikeyan M, Krishnan S, Pandey AK (2006) Harvesting chemical information from the internet using a distributed approach. Chem Extreme J Chem Inf Model 46:452–461

80. Karthikeyan M, Bender, A (2005) Encoding and Decoding Graphical Chemical Structures as Two-Dimensional (PDF417) Barcodes. J Chem Inform Model 45:572–580

81. http:// www. moltable.ncl.res.in

82. Valko AT, Johnson AP (2009) CLiDE Pro: the latest generation of CLiDE, a tool for optical chemical structure recognition. J Chem Inform Model 49:780–787

83. Filippov IV, Nicklaus MC (2009) Optical structure recognition software to recover chemical information OSRA, an open source solution. J Chem Inf Model 49(3):740–743

84. http://infochem.de/products/index.shtml

85. Karthikeyan M, Krishnan S, Pandey AK, Bender A (2008) Distributed chemical computing using Chemstar: an open source Java Remote Method Invocation architecture applied to large scale molecular data from Pubchem. J Chem Info Model 48:691–703

86. Song CM, Bernardo PH, Chai CL, Tong JC (2009) CLEVER: pipeline for designing insilico chemical libraries. J Mol Graph Model 27(5):578–583

87. Huang Z (1998) Extensions to the k-means algorithm for clustering large data sets with categorical values. Data Min Knowl Discov 2:283–304
88. Hoon MJL, Imoto S, Nolan J, Miyano S (2004) Open source clustering software. Bioinforma 20(9):1453–1454
89. Saldanha AJ (2004) JAVA treeview extensible visualization of microarray data. Bioinforma 20:3246–3248
90. http://www.chemaxon.com/products/jklustor/
91. Ullman J (1997) First course in database systems. Prentice-Hall Inc., Simon & Schuster, p 1
92. Mike C SQL Fundamentals
93. http://pubchem.ncbi.nlm.nih.gov/summary/summary.cgi?cid=712

Chapter 2
Chemoinformatics Approach for the Design and Screening of Focused Virtual Libraries

Abstract It is challenging to handle a large volume of molecular data without appropriate tools. Here, we describe the need and the approaches for the development of focussed virtual libraries to design efficient molecules and optimize them for lead generation. The experimental chemists and biologists are more interested in properties of chemicals and their response to biological system in both beneficial and adverse effects context rather than just their structures. In this chapter, the focus is to relate newly designed chemical structures to their predicted activity, property or toxicity. Property prediction tools save time, money and lives of experimental animals. They come in handy while taking informed decisions especially in certain cases involving pharmacodynamic studies of drug molecules in humans where there are inevitable ethical and safety concerns. Property prediction is an important component in virtual screening which is at the heart of drug design and the most important step where chemoinformatics plays a major role. The other fields where structure–activity relation-based principles hold good for virtual screening are agrochemicals and environmental science, specifically the toxicity and biodegradability prediction of pollutant molecules. In this chapter, we will show how to design software tools to handle generation of focussed virtual libraries from a given set of molecules with common features, fragments or bioactivity spectrum.

Keywords Descriptors · Chemical properties · Chemoinformatics · Drug design

2.1 Introduction to Structure–Property Correlations

Chemists are mainly interested in the structure of chemicals to know those properties which can be of some use to us. Physico-chemical properties, bioactivities and toxicity-related data of chemicals available from scientific literature or from experimental results are used for building predictive models applying advanced mathematical methods or machine learning techniques based on the principle of 'similar structures possess similar property' [1–3]. The quality of predictive models basically depends on the selection of relevant molecular descriptors and accuracy of experimental data [4]. Basically, molecular descriptors are the structural features

M. Karthikeyan, R. Vyas, *Practical Chemoinformatics,*
DOI 10.1007/978-81-322-1780-0_2, © Springer India 2014

encoding independent property of interest such as activity, property and toxicity [5]. The relation between structure and property is studied by computing binary fingerprints and descriptors from the molecular graph and its three-dimensional (3D) chemical structure respectively [6].

2.1.1 Descriptors

Descriptors are properties that describe a molecule on the basis of either some physico-chemical property like melting point, boiling point or an algorithm like two-dimensional (2D) fingerprint [7]. There are several types of molecular descriptors and features used for establishing structure–property links. Most commonly used molecular descriptors are constitutional, surface, molecular connectivity, electrostatic, shape, geometry, quantum chemical, physico-chemical, hybrid, etc. which are all intimately related to each other [8]. The constitutional descriptors are the most simple and common ones that just provide information on the chemical composition of molecules [9]. The topological descriptors which encode the surface properties of a molecule are used to ascertain the solubility and permeability of a proposed drug. Electrostatic descriptors such as polarizability, dipole moment and ionization energy predict crystalline density [10]. Geometrical or 3D descriptors based on xyz coordinates provide rich information regarding a molecule's orientation in space and are often more useful than others in predicting biological activity [11]. Quantum chemical descriptors in theory encompass all the electronic and geometrical features of a molecule compared to empirical ones, the only drawback being the computational overload [12]. Some of the quantum chemical descriptors include lowest unoccupied molecular orbital (LUMO) energies, orbital electron density, delocalizability, etc. [13]. Hybrid descriptors such as BCUT [14] WHIM [15] were initially developed for chemical diversity but later found useful as inputs for building predictive models. Another class of descriptors include the binary bit string-based fingerprint descriptors which are employed for similarity searching in databases. The known literature fingerprints, viz. Molecular Design Limited Molecular ACCess System (MDL MACCS) 166-bit keys [16], circular fingerprints[17], Extended Convective Forecast Product (ECFP) [18], FCF2 [19], Unity [20], PubChem fingerprints [21] and TPC [22], have been applied to a wide range of applications including prediction of absorption, distribution, metabolism, excretion and toxicity properties (Fig. 2.1).

From a drug discovery point of view, the most important descriptor among molecular properties is the solubility of a compound [23]. This in turn impacts the oral bioavailability of a drug—an important pharmacokinetic parameter [24]. Solubility is also found to be an important parameter for lipid-based formulation excipients in pharmacy [25]. Another equally relevant descriptor is logP, i.e. the water/octanol partition coefficient [26]. The prior knowledge of these descriptors is considered important during the preclinical trial stage in the drug discovery pipeline. Currently, descriptors for target and ligand are computed simultaneously for predicting side effects in drugs and polypharmacology, an emerging concept in medicine, wherein other therapeutic options are explored for a known marketed drug [27]. Apart from drug design, another field where descriptors play an important role is material science where

Fig. 2.1 Commonly
employed descriptors and fingerprints in structure–property correlation studies

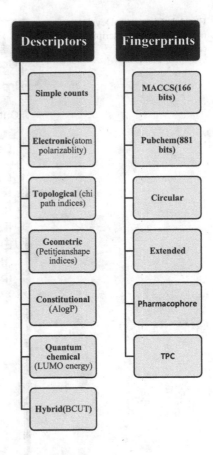

the selection of a right descriptor can lead to improved energetic substances [28]. By evaluating molecular, microscopic and structural descriptors of an adsorbate–adsorbent system, single-component adsorption isotherms can be predicted [29].

In this section, we shall practically see how to compute descriptors using open-source, free, commercial tools for a given set of molecules. The right choice of independent uncorrelated descriptors is the next important step. Genetic algorithm (GA)-based approaches are employed to select the optimal subset of descriptors [30]. Many linear and non-linear models to predict a physico-chemical property or bioactivity can be built using selected descriptors by employing machine learning methods like neural networks which are discussed in detail in the next chapter.

2.1.1.1   Open-Source Tools for Computing Descriptors

Chemistry Development Kit

SMILES notation of a molecule can be input to calculate properties/descriptors using open-source programs. The Chemistry Development Kit (CDK) is a scientific, Lesser General Public License (LGPL)-ed library for bio- and cheminformat-

Fig. 2.2 Chemistry Development Kit (*CDK*) descriptors calculator

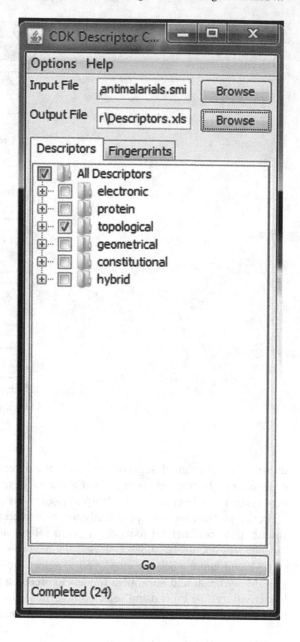

ics and computational chemistry written in Java [31]. A CDK descriptor calculator (v1.3.8) has been developed for cdk1.5 which calculates descriptors and fingerprints given a .smi or .sdf input file [32]. The user can select only the type of descriptor; the program currently uses a default parameter setting for each descriptor (Fig. 2.2).

JOElib

It is an integrated chemoinformatics package governed by the GNU general public license [33]. The Java libraries are available at its homepage site. The descriptors include simple atom group counts which are good enough to build primitive quantitative structure–property relationships (QSPR) models but for predicting complex biological properties transformed descriptors should be computed. One can also write their own descriptor and classes into the program.

Source Code for Computing JOELib Descriptors from Simplified Molecular-Input Line-Entry System format of Any Molecule

```
public String[] getData(String smi) {
     BasicDescriptors bd = new BasicDescriptors();
     JOEMol mol = new JOEMol(IOTypeHolder.instance().getIOType("SMILES"),
IOTypeHolder.instance().getIOType("SDF"));
     String[] out = new String[2];
     try {
         JOESmilesParser.smiToMol(mol, smi, "mol_name");
         double logP = 0;
         int premiscuious = 0;
         bd.computeDescriptors(mol, logP, premiscuious);
         DecimalFormat df1 = new DecimalFormat("####.####");
         LogP lp = new LogP();
         bd.logP = lp.getDoubleValue(mol);
         out[0] = "";
         out[0] += "HBD:" + bd.hbd + ";LogP:" + df1.format(bd.logP) + ";M.Wt:" +
df1.format(bd.mw) + ";Promiscuous:" + bd.promiscuous + ";TPSA:" +
df1.format(bd.tPSA) + ";Basic Score:" + bd.basicScore() + ";HBA:" + bd.hba + ";DL
Failures:" + bd.drugLikeFailures() + ";LL Failures:";
         out[0] += bd.leadLikeFailures() + ";" + bd.basicScore() + ";PDL:" +
df1.format(bd.PDL()) + ";PLL:" + df1.format(bd.PLL()) + ";CFMS Penalties:" +
bd.CFMSpenalties() + "';";
         out[1] = bd.stringSSKey3DS;
         out[0] += ";numberOfBadAtoms :" + bd.numberOfBadAtoms;
         out[0] += ";numberOfCF3 :" + bd.numberOfCF3;
         out[0] += ";numberOfN :" + bd.numberOfN;
         out[0] += ";numberOfNO2 :" + bd.numberOfNO2;
         out[0] += ";numberOfO :" + bd.numberOfO;
         out[0] += ";numberOfS :" + bd.numberOfS;
         out[0] += ";numberOfSO2 :" + bd.numberOfSO2;
         out[0] += ";numberOfX: " + bd.numberOfX;
         String[] rp = bd.reactivePatterns;
         for (int i = 0; i < rp.length; i++) {
             out[0] += ";numberOf RP" + i + ":" + rp[i];
         }
         String[] wp = bd.warheadPatterns;
         for (int i = 0; i < wp.length; i++) {
             out[0] += ";numberOf WHP" + i + ":" + wp[i];
         }
         getSMPatterns sm = new getSMPatterns();
         String[] out1 = sm.getToxicophoreFP(smi); //toxicophoreFingerprints
         for (int i = 0; i < out1.length; i++) {
             out[0] += ";toxph FP:" + i + ":" + out1[i];
         }
         String[] out2 = sm.getChemClassFP(smi);  //ChemicalClassFP
         for (int i = 0; i < out2.length; i++) {
             out[0] += ";chem FP:" + i + ":" + out2[i];
         }
     } catch (Exception e) {
         System.out.println(e);
     }

     return out;

}
```

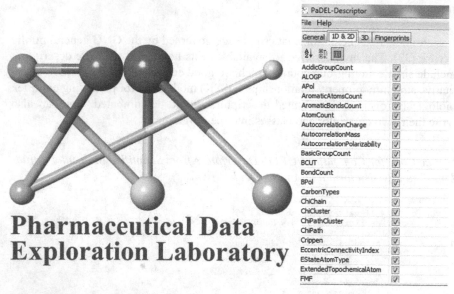

Fig. 2.3 PaDEL descriptor calculator graphical user interface (*GUI*)

PaDEL

It is an open-source program that computes 797 descriptors and 10 types of fingerprints [34]. It uses the CDK library for computing descriptors; however, some new descriptors have been added, mainly electrotopological state descriptors [35]. Both graphical user interface (GUI) and command line options are available. The advantage of this software is the large number of file formats it supports, around 90 in number. Further, it surpasses the CDK calculator with regard to its speed due to its multithreaded nature (Fig. 2.3).

2.1.1.2 Free Programs

PowerMV is a descriptor generation and compound annotation tool designed biologists and statisticians for quickly screening their assay results and gaining some knowledge regarding their potential biological mechanism [36]. Four descriptor sets are used, four bit string and two continuous, which are used for nearest neighbour searching in annotated databases. The package is written in Visual C and C++and runs on a .NET framework unlike previous Java-based programs. The program provides users with two versions: basic and affiliate with greater graphics and better descriptors in the latter [37]. One can build classification and regression models through graphical interface to the R program.

2.1.1.3   Tools Requiring An Academic License

Calculator plug-in in Marvin Beans from ChemAxon is used for calculating a number of descriptors and is available via an academic request [38]. It can be accessed from Marvin Sketch and Marvin view modules. For efficiency, it is advisable to run it using *cxcalc* command in batch mode from command prompt. A number of diverse descriptors can be computed in a short time.

A Practice Tutorial

Here, we compute some selected properties for a .smi file containing 100 molecules belonging to the well-known Ames data set [39]. Download this file and put in the Marvin Beans directory. We begin by calculating simple but powerful atomic descriptors like atom counts and atomic composition. The cxcalc commands are available in the original directory where ChemAxon is installed and then go to the subdirectory Marvin Beans docs users cxcalc-calculations.html. First, navigate to the directory containing Marvin Beans bin folder in command prompt and type cxcalc -h to list the commands. Then, type the commands cxcalc atomcount -z 7 Ames100. smi and then cxcalc composition -S true Ames100.smi to compute the atom counts and atomic composition for all the 100 molecules in the data set. Similarly, type cxcalc atomicpolarizability test Ames100.smi to calculate the polarizability of each atom in all 100 molecules (Fig. 2.4).

We can also compute 3D descriptors using the 'cxcalc' option. Draw a structure of aspirin molecule (acetyl salicylic acid) in Marvin Sketch and save it as .smi in the Marvin Beans folder. In the command window, type cxcalc stereoisomers -v true aspirin.mol to generate the stereoisomer of the molecule. Similarly, the command cxcalc lowestenergyconformer -f mrv test aspirin.mol calculates the lowest energy conformer of aspirin (Fig. 2.5).

Molecular graph-based descriptors can also be calculated using the cxcalc command. Here, let us compute Randic index [40] and Wiener index [41] which are important molecular connectivity descriptors. Randic index, also called bond index, is the sum of bond contributions in a molecule and Wiener path is a topologic index describing the shortest path between all pairs of vertices. The syntax of the commands is cxcalc randicindex test ames100.smi and cxcalc wienerindex test ames100.smi (Fig. 2.6).

Data processed in one program can be piped into another using the | vertical line command. Let us compute the logP values for 100 molecules in Ames data set and then pipe the output data to Marvin view to view the table alongside. The command to do so is cxcalc -S -t myLOGP logP -a 0.15 -k 0.05 test ames100.smi | mview— (Fig. 2.7).

Log p is the water/octanol partition coefficient [42]; there is another descriptor called logD [43] which is a distribution coefficient especially useful for determining lipophilicity of ionizable compounds as it accounts for pH dependence of molecules in aqueous solution.

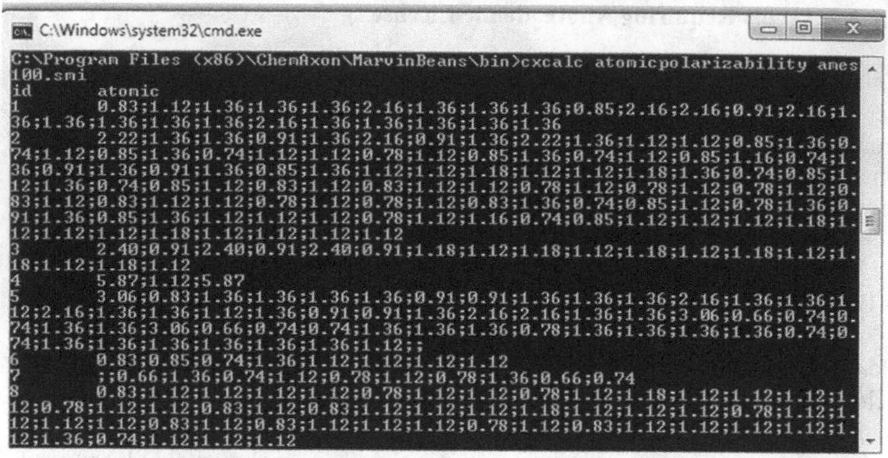

```
C:\Windows\system32\cmd.exe

C:\Program Files (x86)\ChemAxon\MarvinBeans\bin>cxcalc atomicpolarizability anes
100.smi
id      atomic
1       0.83;1.12;1.36;1.36;1.36;2.16;1.36;1.36;1.36;0.85;2.16;2.16;0.91;2.16;1.
36;1.36;1.36;1.36;1.36;2.16;1.36;1.36;1.36;1.36;1.36
2       2.22;1.36;1.36;0.91;1.36;2.16;0.91;1.36;2.22;1.36;1.12;1.12;0.85;1.36;0.
74;1.12;0.85;1.36;0.74;1.12;1.12;0.78;1.12;0.85;1.36;0.74;1.12;0.85;1.16;0.74;1.
36;0.91;1.36;0.91;1.36;0.85;1.36;1.12;1.12;1.18;1.12;1.12;1.18;1.36;0.74;0.85;1.
12;1.36;0.74;0.85;1.12;0.83;1.12;0.83;1.12;1.12;0.78;1.12;0.78;1.12;0.78;1.12;0.
83;1.12;0.83;1.12;1.12;0.78;1.12;0.78;1.12;0.83;1.36;0.74;0.85;1.12;0.78;1.36;0.
91;1.36;0.85;1.36;1.12;1.12;1.12;0.78;1.12;1.16;0.74;0.85;1.12;1.12;1.12;1.18;1.
12;1.12;1.12;1.18;1.12;1.12;1.12;1.12
3       2.40;0.91;2.40;0.91;2.40;0.91;1.18;1.12;1.18;1.12;1.18;1.12;1.18;1.12;1.
18;1.12;1.18;1.12
4       5.87;1.12;5.87
5       3.06;0.83;1.36;1.36;1.36;1.36;0.91;0.91;1.36;1.36;1.36;2.16;1.36;1.36;1.
12;2.16;1.36;1.36;1.12;1.36;0.91;0.91;1.36;2.16;2.16;1.36;1.36;3.06;0.66;0.74;0.
74;1.36;3.06;0.66;0.74;0.74;1.36;1.36;1.36;1.36;0.78;1.36;1.36;1.36;1.36;0.74;0.
74;1.36;1.36;1.36;1.36;1.36;1.36;1.12;;
6       0.83;0.85;0.74;1.36;1.12;1.12;1.12;1.12
7       ;;0.66;1.36;0.74;1.12;0.78;1.12;0.78;1.36;0.66;0.74
8       0.83;1.12;1.12;1.12;1.12;0.78;1.12;1.12;0.78;1.12;1.18;1.12;1.12;1.12;1.
12;0.78;1.12;1.12;0.83;1.12;0.83;1.12;1.12;1.12;1.18;1.12;1.12;0.78;1.12;1.
12;1.12;0.83;1.12;0.83;1.12;1.12;1.12;0.78;1.12;0.83;1.12;1.12;1.12;1.12;1.
12;1.36;0.74;1.12;1.12;1.12
```

Fig. 2.4 Atomic polarizabilities for 100 molecules using the cxcalc program in Marvin Beans

Code for Reading a Molecule from a Structure Data File and Printing LogD Values in a Given pH Range

```
plugin.setMolecule(mol);
plugin.run();
//get and print logD values
double[ ] pHs = plugin.getpHs();
double[ ]logDs=plugin.getlogDs();
for( int i=0; i<logDs.length; i++) {
double pH =pHs[i];
double logD = logDs[i];
System.out.println(pH+", "+logD);
}
```

2.1.1.4 Commercial Software to Calculate Molecular Properties

OpenEye

OpenEye company provides software to the pharmaceutical industry for molecular modelling and chemoinformatics. Their Shape TK module facilitates the calculation of molecular descriptors for shape volume overlap between molecules and spatial similarity of chemical groups [44].

Schrodinger

The QikProp module computes pharmaceutically relevant descriptors for a large data set containing million compounds in an hour in batch mode [45]. It is a quick,

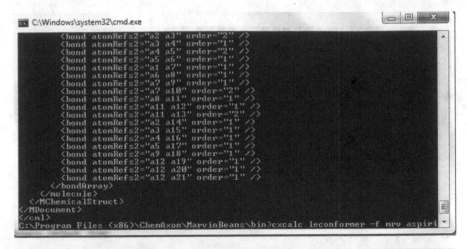

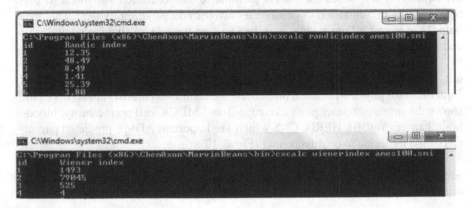

Fig. 2.5 Computed stereoisomer and lowest energy conformer of aspirin using cxcalc command

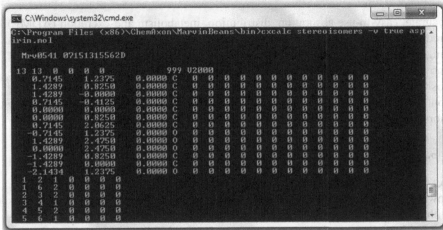

Fig. 2.6 Randic and Weiner index values computed for the data set

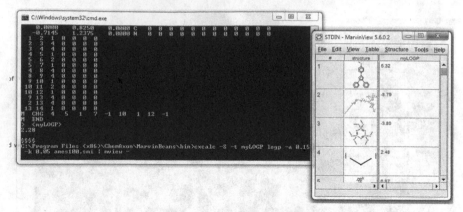

Fig. 2.7 LogP data for 100 molecules piped to Marvin View to visualize the tabulated results

accurate, easy-to-use absorption, distribution, metabolism and excretion (ADME) prediction program designed by Professor William L. Jorgensen [46]. It provides ranges for comparing a particular molecule's properties with those of 95 % of known drugs. It can flag 30 types of reactive functional groups that may cause false positives in high-throughput screening (HTS) assays. QikProp input must be a file containing the 3D structure (x, y, and z coordinates and atomic numbers) of one or more molecules.

A Practice Tutorial

Let us compute the ADME properties of the previous Ames100 data set. First, download the data set from www.chemoinformatics.org. It contains 100 molecules with the binary mutagenicity classification data. We will compute QikProp descriptors for them. Before submitting to QikProp, it is advisable to prepare the molecules using the LigPrep module in Schrodinger. LigPrep automatically converts them to 3D structures; also check for correct tautomeric and ionization variations. It performs energy minimization to generate a customized ligand library [47]. The .mae output file from LigPrep is input into the QikProp module by clicking applications and submitting the job (Figs. 2.8 and 2.9).

The output from the QikProp is obtained in four files, viz. qikpropames100. out, qikpropames100.mae, qikpropames100.qpsa and qikpropames100.csv. Apart from the usual physico-chemical properties, the comma-separated values (CSV) file shows the important descriptors like caco-2 and MDCK cell permeability, blood–brain barrier (logBB), HERG, CNS which are important ADME predicted parameters for a molecule to qualify as drug (Fig. 2.10).

Alternatively, a simple python script can be downloaded from the Schrodinger Script Center for generating molecular descriptors like topological, Molecular Orbital PACkage (MOPAC) and QuikProp (Script name: molecular_descriptors.py

Fig. 2.8 LigPrep input screen

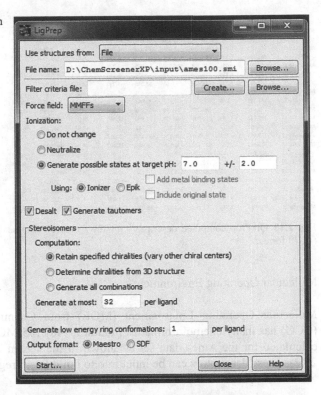

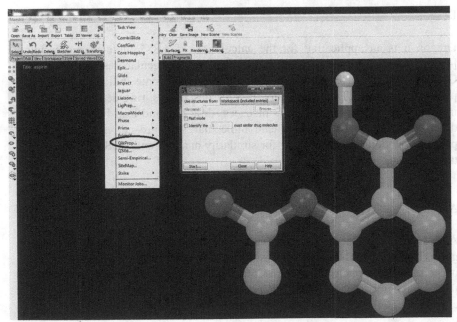

Fig. 2.9 QikProp input screen

	mol_MW	dipole	SASA	FOSA	FISA	ClQPlogs	QPlogHERG	QPPCaco	QPlogBB	CNS	QPPMDCK	QPlogKp	IP(eV)	EA(eV)
Molecule 1	267.836	2.321	238.882	65.019	0	-1.162	-2.341	9906.038	0.322	1	10000	-1.52	9.468	1.47
Molecule 2	786.845	6.541	1170.359	230.926	365.42	-10.133	-5.665	0.218	-5.337	-2	0.091	-4.993	8.508	1.306
Molecule 3	150.088	3.698	307.816	32.383	275.433	-0.592	1.091	1.553	-1.822	-2	0.737	-6.116	11.184	-0.352
Molecule 4	748.993	9.332	1002.147	927.762	74.385	-3.276	-6.089	121.425	-0.122	0	61.995	-5.863	8.673	-0.888
Molecule 5	748.993	6.218	955.204	860.929	94.275	-3.276	-5.603	78.649	-0.257	0	38.769	-6.23	8.644	-1.104
Molecule 6	363.419	3.858	410.036	0.045	93.193	-5.792	-0.894	327.908	0.378	0	10000	-2.853	11.627	1.089
Molecule 7	363.419	3.249	411.454	0.05	93.515	-5.792	-0.909	325.607	0.377	0	10000	-2.859	11.566	1.009
Molecule 8	363.419	3.484	420.233	0	93.591	-5.792	-1.207	325.072	0.368	0	10000	-2.86	11.493	1.089
Molecule 9	363.419	3.92	405.132	0.052	93.449	-5.792	-0.678	326.081	0.384	0	9546.993	-2.858	11.528	1.097
Molecule 10	572.427	13.989	797.008	145.86	469.231	-2.578	-4.041	0.089	-5.275	-2	0.027	-8.286	8.508	0.222
Molecule 11	739.693	6.539	689.943	340.723	35.904	-12.462	-4.382	4522.998	0.257	1	10000	-1.029	10.813	1.148
Molecule 12	739.693	6.977	657.118	342.326	37.711	-12.462	-3.969	4347.944	0.172	1	10000	-1.062	10.818	1.053
Molecule 13	739.693	6.239	707.365	338.973	37.445	-12.462	-4.603	4373.33	0.272	1	10000	-1.057	10.886	1.256
Molecule 14	739.693	7.508	682.294	330.645	36.486	-12.462	-4.245	4465.872	0.263	1	10000	-1.04	10.79	1.108
Molecule 15	739.693	6.747	655.209	305.491	37.185	-12.462	-3.864	4398.245	0.269	1	10000	-1.053	10.872	1.264
Molecule 16	739.693	8.7	720.392	348.038	36.474	-12.462	-4.795	4467.053	0.283	1	10000	-1.04	10.907	1.148
Molecule 17	739.693	6.383	680.288	340.109	37.138	-12.462	-4.236	4402.669	0.227	1	10000	-1.052	10.898	1.225
Molecule 18	739.693	6.125	660.244	316.383	37.054	-12.462	-3.975	4410.81	0.251	1	10000	-1.05	10.908	1.228
Molecule 19	739.693	7.103	697.749	339.842	35.843	-12.462	-4.521	4528.985	0.271	1	10000	-1.028	10.86	1.262
Molecule 20	739.693	6.264	718.947	350.061	35.806	-12.462	-4.767	4532.656	0.285	1	10000	-1.027	10.809	1.206
Molecule 21	739.693	7.036	702.693	348.322	35.869	-12.462	-4.587	4526.436	0.259	1	10000	-1.028	10.844	1.268

Fig. 2.10 QikProp computed descriptors for the Ames100 data set in comma-separated values (*CSV*) format

Molecular Operating Environment

Molecular Operating Environment (MOE) from Chemical Computing Group (CCG) has many chemoinformatics modules; [48] 185 MOE 2D descriptors were calculated for the Ames data set as shown in the screen capture (Figs. 2.11 and 2.12). These descriptors can be input into to build linear regression models.

Dragon

Dragon 6 is an application for the calculation of 4,885 molecular descriptors [49]. The latest version of Dragon includes new molecular descriptors such as CATS 2D, Klein TDB autocorrelations, atom-type E-state indices, extended to-pochemical atom (ETA) descriptors, P_VSA descriptors, ring descriptors, several indices from different 2D and 3D matrices, drug-like and lead-like filters. These descriptors can be used to evaluate molecular structure–activity or structure–property relationships, as well as for similarity analysis and HTS of molecule databases.

Accelerys

ADME and molecular mechanics descriptors can be calculated using Accelerys program. Their TOPKAT module is an established *in silico* method for assessing toxicity prediction of organic compounds [50]. TOPKAT can help assess environmental fate, ecotoxicity, toxicity, mutagenicity, and reproductive/developmental toxicity of chemicals. TOPKAT technology is currently used to optimize therapeutic ratios of

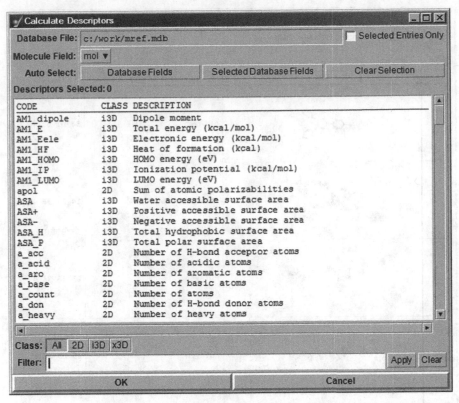

Fig. 2.11 Descriptors list in Molecular Operating Environment (Chemical Computing Group) *MOE(CCG)*

lead compounds, prioritise promising compounds for further development/investment, evaluate intermediates, metabolites and pollutants screen compounds generated via HTS systems, assess pharmaceutical, commercial, industrial and agricultural chemical products for potential safety problems and set dose ranges for animal assays.

2.1.1.5   In-House-Developed Open-Source Tool

Large-scale distributed computing of chemical properties has been carried out using ChemStar, wherein the Topological Polar Surface Area (TPSA) property of 18 million compounds was studied using Java Remote Method Invocation (JAVA RMI) [51].

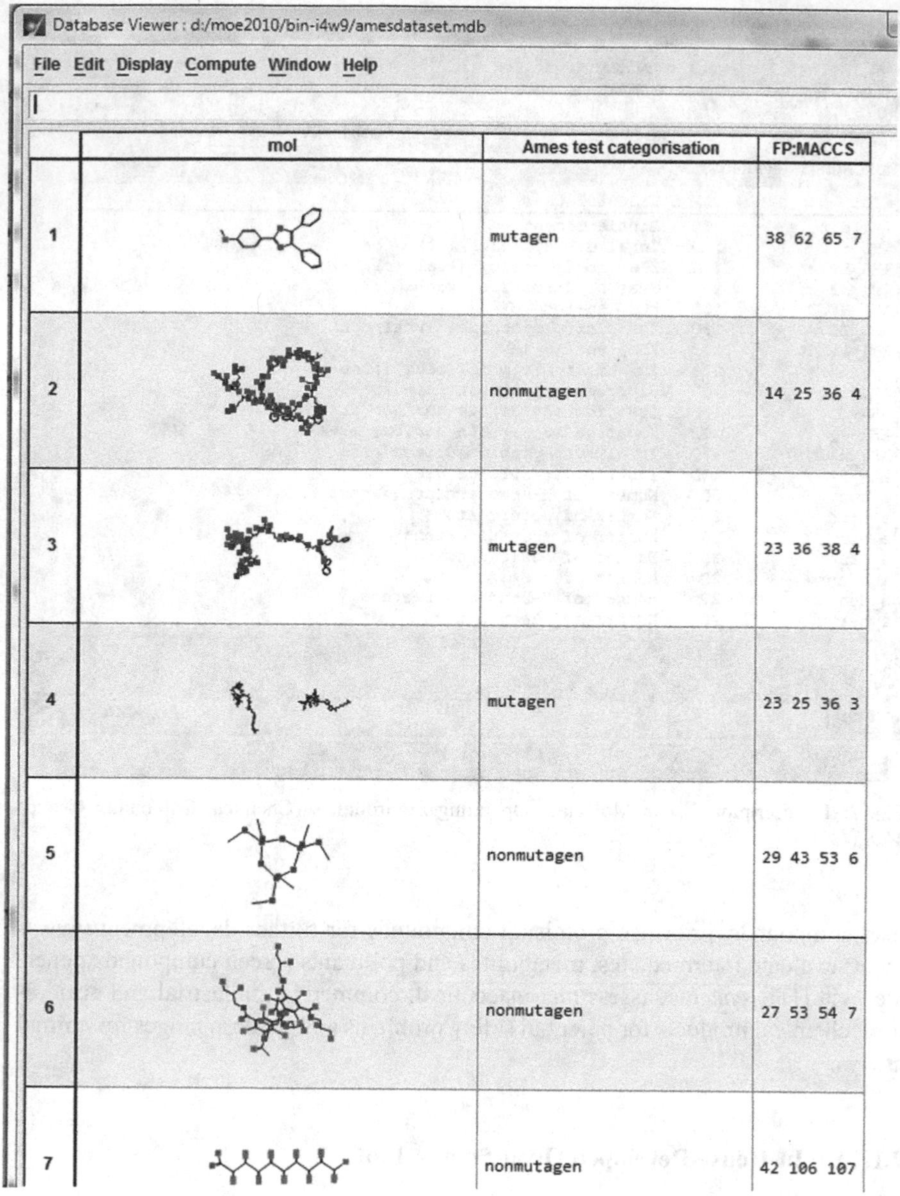
Database Viewer : d:/moe2010/bin-i4w9/amesdataset.mdb
File Edit Display Compute Window Help
mol
Ames test categorisation
FP:MACCS
1
mutagen
38 62 65 7
2
nonmutagen
14 25 36 4
3
mutagen
23 36 38 4
4
mutagen
23 25 36 3
5
nonmutagen
29 43 53 6
6
nonmutagen
27 53 54 7
7
nonmutagen
42 106 107

Fig. 2.12 Molecular Design Limited Molecular ACCess System (*MACCS*) fingerprints computed for Ames data set

Code for Distributed Computing Of Molecular Properties Using ChemStar[1]

```
Class:
Read Input file(String fname){

Distribute the tasks to Clients

Client Components (Parallel mode)

-Get List of Calculator Plugins (ChemAxon / PADEL / CDK / JOELib)
-LogP
-TPSA
-MWT
-HBA
-HBD
-WeinerPath
-Volume
-ADMET
-Toxicophores
-Chemophores
-Pharmacophores
-MACCS Keys
-nAtoms (C, H, N,S,O,Cl,Br,I,N,P)

Send the results to Server

}

class AppendFileStream extends OutputStream
{

    public AppendFileStream(String s)
        throws IOException
    {
        fd = new RandomAccessFile(s, "rw");
        fd.seek(fd.length());
    }

    public void close()
        throws IOException
    {
        fd.close();
    }

    public void write(byte abyte0[])
        throws IOException
    {
        fd.write(abyte0);
    }

    public void write(byte abyte0[], int i, int j)
        throws IOException
    {
        fd.write(abyte0, i, j);
    }

    public void write(int i)
        throws IOException
    {
        fd.write(i);
    }

    RandomAccessFile fd;
}
```

[1] Interested readers are encouraged to download the supporting materials related to ChemStar application (JCIM' 2008, ACS).

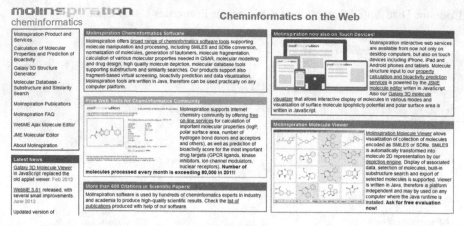

Fig. 2.13 Home page of molinspiration server on the web

2.1.2   Online Property Prediction Tools

All of them mostly employ any of the machine learning-based quantity structure–activity relationship (QSAR)/QSPR methods for property prediction.

2.1.2.1   Molinspiration

This site provides a range of tools for structure drawing, property prediction, etc. [52], Fig. 2.13.

2.1.2.2   Prediction of Activity Spectra for Substances

The acronym PASS stands for prediction of activity spectra for substances [53]. Upon entering a structural formula of a chemical substance, the program computes the potential biological activities of this compound. To execute the prediction, PASS requires a knowledge base about structure–activity relationships (SAR) for compounds with known biological activities. This is provided by SAR Base, containing the analysis results obtained with an in-house training set of more than 250,000 compounds with known biological activities. This training set is continuously curated and expanded. SAR Base can also be replaced by an exclusive knowledge base, which can be created using in-house data. The knowledge base together with the user-defined constraints of biological activities of interest and relevant parameters provides PASS the starting point for the computational prediction (Fig. 2.14).

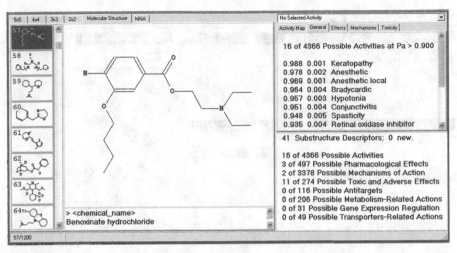

Fig. 2.14 Prediction of activity spectra for substances (*PASS*) property prediction server

2.1.2.3 AquaSol

A web-based predictor, AquaSol, is available online through the ChemDB portal that can be applied to the problem of predicting aqueous solubility [54]. Molpro, another module in the portal, predicts molecular properties other than 3D structures.

2.1.2.4 Molecular descriptor family prediction SAR

An algorithm for extracting useful information from the topological and geometrical representation of chemical compounds was developed and integrated to calculate molecular descriptor family (MDF) members [55]. The activity is predicted based on a learning set, a preciously obtained MDF SAR model and a molecule submitted as HIN file by the user.

2.1.2.5 preADMET

It is a commercial website used to compute 2,000 descriptors including absorption, distribution, metabolism, elimination and toxicity (ADMET)-relevant properties like caco-2 cell permeability, blood–brain barrier, human intestinal absorption, etc. [56]. It also comes with a drawing tool and library builder.

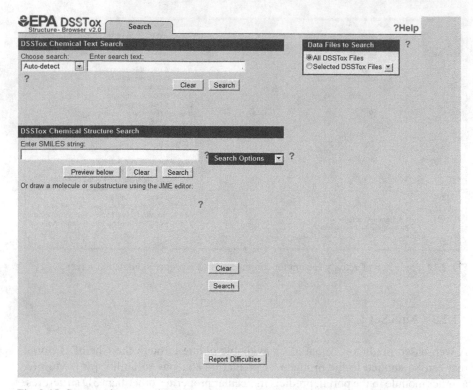

Fig. 2.15 Structure browser of Distributed Structure-Searchable Toxicity (*DSSTox*)

2.1.2.6   Distributed Structure-Searchable Toxicity Prediction Server

It is hosted by Environmental Protection Agency (EPA) USA to predict the toxicity of compounds [57]. It encourages and uses the structure data file (sdf) format. It has a browser developed from open-source tools to search its data files. The files can be downloaded into any chemical relational database for chemical analog searching to enable model building (Fig. 2.15).

2.1.2.7   Estimation Programs Interface Suite

The Estimation Programs Interface (EPI) suite is a free package to compute descriptors specifically to predict the biodegradability of compounds [58], Fig. 2.16.

The EPI Suite developed by EPA is a physical/chemical property and environmental fate estimation program. EPI Suite uses a set of several estimation programs like KOWWIN, AOPWIN, HENRYWIN, MPBPWIN, BIOWIN, BioHCwin, KOCWIN, WSKOWWIN, WATERNT, BCFBAF, HYDROWIN, KOAWIN and AEROWIN, WVOLWIN, STPWIN, LEV3EPI and ECOSAR. Every module in this program and similar programs has its own level of approximation and accuracy.

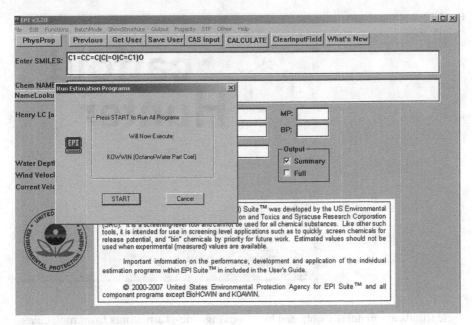

Fig. 2.16 EPI user interface

2.1.3 *Virtual Library Generation (Enumeration)*

The concept of designing virtual libraries to enhance the diversity of compounds for efficient lead generation is well known [59]. A virtual library is composed of scaffold, linkers and functional groups. First, let us see what is a scaffold and what are the known scaffold generation tools.

2.1.3.1 Scaffold

The term 'scaffold' is used broadly in chemistry; the precise meaning of the word is context- and chemist-dependent. Bemis and Murcko outlined a popular method for computationally deriving scaffolds from molecules by removing side-chain atoms [60]. Atoms in ring systems or linking ring systems, and sp^2 atoms directly bonded to these atoms, were preserved. Alternative scaffold definitions rely on abstraction or decomposing the framework into simpler substructural elements. For example, a molecular framework can be interpreted as a graph containing nodes and edges representing atom and bond types, respectively. Removing atom and bond labels or agglomerating nodes by chemotype yields a hierarchy of reduced graphs, or molecular equivalence classes, that represent sets of related molecules. Likewise, a framework can be further decomposed into individual rings (or the core ring assembly) using chemically intuitive rules; the rings can individually or jointly be considered as scaffolds derived from the original compound.

Fig. 2.17 Scaffold Hunter
start-up screen

Scaffolds are generally obtained by removing side-chain atoms from molecules with the definition of both 'side chains' and the equivalence of atoms and rings within the scaffolds being dependent on the particular implementation of the algorithm. Scaffolds constitute the major 'denominator' in drug design, as evident from approaches such as 'scaffold hopping', their link to bioactivity patterns and the fact that scaffold enumerations (Markush structures) are routinely used for patenting chemical series in the pharmaceutical context. From this, it becomes apparent that the scaffold is a truly relevant entity in synthetic organic as well as medicinal chemistry.

Open-Source Tools for Scaffold Generation

Scaffold Hunter

Scaffold Hunter is a Java-based open-source tool for the visual analysis of data sets with a focus on data from the life sciences, aiming at an intuitive access to large and complex data sets [61]. The tool offers a variety of views, e.g. graph, dendrogram and plot view, as well as analysis methods, e.g. for clustering and classification. Scaffold Hunter has its origin in drug discovery, which is still one of the main application areas and is evolved into a reusable open-source platform for a wider range of applications. The tool offers flexible plug-in and data integration mechanisms to allow adaption to new fields and data sets, e.g. from medical image retrieval. Scaffold Hunter is used worldwide in research, both academic and commercial, (Fig. 2.17).

OpenEye

BROOD is a software application designed to help project teams in drug discovery explore chemical and property space around their hit or lead molecule [62]. BROOD generates analogs of the lead by replacing selected fragments in the molecule with fragments that have similar shape and electrostatics, yet with selectively modified molecular properties. BROOD fragment searching has multiple applications, including lead hopping, side-chain enumeration, patent breaking, fragment merging, property manipulation and patent protection by SAR expansion.

Code Optimized for Scaffold Generation Using Free And Open-Source Tools

```
/*
 * To change this template, choose Tools | Templates
 * and open the template in the editor.
 */
package cheminfbook;

import chemaxon.struc.Molecule;
import chemaxon.util.MolHandler;
import chemaxon.sss.search.MolSearch;
import chemaxon.formats.MolImporter;
import java.util.Vector;
import java.io.*;
import joelib.io.*;
import joelib.smiles.*;
import joelib.molecule.JOEMol;
import joelib.molecule.JOEAtom;
import joelib.util.iterator.AtomIterator;
import joelib.molecule.JOEBond;
import joelib.util.iterator.BondIterator;
import java.util.*;
import chemaxon.util.MolHandler;
import chemaxon.struc.Molecule;

/**
 *
 * @author M Karthikeyan and Renu Vyas
 */
public class Cheminfbook {

    /**
     * @param args the command line arguments
     */
    Cheminfbook() {
    }

    public static void main(String[] args) {
        // TODO code application logic here
        Cheminfbook cb = new Cheminfbook();
        try {
            String smi = "C1C(Br)C(OC)CC(Cl)C1C2=C(C=C)C=CC=C2C3N(C)C3";
            Molecule m = MolImporter.importMol(smi);
            // m.clean(3, null);
            // System.out.println(m.toFormat("
            String[] out = cb.getScaffold(smi, true, true);
            System.out.println(smi + ">>" + out[0]);
        } catch (Exception e) {
            System.out.println(e);
        }
```

```
    }

    public static JOEMol ReadSMILES(String smiles, IOType inType, IOType outType) {
        JOEMol mol = new JOEMol(inType, outType);
        try {
            JOESmilesParser.smiToMol(mol, smiles, ".");
        } catch (Exception e) {
            System.out.println(e);
        }
        mol.addHydrogens();

        return mol;
    }

    //== Module to generate scaffold from SMILES format ==//
    public static String[] getScaffold(String smiles, boolean
removeAtomAndBondTypes, boolean c_atom) {

        String[] output = new String[5];
        output[0] = "";
        output[3] = "";
        int i = 0;
        JOEMol mol = ReadSMILES(smiles,
IOTypeHolder.instance().getIOType("SMILES"),
IOTypeHolder.instance().getIOType("SDF"));
        JOEMol framework = (JOEMol) mol.clone();
        JOEMol RGp = new JOEMol();
        int max = 100;
        String[] del_bond = new String[max];
        framework.deleteHydrogens();
        JOEAtom atom;
        JOEBond bond;
        int a_cnt = framework.numAtoms();
        int b_cnt = framework.numBonds();
        String[] at_inf = new String[a_cnt];
        int db_cnt = 0;
        int at_cnt = 0;
        int[][] da_inf = new int[b_cnt][5];
        String[][] db_inf = new String[b_cnt][4];
        for (int z = 0; z < b_cnt; z++) {       //b_cnt
            bond = mol.getBond(z);
            da_inf[z][0] = bond.getBeginAtomIdx();
            da_inf[z][1] = bond.getEndAtomIdx();
            da_inf[z][2] = bond.getBondOrder();
            db_inf[z][0] = bond.getBeginAtom().toString();
            db_inf[z][1] = bond.getEndAtom().toString();
        }
        AtomIterator ait;
        JOEAtom h_atom = new JOEAtom();
        h_atom.setAtomicNum(1);
        boolean atomDeleted;
        String s = "";
        int d = 0;
        do {
            atomDeleted = false;
            ait = framework.atomIterator();
            while (ait.hasNext()) {
                StringBuffer sb = new StringBuffer();
                atom = ait.nextAtom();
                boolean m = atom.isInRing();
                atom.getCIdx();
                Vector vectBonds = atom.getBonds();
                if (m) {
                    JOEBond r_bond = (JOEBond) vectBonds.firstElement();
                    JOEAtom ra1 = r_bond.getBeginAtom();
                    JOEAtom ra2 = r_bond.getEndAtom();
                } else {
                    JOEBond nr_bond = (JOEBond) vectBonds.firstElement();
```

```
                        JOEAtom ra1 = nr_bond.getBeginAtom();
                        JOEAtom ra2 =
                }

                if (vectBonds.size() == 1 && d == 0) {
                        bond = (JOEBond) vectBonds.firstElement();
                        if (!(!removeAtomAndBondTypes && atom.isOxygen() &&
bond.isCarbonyl())) {
                                atomDeleted = true;
                                JOEAtom a1 = bond.getBeginAtom();
                                JOEAtom a2 = bond.getEndAtom();
                                int t1 = a1.getIdx();
                                int t2 = a2.getIdx();
                                int c1 = a1.getCIdx();
                                int c2 = a2.getCIdx();
                                if (a2.isInRing()) {
                                    da_inf[i][3] = t2;
                                    da_inf[i][4] = t1;
                                    db_inf[i][2] = a2.toString();
                                    db_inf[i][3] = a1.toString();
                                    at_inf[at_cnt] = a1.getType() + "_" + a2.getType();
                                    joelib.util.types.IntInt a = new
joelib.util.types.IntInt();
                                    a.i1 = a1.getIdx();
                                    a.i2 = a2.getIdx();
                                    RGp.beginModify();
                                    a1.setFormalCharge(0);
                                    RGp.addAtom(a1);
                                    a2.setFormalCharge(0);
                                    RGp.addAtom(a2);
                                    RGp.addBond(bond);
                                    RGp.endModify();
                                    d++;
                                    System.out.println("a2 " + framework + " d " + d);
                                } else {
                                    da_inf[i][3] = t2;
                                    da_inf[i][4] = t1;
                                    db_inf[i][2] = a2.toString();
                                    db_inf[i][3] = a1.toString();
                                    at_inf[at_cnt] = a1.getType() + "_" + a2.getType();
                                    joelib.util.types.IntInt a = new
joelib.util.types.IntInt();
                                    a.i1 = a1.getIdx();
                                    a.i2 = a2.getIdx();
                                    RGp.beginModify();
                                    a1.setFormalCharge(0);
                                    RGp.addAtom(a1);
                                    a2.setFormalCharge
                                    RGp.addAtom(a2);
                                    RGp.addBond(bond);
                                    RGp.endModify();
                                    framework.deleteBond(bond);
                                    framework.deleteAtom(atom);
                                }
                                i++;
                        }
                }
        }
} while (atomDeleted && i < max);
JOEMol e_scaff = (JOEMol) framework.clone();
output[0] = e_scaff.toString(IOTypeHolder.instance().getIOType("SMILES"));
if (removeAtomAndBondTypes) {
    BondIterator bit = framework.bondIterator();
    JOEAtom atom1;
    JOEAtom atom2;
    int index;
    while (bit.hasNext()) {
        bond = bit.nextBond();
```

```
        atom1 = bond.getBeginAtom();
        atom2 = bond.getEndAtom();
        index = bond.getIdx();
        bond.set(index, atom1, atom2, 1, 0);
    }
    ait = framework.atomIterator();
    while (ait.hasNext()) {
        atom = ait.nextAtom();
        atom.setFormalCharge(0);
        atom.unsetStereo();
        if (!atom.isCarbon() && c_atom) {
            atom.setAtomicNum(6);
            boolean m = atom.isInRing();
            atom.getIdx();
        } else if (!atom.isCarbon() && !c_atom) {
            JOEMol f_scaff = (JOEMol) framework.clone();
        }
    }
    if (c_atom) {
        output[1] =
framework.toString(IOTypeHolder.instance().getIOType("SMILES"));
    } else if (!c_atom) {
        output[1] =
framework.toString(IOTypeHolder.instance().getIOType("SMILES"));
    }
}
    framework.stripSalts();
    mol.deleteHydrogens();
    output[2] = "";
    for (int l = 1; l < RGp.numAtoms() + 1; l += 2) {
        int t1 = (l - 1) / 2;
        output[2] += db_inf[t1][2] + "," + db_inf[t1][3] + "," + da_inf[t1][3]
+ "," + da_inf[t1][4] + "\n";
    }
    output[3] = "";
    for (int l = 0; l < mol.numBonds(); l++) {
        output[3] += da_inf[l][0] + "," + da_inf[l][1] + "," + db_inf[l][0] +
"," + db_inf[l][1] + "," + da_inf[l][2] + "\n";
    }
    output[4] = (String)
RGp.toString(IOTypeHolder.instance().getIOType("SMILES"));
    return output;
    }
}
```

The above code was used to extract scaffold B from molecule A.

Commercial Tools

ReCore

ReCore replaces a given core: Given a predefined central unit of a molecule (the core), fragments are searched in a 3D database for the best-possible replacement—

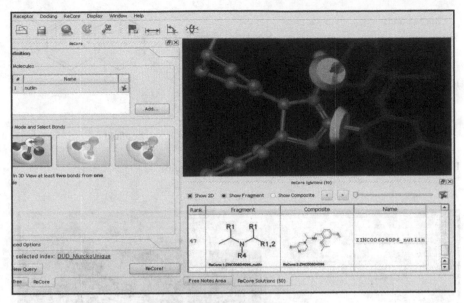

Fig. 2.18 Cutting points in a molecule defined using ReCore

while keeping all connected residues, i.e. the rest of the query compound in place [63]. Additionally, user-defined 'pharmacophore' constraints can be employed to restrict solutions. For further details, the reader is encouraged to download the manual from the website address http://www.biosolveit.de/ReCore/ (Fig. 2.18).

Molecular Operating Environment Chemical Computing Group

Scaffold Replacement (or scaffold hopping) is an approach used to discover new chemical classes by replacing a portion of a known compound (the scaffold), while preserving the remaining chemical groups [64]. This application is built upon MOE's pharmacophore modelling tools. It generates novel structures from all or part of a ligand (possibly bound to a receptor). Three types of operations are supported:

1. Scaffold Replacement: replace a portion of the ligand with a linker
2. Link Multiple Fragments: connect separate fragments with a linker
3. Add Group to Ligand: extend the ligand with a linker

The user indicates the atoms or bonds to be replaced or extended and can specify QuaSAR Descriptor, Model file and/or pharmacophore query filters to limit the results. For example, a pharmacophore query can be used to enforce specific interactions (or restrictions) on the generated structures when growing in a receptor pocket. Scaffold Replacement can be used as part of a ligand-based or structure-based discovery methodology.

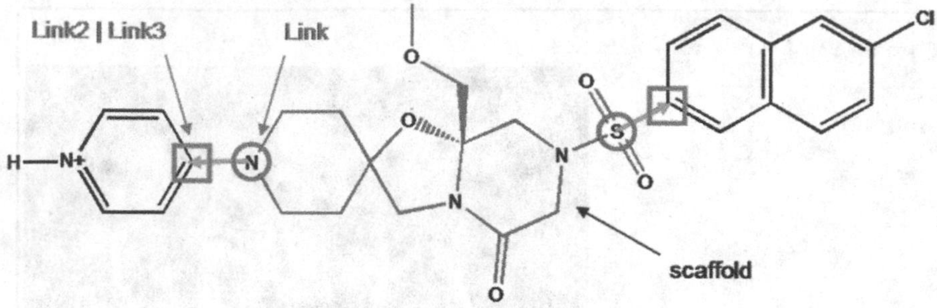

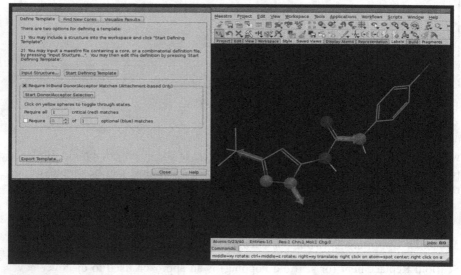

Fig. 2.19 Select red atoms for Replace Scaffold (Select Scaffold)

Fig. 2.20 Core hopping module of Schrodinger

Using 'Replace Scaffold (Select Scaffold)' and selecting the atoms indicated in red results in two connection points (indicated by arrows). The R-groups are indicated in black (Figs. 2.19 and 2.20).

Schrodinger

The steps for the two core hopping strategies are given below [65]:

- Start with template with attachment bonds
- … and with protocore with many possible attachments
- Find ways for protocore to align with template
  - Two alignments are shown in Fig. 2.21
- Add template's R groups to the new core

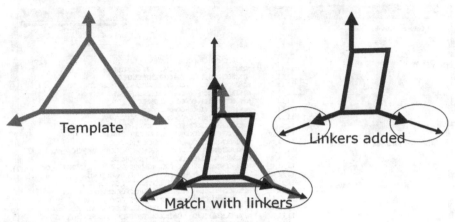

Fig. 2.21 Core hopping methods

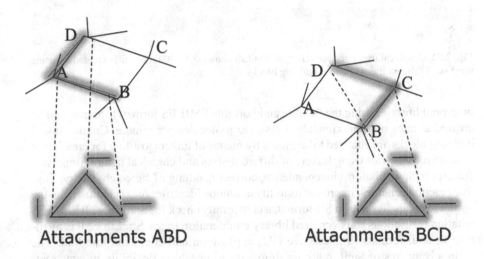

Show 1 linker (maximum) per attachment; default is 2

- All combinations of linkers in all attachment bonds are tested
    - Example shows that two attachment points used linkers
- Suite 2012: a variety of linkers available (2011: methylene)

2.1.3.2 Open-Source Tools for Virtual Library Synthesis

SmiLib

SmiLib is a Java-based combinatorial library enumeration tool developed by Andreas Schuller [66]. SmiLib v2.0 offers the possibility to construct very large com-

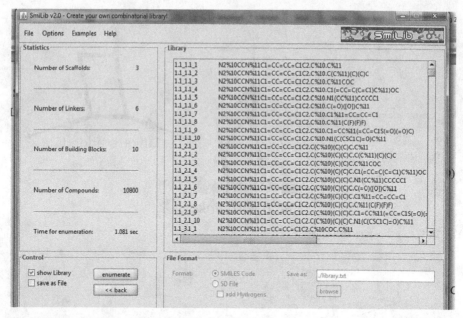

Fig. 2.22 Graphical user interface (*GUI*) of SmiLib showing 10,800 molecules created by using three scaffolds, six linkers and ten building blocks

binatorial libraries using the flexible and portable SMILES format. Libraries can be created at rates of approximately 8,700,000 molecules per minute. Combinatorial building blocks are attached to scaffolds by means of linkers to allow for creation of customized libraries using linkers of different sizes and chemical nature. Important features include platform independence, correct handling of stereo chemistry, flexible reaction schemes, improved usability, a unique identifier for each molecule, the option to create libraries in SD format, a conformity check for SmiLib v2.0 SMILES notation restrictions and decreased library enumeration times. SmiLib v2.0 is available in both formats as interactive GUI application and command line tool. The main advantages of SmiLib are its simplicity to use, high flexibility in constructing combinatorial libraries (exact subset of molecules for virtual synthesis can be specified) and high speed of library construction [67]. The SMILES format is used as both input and output format. SmiLib uses a special syntax for ring closures, i.e. any two-digit number preceded by a percentage sign. For example, 'C%10.C%10' ≡ 'C1.C1' ≡ 'CC' (Ethane C2H6). In addition to normal SMILES, [R1], [R2], [R3], etc. are used as labels for sites of variability and [A] is used as a label for attachment sites. An attachment site is part of the molecule, which is to be attached to a scaffold or a linker. It is a platform-independent program written in Java; SmiLib is run with help of the Java virtual machine with 'java –jar SmiLib.jar'. It requires three American Standard Code for Information Interchange (ASCII) files containing all scaffold, linker and building block molecule fragments in SMILES format (command line parameters -s<scaffolds.smi>, -l<linkers.smi>, -b<buildingblocks. smi>). A reaction scheme file for the enumeration of a combinatorial library can be specified with the option '-r<reaction_scheme>' (Fig. 2.22).

Molecular Operating Environment Chemical Computing Group

In MOE, a proprietary software is also supplied with a combinatorial library generation tool [68]. A combinatorial library is specified by:

- A database of scaffold molecules or a single scaffold molecule
- Databases of functional groups
- Connection information specifying where the functional groups attach on each scaffold

Attachment Points A single combinatorial product is constructed by attaching *R-groups* to a scaffold at marked *attachment points,* called *ports.* The entire combinatorial library is enumerated by exhaustively cycling through all combinations of R-groups at every attachment point on every scaffold. The virtual library is written to an output database. Attachment points are terminal atoms named 'An', where n is a positive integer. In the QuaSAR-CombiGen panel, n is limited to the range [0 … 9]. When using the scientific vector language (SVL) command QuaSAR_CombiGen, however, n can be in the range [0 … 999]. If the terminal atom is attached to the main molecule by a higher-order bond, substitution will be made through a bond of the same order. Note that the bond order at the scaffold attachment point must agree with that at the R-group attachment point: Either at least one of the bond orders must be 1 (single bond) or both must be of the same order. Fragment molecules are created by appropriately naming atoms at the desired points of substitution. One can use the Builder to perform this operation and the Clip R-Groups application in a database can be used to create fragments with named attachment points.

Attachment points must be specified on both the R-group and the scaffold molecule (Fig. 2.23).

2.1.4 Virtual Screening

Bio- and chemoinformatics are crucial for the success of virtual screening of compound libraries which is an alternative and complementary approach to HTS in the lead discovery process [69]. A combination of drug-derived building blocks and a restricted set of reaction schemes is the key for the automatic development of novel, synthetically feasible structures that can be docked into the active site of a drug target for lead identification using computers which is the essence of virtual screening [70]. The virtual screening of combinatorial libraries is used to rationally select compounds for biological in vitro testing from databases of hundreds of thousands of compounds. In addition to structural descriptors, such as fingerprints and pharmacophores, the application of relatively simple structural descriptors traditionally used in quantitative structure–activity studies offers speed and efficiency for rapidly measuring the molecular diversity of such collections capable of screening large data sets of organic compounds for potential ligands. The methods described in this section are used for computationally prioritising candidate molecular libraries for synthesis and screening by using certain filters. These statistical methods are powerful because they provide a simple way to estimate the properties of the overall

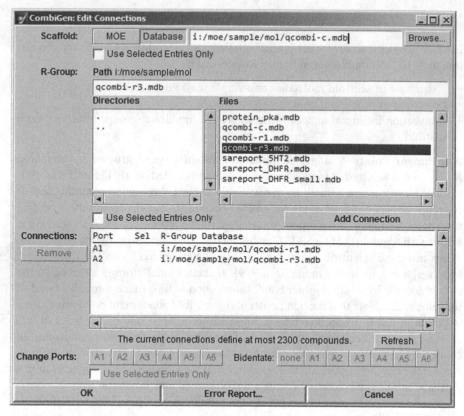

Fig. 2.23 Virtual library synthesis in Molecular Operating Environment (*MOE*) using CombiGen

library without explicitly enumerating all of the possible products. Current virtual screening applications focus not only on biological activity but also on other relevant properties of drug candidates, like ADME. In the first step of virtual screening, the prediction algorithm must be very fast because typically several millions of compounds have to be processed to generate hit lists of molecules which can be further subjected to actual experimental confirmation in laboratory.

A typical virtual screening workflow in a drug design experiment involves the following steps:

1. Scaffold extraction from a data set of molecules.
2. Use these scaffolds as seeds to enumerate a virtual library by supplying linkers and functional groups.
3. Apply any of the filters below either independently or in combination depending upon prior knowledge (Rule of five(RO5) Lipinski Pharmacophore model QSAR Docking Select Hits or no hits) (Fig. 2.24).

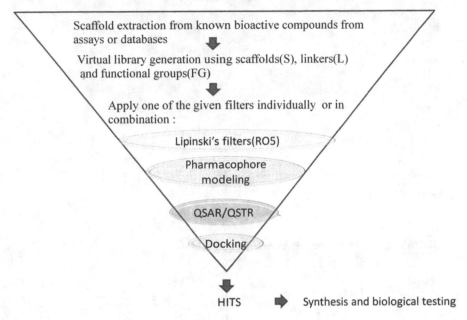

Fig. 2.24 A general virtual screening protocol

2.1.4.1 Free Virtual Library Screening Platforms

Screening Assistant 2

Screening Assistant 2 (SA2) is a modular software dedicated to perform various simple and advanced chemoinformatics analysis around chemical libraries [69], Fig. 2.25.

SA2 is a free and open-source Java software dedicated to the storage and the analysis of small to very large chemical libraries. SA2 stores unique chemical structures using a MySQL database and associates to the molecules various standard precomputed descriptors as well as user-defined properties/descriptors that can be imported in a flexible way. Various standard and advanced chemoinformatics methods have been implemented, including chemical space visualization/creation, substructure and similarity searches, diverse subset extraction and diversity indices calculation. Its modular architecture, based on the NetBeans Platform, eases the addition of new functionalities to the software. The program and source code are freely available (GPL), The system is programmed in Java and data are managed by a MySQL server. The software allows to calculate drug-like and lead-like properties, as well as to study the libraries in terms of uniqueness, of internal duplicates, diversity and frameworks (http://www.univ-rleans.fr/icoa/screeningassistant/). The software is available on Sourceforge: http://sourceforge.net/projects/screenassistant.

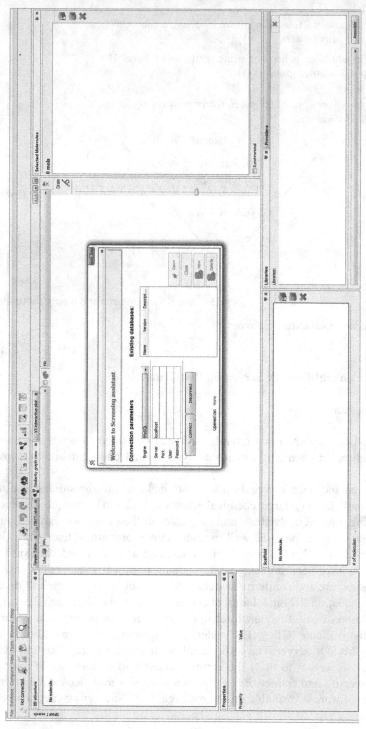

Fig. 2.25 The welcome screen of Screening Assistant 2

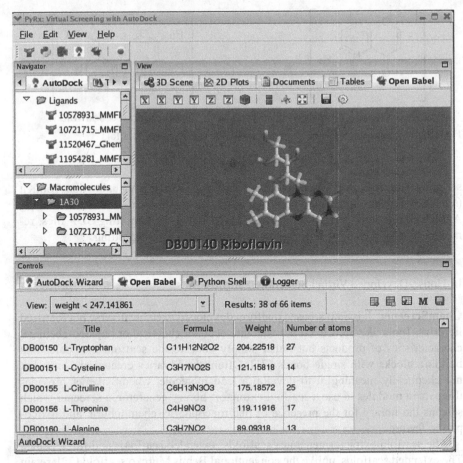

Fig. 2.26 A screenshot of PyRx platform

• PyRx

PyRx is a free and open-source software for computer-aided drug design distributed under Simplified BSD License [70]. PyRx is a Virtual Screening software for Computational Drug Discovery that can be used to screen libraries of compounds against potential drug targets. PyRx enables medicinal chemists to run virtual screening from any platform and helps users in every step of this process—from data preparation to job submission and analysis of the results. PyRx includes a docking wizard with easy-to-use user interface which makes it a valuable tool for computer-aided drug design. PyRx also includes chemical spreadsheet-like functionality and powerful visualization engine that are essential for rational drug design (Fig. 2.26).

A number of open-source software are used such as AutoDock 4 and AutoDock Vina for docking AutoDockTools to generate input files, Python as a programming/

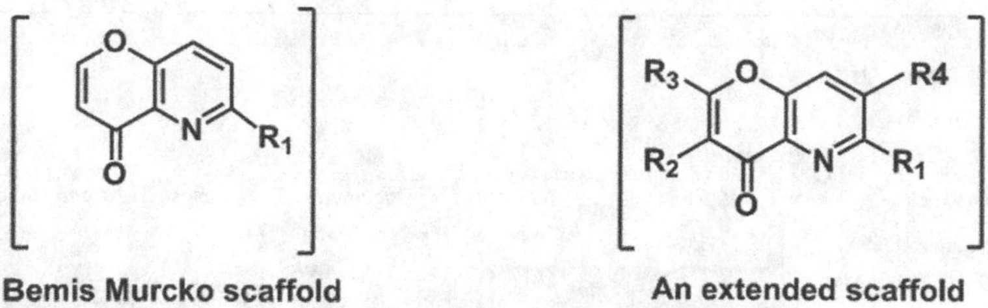

Bemis Murcko scaffold　　　　　　　　　**An extended scaffold**

Fig. 2.27 A comparison of Bemis Murcko scaffold and ChemScreener scaffold

scripting language, wxPython for cross-platform GUI, the Visualization ToolKit (VTK) by Kitware Inc, Enthought Tool Suite, including Opal Toolkit for running AutoDock remotely using web services, Open Babel for importing SDF files, removing salts and energy minimization and matplotlib for 2D plotting.

In-House-Developed Virtual Screening Platform

ChemScreener It is a Java-based platform developed to create diverse but focussed libraries. Tools like SmiLib do not take into account chemical or physico-chemical characteristics of products but rather simply concatenate scaffold molecules and building blocks with single bonds [71]. Often, the libraries created are huge but not chemically meaningful to develop a lead molecule. ChemScreener provides three main modules to use a scaffold extractor: library generator, a screener which screens the library for the presence of pharmacophoric, chemophoric and toxicophoric features.

The scaffold extractor generates scaffolds, extended scaffolds and frameworks. The extended scaffolds, unlike the conventional Bemis Murcko scaffold [72], retain connection information and are used in focussed library synthesis (Fig. 2.27).

In the medicinal chemistry literature, a number of substructures have been identified as toxicophores, such as some aromatic amines, azides, diazo structures, triazenes, aromatic azo moieties, aromatic hydroxylamines, aliphatic halides, etc. 'Chemophores' refer to substructural groups which are too reactive or inert or synthetically inaccessible, which would lead to practically irrelevant molecules. Medicinal chemists design compounds on the basis of chemophoric features; for instance, the presence of OMe group in a molecule is generally known to enhance its bioactivity, alkyl groups are introduced to increase selectivity, a fluoro group for metabolic stabilization whereas a nitro group will impact the activity in an adverse way which implies later side effects in drug efficacy. Toxicophores were collected from literature databases such as RTECS [73], NIOSHTIC [74], EPA and pharmacophores and chemophores were extracted from literature. This program provides an alert indicating the number of chemophore, toxicophore and pharmacophore matches to assist in fine-tuning the library generated. The virtual library can also be screened on the basis of binding affinity-based filters provided the target structure

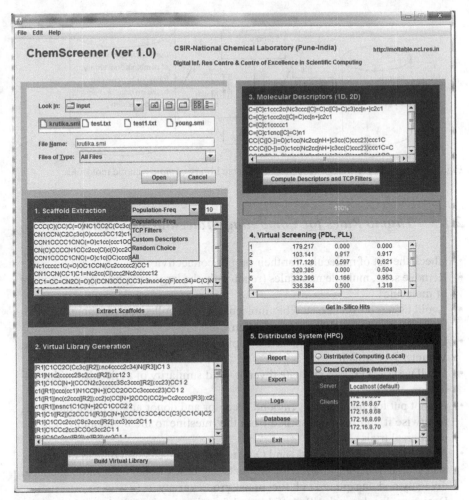

Fig. 2.28 Homepage of in-house-developed ChemScreener virtual screening platform

or a good homology model is available as ChemScreener can be complemented with docking-based screening tools such as Autodock 4.0 (Fig. 2.28).

A virtual library of 150 million antipsychotic molecules of 2 GB file size was generated from four seed scaffolds using the ChemScreener program which is currently not possible with the existing software. It could also reveal significant bioactivity data patterns from scaffold extraction in big databases like PubChem[75] and ChEMBL [76], (Fig. 2.29).

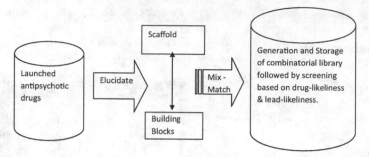

Fig. 2.29 Flowchart to create a focussed and diverse library of antipsychotic molecules

2.1.5   Thumb Rules for Computing Molecular Properties

- Check the list of molecules for their proper connectivity
- Remove salts, multiple molecules (retain only the large molecule from a mixture of molecule)
- Avoid using too small or too big molecules in the collection (as input for automatic focussed virtual library generation)
- Compute basic descriptors related to Lipinski's rule of five in addition to TPSA, Volume, Weinerpath
- Do 2D and 3D PCA to evaluate diversity and similarity of molecules in the collection
- Do not put too many hydrogen bond acceptor and donor atoms into a molecule, otherwise it will not be absorbed from the intestine to the blood and fail in the preclinical trials
- Apart from the usual Lipinski and Oprea criteria for selection of lead molecules, also search in natural and marine products databases which offer more chemical diversity and unexplored rich functional group variety
- Design molecules with NP scaffolds and functional groups for better bioactivity (based on early reports) and more scope for patenting

2.1.6   Do it Yourself

1. Use the relevant code given in the text to extract scaffolds from SMILES of top ten drugs in the market
2. Retrieve ten drug molecules from drug bank database and ten known pesticides, calculate Lipinski's drug-like properties, ADMET and biodegradability parameters using any of the free online tools. Comment on the results
3. Generate a virtual library using SmiLib from molecules belonging to anti-anginal compounds

2.1.7   Questions

1. Write a brief essay on the known property prediction tools in chemoinformatics.
2. How is a virtual library constructed? What are the methods known to screen a virtual library?
3. How do you obtain a diverse but focussed virtual library for a class of therapeutic compounds?
4. Define scaffold hopping. Highlight the tools which can be used for scaffold hopping.
5. What is a scaffold? Elaborate on the known methods of scaffold extraction.

References

1. Leo A, Hansch C, Church C (1969) Comparison of parameters currently used in the study of structure-activity relationships. J Med Chem 12:766–771
2. Admason GW, Bawdon D (1976) An empirical method of structure-activity correlation for polysubstituted cyclic compounds using wiswesser line notation. J Chem Inf Comput Sci 16(3):161–165
3. Choplin, F (1990) Computers and the medicinal chemist. In: Hansch C, Sammes PG, Taylor JB (eds) Comprehensive Medicinal Chemistry Pergamon Press, UK 4:33–58
4. Tropsha A, Gramatica P, Gombar V (2003) The importance of being earnest: validation is the absolute essential for successful application and interpretation of QSPR models. Mol Inform 22(1):69–77
5. http://www.moleculardescriptors.eu/
6. Seybold PG, May M, Bagel UA (1987) Molecular structure property relationships. J Chem Educ 64(7):575
7. Todeschini R, Consonni V (2009) Molecular descriptors for chemoinformatics, vol 2. Wiley-VCH
8. Karelson M (2000) Molecular descriptors in QSAR/QSPR. Wiley
9. http://www.vcclab.org/lab/indexhlp/consdes.html
10. http://www.codessa-ro.com/descriptors/electrostatic/index.htm
11. Balaban AT (1997) From chemical topology to three dimensional geometry. Plenum Press, New York, 1–24
12. Karelson M, Lobanov V, Katritzky AR (1996) Quantum chemical descriptors in QSAR/QSPR studies. Chem Rev 96:1027–1043
13. Enoch SJ (2010)The use of quantum mechanics derived descriptors in computational toxicology. In: Puzyn T et al (ed) Challenges and advances in computational chemistry and physics, vol 8. Springer Science pp 24–27
14. Stanton D (1999) Evaluation and use of BCUT descriptors in QSAR and QSPR studies. J Chem Inf Comput Sci 39(1):11–20
15. Ma SL, Joung JY, Lee S, Cho KH, No KT (2012) PXR ligand classification model with SFED weighted WHIM and CoMMA descriptors. SAR QSAR Environ Res 23(5–6):485–504
16. http://rdkit.org/docs/api/rdkit.Chem.MACCSkeys-pysrc.html
17. Todeschini R, Bettiol C, Giurin G, Gramatica P, Miana P, Argese E (1996) Modeling and prediction by using WHIM descriptors in QSAR studies: submitochondrial particles(SMP) as toxicity biosensors of chlorophenols. Chemosphere 33:71–79

18. Hinselmann G, Rosenbaum L, Jahn A, Fechner N, Zell AJ (2011) Compound Mapper: an open source JAVA library and command line tool for chemical fingerprints. J Chemoinformatics 3:3
19. Rogers D, Mathew H(2010) Extended connectivity fingerprints. J Chem Inf Model 50(5):742–754
20. Bender A, Hamse Y, Mussa HY, Glen C (2010) Similarity searching of chemical databases using atom environment descriptors (Molprint 2D) evaluation of performance. J Chem Inf Comput Sci 44:1708–1718
21. Deursen R, Blum Lorenz CB, Reymond JL (2010) A searchable map of PubChem. J Chem Inf Model 50(11):1924–1934
22. Chemscreener unpublished results
23. Jorgenson WL, Duffy EM (2002) Prediction of drug solubility from structure. Adv Drug Deliv Rev 54:355–366
24. Livingstone DJ, Waterbeemd VD, Han I (2009) In silico prediction of human oral bioavailability. Method Prin Med Chem 40:433–451
25. Persson LC, Porter CJ, Charman WN, Bergstrom CA (2013) Computational prediction of drug solubility in lipid based formulation excipients. Pharm Res PMID:23771564
26. Faller B, Ertl P (2007) Computational approaches to determine drug solubility. Adv Drug Deliv Rev 59:533–545
27. Cortes-Cabrera A, Morris GM, Finn PW, Morreale A, Gago F (2013) Comparison of ultra fast 2D and 3D descriptors for side effect prediction and network analysis in polypharmacology. Br J Pharmacol. doi:10.1111/bph.12294
28. Rice BM, Byrd EF (2013) Evaluation of electrostatic descriptors for crystalline density. Langmuir
29. Garcia EJ, Pellitero PJ, Jallut C, Pirngruber GD (2013) Modeling adsorption properties on the basis of microscopic, molecular structural descriptors for non polar adsorbents. J Chem Inf Model
30. Wegner JK, Zell A (2003) Prediction of aqueous solubility and partition coefficient optimized by genetic algorithm based descriptors selection method. J Chem Inf Comput Sci 43(3):1077–1084
31. Steinbeck C, Hoppe C, Kuhn S, Matteo F, Guha R, Willighagen EL (2006) Recent development of the CDK (Chemistry Development Kit) an open source JAVA library for chemo and bioinformatics. Curr Pharm Design 12(17):2111–2120
32. http://www.rguha.net/code/java/cdkdesc.html
33. Steinbeck C (2008) Open toolkits and applications for chemoinformatics teaching Abstracts of Papers, 235th ACS National Meeting, New Orleans, LA, United States, April 6–10
34. http://padel.nus.edu.sg/software/padeldescriptor/
35. Yap CW (2011) Padel descriptor an open source software to calculate molecular descriptors and fingerprints. J Comput Chem 32(7):1466–1474
36. http://nisla05.niss.org/PowerMV/?q=PowerMV
37. Liu K, Feng J, Young SS (2005) A software environment for molecular viewing, descriptor generation, data analysis and hit evaluation. J Chem Inf Model 45(2):515–522
38. http://www.chemaxon.com/marvin/help/calculations/calculator-plugins.html
39. http://cheminformatics.org/datasets/
40. Xueliang L, Yongtang S, Wang L (2012) On a relation between randic index and algebraic connectivity. Match 68(3):843–839
41. Ivanciuc O, Ivanciuc T, Douglas KJ, William SA, Balaban T (2001) Wiener index extension by counting even/odd graph distances. J Chem Inf Model 41(3):536–549
42. Benet LZ, Broccatelli F, Oprea TI (2011) BDDCS applied to over 900 drugs. AAPS J 13(4):519–547
43. Lu D, Chambers P, Wipf P, Xie X-Q, Englert D, Weber S (2012) Lipophilicity screening of novel drug like compounds and comparison to clogp. J Chromatogr A 1258:161–167
44. http://www.eyesopen.com/oechem-tk
45. QikProp (2012) version 3.5, Schrödinger, LLC, New York

46. Kerns E, Li D (2010) Drug like properties, concepts, structure design and methods. Academic Press
47. LigPrep (2012) version 2.5, Schrödinger, LLC, New York
48. Molecular Operating Environment (MOE) (2012)10; Chemical Computing Group Inc., 1010 Montreal, QC, Canada, H3A 2R7, 2012
49. Gerardo CMM, Yovani MP, Khan MTH, Arjumand A, Khan KM, Torrens F, Rotondo R (2007) Dragon method for finding novel tyrosinase inhibitors biosilico identification and experimental in vitro assays. Eur J Med Chem 42(11–12):1370–1381
50. http://accelrys.com/products/discovery-studio/admet.html
51. Karthikeyan M, Krishnan S, Pandey AK, Bender A, Tropsha A (2008) Distributed chemical computing using ChemStar: An open source java remote method invocation architecture applied to large scale molecular data from pubchem. J Chem Inf Model 48(4):691–703
52. http://www.molinspiration.com/
53. http://www.pharmaexpert.ru/passonline/
54. Lusci A, Pollastri G, Baldi P (2013) Deep architectures and deep learning in Chemoinformatics: the prediction of aqueous solubility for drug like molecules. *J Chem Inf Model* 53(7):1563–1575
55. Sorana BD, Lorentz J (2011) Predictivity approach for quantitative structure prediction models: application for blood barrier permeation for diverse drug like compounds. Int J Mol Sci 12(7):4348–4386
56. www.preadmet.bmdrc.org/
57. http://www.epa.gov/ncct/dsstox/
58. http://www.epa.gov/opptintr/exposure/pubs/episuite.htm
59. Ulrich A, Koch C, Speitling M, Hansske FG (2002) Modern methods to produce natural-product libraries. Curr Opin Chem Biol 6(4):453–458
60. Bemis GW, Murcko MA (1999) Properties of known drugs, 2: Side chains. J Med Chem 42(25):5095–5099
61. Wetzel S, Karsten K, Renner S, Rauh D, Oprea TI, Mutzel P, Waldmann H (2009) Interactive exploration of chemical space with scaffold hunter. Nat Chem Biol 5(9):696
62. http://www.eyesopen.com/brood
63. Van Drie JH (2009) ReCore. J Am Chem Soc 131(4):1617
64. http://www.chemcomp.com/journal/newscaffold.htm
65. Core Hopping (2011), version 1.1, Schrödinger, LLC, New York
66. Schuller A, Hahnke V, Schneider G (2007) SmiLib v2.0: A Java-Based Tool for Rapid Combinatorial Library Enumeration. QSAR Comb Sci 3:407–410
67. http://gecco.org.chemie.uni-frankfurt.de/smilib/
68. http://www.chemcomp.com/MOE-Cheminformatics_and_QSAR.htm#CombinatorialLibraryDesign
69. Tropsha A (2008) Integrated chemo and bioinformatics approaches to virtual screening. In: Tropsha A, Varnek A (ed) Chemoinformatics approaches to virtual screening. SC Publishing, pp 295–325
70. Perola E, Xu K, Kollmeyer TM, Kaufmann SH, Prendergast FG, Pang Y-P (2000) Successful virtual screening of a chemical database for farnesyl transferase inhibitor leads. J Med Chem 43(3):401–408
71. Oprea TI (2002) Virtual screening in lead discovery a viewpoint. Molecules 7:51–62
72. Unpublished results
73. http://www.cdc.gov/niosh/rtecs/default.html
74. http://www2a.cdc.gov/nioshtic-2/
75. http://pubchem.ncbi.nlm.nih.gov/
76. https://www.ebi.ac.uk/chembl/

Chapter 3
Machine Learning Methods in Chemoinformatics for Drug Discovery

Abstract It is well known that the structure of a molecule is responsible for its biological activity or physicochemical property. Here, we describe the role of machine learning (ML)/statistical methods for building reliable, predictive models in chemoinformatics. The ML methods are broadly divided into clustering, classification and regression techniques. However, the statistical/mathematical techniques which are part of the ML tools, such as artificial neural networks, hidden Markov models, support vector machine, decision tree learning, Random Forest and Naive Bayes and belief networks, are best suited for drug discovery and play an important role in lead identification and lead optimization steps. This chapter provides stepwise procedures for building ML-based classification and regression models using state-of-art open-source and proprietary tools. A few case studies using benchmark data sets have been carried out to demonstrate the efficacy of the ML-based classification for drug designing.

Keywords Machine learning · Neural networks · SVM · SVR · Genetic programming · Chemoinformatics · Drug design

3.1 Introduction

Statistical and machine learning (ML) methods have often been employed in chemoinformatics especially for drug design studies. While there is some amount of overlap between both the domains, there are many subtle differences, the most important one being that while the former methods are used for drawing inference from the data the latter are used for building predictive models from the data [1]. A list of commonly used statistical and ML-based methods used in drug design context is provided here (Fig. 3.1).

As statistics is a very vast domain, here in this chapter, we will focus mainly on the ML-based methods and tools with suitable worked-out examples using real data sets. Experimental chemists and biologists are interested in the properties of the chemicals and their response to biological systems in both beneficial and adverse effects contexts. Several research groups across the world have compared chemical

M. Karthikeyan, R. Vyas, *Practical Chemoinformatics,*
DOI 10.1007/978-81-322-1780-0_3, © Springer India 2014

133

Fig. 3.1 Commonly employed statistical and machine learning methods in chemoinformatics

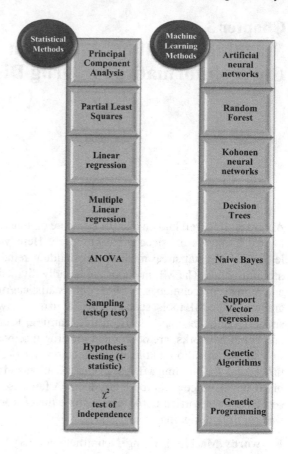

and drug databases to identify the molecular descriptors that can be used to classify molecules as drugs/nondrugs and toxins/nontoxins [2].

3.2 Machine Learning Models for Predictive Studies

In the context of drug design, biological activity is a function of the descriptor or property, so the general form of a ML model can be given as:

$$y = f(x, d)$$

where x represents an N-dimensional vector ($X = [X_1, X_2, \ldots, X_n]^T$) of descriptors (model inputs).

x refers to model parameters and y denotes model output describing activity/property/toxicity (Fig. 3.2).

The main task of ML models in drug design context is to distinguish between active and inactive molecules in a given database. There are generally two types of models that can be developed, viz. continuous and binary, depending upon the type

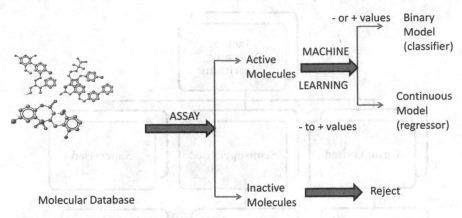

Fig. 3.2 Machine learning for drug design experiments

of bioassay data available. In a continuous model, it is possible to predict the model output in a range using a regressor, whereas in a binary model built using a classifier, the outcome would be either 'yes' or 'no'.

The major ML-based predictive models in drug design comprise the following four categories:

1. Quantitative Structure–Activity Relationship (QSAR) Models

The bioactivities generally modelled are half maximal inhibitory concentration (IC_{50}), minimum inhibitory concentration (MIC) and half maximal effective concentration (EC_{50}) obtained in biological assays; statistical methods used in QSAR studies are principal component analysis, partial least squares, Kohonen neural network, artificial neural network, etc. [3].

2. Quantitative Structure–Property Relationship (QSPR) Models

QSPR models are built generally for correlating some properties of the molecule like melting point, boiling point, λ_{max} solubility, etc. [4].

3. Quantitative Structure–Toxicity Relationship (QSTR) Models

The LD_{50} and TD_{50} are, respectively, the lethal and toxic median dose parameters important for medicinal purposes, and hence many efforts have been devoted to build predictive models. Toxicity is another important parameter which needs to be assessed from molecular structures [5].

4. Quantitative Structure–Biodegradability Relationship (QSBR) Models

Structures are also correlated with the environmental biodegradability of a molecule. Thus, in view of increasing environmental legislation [6], QSBR models play an important role in predicting the biodegradability of a molecule.

The applicability domain is one of the most important factors which should be taken into consideration while building mathematical models or while applying the prebuilt models for predictive studies [7]. Explaining outliers in the training set, test set and predicted set is one of the requirements in modern structure–property–

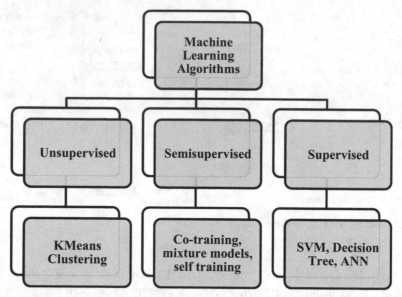

Fig. 3.3 Types of machine learning approaches and their methods

activity relationship studies [8]. The benchmarked data sets for binary (active, inactive) and continuous outputs pertaining to QSAR and QSPR studies are available at http://www.chemoinformatics.org site and uci ML repository for downloading.

3.3   Machine Learning Methods

ML is a branch of artificial intelligence, which is concerned with the construction and study of computational systems that can learn from data [9]. A ML system could be trained based on properties and features and on the basis of that information, predictions can be done. The aim of ML is to teach a machine to learn from experiences, i.e. to feed it with a set of example objects and, based on the information content thereof, to build a classifier or a predictive model [10] (Fig. 3.3).

The ML-based classifiers can be divided into the following types:

1. Supervised Learning Algorithms [11]

They mainly consist of a training data set and analyse this training data to learn relationships between data elements to produce an inferred function. They involve algorithms such as Bayesian statistics, decision tree (DT) learning, support vector machine (SVM), random forest (RF) and nearest neighbour algorithms.

2. Unsupervised Learning Algorithms [12]

In this type of algorithm, there is no 'supervising' (as in supervised learning) label data in the training set to figure out the hidden structure within the unlabelled data set. This mainly involves clustering techniques like K-means, mixture model and hierarchical clustering.

3. Semi-supervised Algorithms [13]

This class of ML algorithms uses both labelled and unlabelled data sets and falls between supervised learning and unsupervised learning algorithms.

In drug discovery, new drugs are designed to interact with the disease/disorder-related or disorder-related molecules and to avoid interaction with the other molecules vital for normal functioning in the human body. Computer-aided screening of drugs heavily relies on various filters, whose aim is to retain drug-like compounds and discard those unlikely to be the drugs. These stepwise filtering processes increase the complexity and specificity of filters. Most of the algorithms behind these computer-aided filters are ANN based since ANNs are relatively easy to use, efficient and versatile tools. They also possess some drawbacks associated with this prediction method [14]. Among them are (1) the 'black-box' character of ANN, which may hamper the interpretation of derived models and fine-tuning; (2) the risk of overfitting (i.e. ability to fit to training data noise rather than to true data structure, thereby resulting in poor generalization); and (3) a relatively long training time.

ANNs, support vector regression (SVR) and genetic programming are exclusively data-driven modelling formalisms that enable a computing machine to capture (learn) relationships existing between input and output variables of an example data set [15]. The k-nearest neighbour (kNN) algorithm is a nonparametric supervised learning algorithm with the underlying principle that the data instances belonging to the same class should lie closer to the feature space [16]. The Naive Bayes (NB) classifier is a simple inductive-learning probabilistic classifier based on the Bayes' theorem with strong (naive) independence assumptions based on conditional probabilities [17]. DTs are simple predictive models generated by the algorithms that identify various ways of splitting a data set into branch-like structures which form an inverted DT originating from a root node at the top of the tree [18].

The strong ML classifiers such as SVM and RF can be used in drug designing [19]. The RF paradigm belongs to a class of methods known as 'ensemble learning' that generates a number of classifier models and aggregates their results [20].

It is another method for classification and regression which operates by constructing a DT. The framework of an RF method consists of several parts which can be mixed and matched to create a large number of specific models. It grows many DTs, to classify a new object from an input vector and to put the input vector in each of the trees in the forest. Each tree provides a classification and 'votes' are assigned to each class and RF finally chooses the classification possessing the maximum votes [21].

The support vector machine (SVM) is a statistical learning theory-based non-probabilistic binary linear classifier and its analogue termed support vector regression (SVR) performs regression and density estimation [22]. Given an example set consisting of data belonging to two categories, the SVM's supervised training algorithm learns the underlying binary classification, and, post training, the SVM is capable of assigning their correct classifications to the new data. Typically, SVM

Input Space   Feature Space

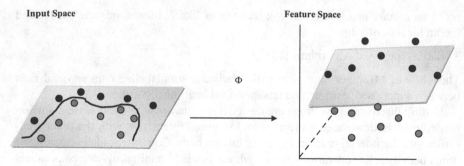

Fig. 3.4 Schematic showing SVM-based binary classification by mapping the original data into high-dimensional feature space

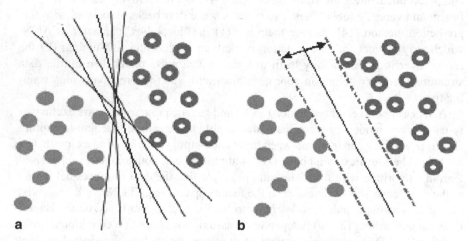

a   b

Fig. 3.5 **a** A schematic of possible hyperplanes for linearly separable data. **b** Optimum hyperplane located by SVM and the corresponding support vectors

(SVR) maximizes the prediction accuracy of the classifier (regression) model while simultaneously escaping from data overfitting. In SVM, the inputs are first nonlinearly mapped into a high-dimensional feature space (Φ) wherein they are classified using a linear hyperplane (Fig. 3.4).

Thus, the SVM is a linear method in a high-dimensional feature space, which is nonlinearly related to the input space. Though the linear algorithm works in the high-dimensional feature space, in practice it does not involve any computations in that space, since through the usage of the 'kernel trick' all necessary computations are performed directly in the input space [23].

Consider a two-class data set that is linearly separable as shown in Fig. 3.5. The SVM constructs an N-dimensional hyperplane (or a set of hyperplanes in a high-dimensional space), to optimally separate data into two categories. From among various alternatives, it locates the hyperplane in a manner such that a good separation is realized between the two classes. This is achieved by placing the hyperplane at the largest distance from the nearest training data point belonging to any class.

In effect, the maximum margin, i.e. optimal hyperplane, is the one that gives the greatest separation between the classes. The data points that are closest to the optimal hyperplane are called 'support vectors (SV)'. In each class, there exists at least one SV; very often there are multiple SVs. The optimal hyperplane is uniquely defined by a set of SVs. As a result, all other training data points can be ignored.

The SVM formulation follows structural risk minimization (SRM) principle, as opposed to the empirical risk minimization (ERM) approach commonly employed within statistical ML methods and also in training ANNs. In SRM, an upper bound on the generalization error is minimized as opposed to the ERM, which minimizes the prediction error on the training data. This equips the SVM with a greater potential to generalize the classifier function learnt during its training phase for making good classification predictions for the new data. An in-depth discussion of SVM and SVR can be found in a number of important publications [24, 25].

The SVM possesses some desirable characteristics such as good generalization ability of the classifier function, robustness of the solution, sparseness of the classifier and an automatic control of the solution complexity. Moreover, the formalism provides an explicit knowledge of the data points (termed 'support vectors'), which are important in defining the classifier function. This feature allows an interpretation of the SVM-based classifier model in terms of the training data. Robustness of SVM is achieved by considering absolute, instead of quadratic, values of the errors. As a consequence, the influence of outliers is less pronounced [26].

3.4 Open-Source Tools for Building Models for Drug Design

There exist a number of software suites/packages to implement ML. The best part is many of them are available as open-source tools. Few of the important ML suites/packages are discussed here.

3.4.1 Library for Support Vector Machines (LibSVM)

It is an integrated software for SVM classification, regression and distribution estimation [27]. It also supports multi-classification. A LibSVM mainly includes:

- SVM formulation
- Efficient multi-classification
- Cross-validation for model selection
- Probability estimates
- Various kernels

A LibSVM package mainly includes the following:

1. Main directory: Core C/C++ programs and sample data. The files svm and cpp implement training and testing algorithms.

2. Tool sub-directory: Includes tools for checking and selecting SVM parameters.
3. Other sub-directories contain pre-built binary files and interfaces to other languages.

There are some other useful utilities in the LibSVM package, which are as follows:

In the LibSVM package, svm-scale is a tool for scaling input data file and svm-train comprises certain parameters depending on which the data are classified which mainly involve:

a. s svm_type: Set type of SVM (default 0), where 0 and 1 are for multi-class classification, 2 is for one-class SVM, 3 is for regression and 4 is for nu-SVR (regression).
b. t kernel_type: Set type of kernel function (default 2), where 0 is for linear, 1 for polynomial, 2 for radial basis, 3 for sigmoid and 4 for precomputed kernel.
c. d degree: Set degree in kernel function (default 3).
d. g gamma: Set gamma in kernel function.
e. b probability_estimates: Whether to train a support vector classification (SVC) or SVR model for probability estimates, 0 or 1.
f. wi weight: Set the parameter C of class i to weight *C, for C-SVC.
g. v n: n-fold cross-validation mode.
h. q quiet mode (no outputs)

The steps for building a radial basis function (RBF) kernel-based SVM model using LibSVM are enumerated here (Fig. 3.6).

When a data set contains a large (inputs) number of features in it, it is possible that many of those features are noisy or they do not contribute significantly towards the classification of the data. It thus becomes important to extract only the relevant features and remove the noisy ones. For example, in drug designing, many descriptors (which are considered as features in SVM) may not contribute towards classifiers' ability to distinguish between drugs and nondrugs. Those descriptors can be removed from the data set.

For the extraction of influential features priority-wise, i.e. ranking of features according to their contribution towards the classification, a technique called Information Gain or Infogain is used for SVM [28]. InfoGain is a Waikato Environment for Knowledge Analysis (WeKa) [29] implementation, which is a measure of the contribution of a particular feature to the model. To run a particular set of data in SVM, a particular file format is required, which is called as LibSVM format. This format can be obtained by converting a comma-separated value (CSV) file by implementing a code in Matrix Laboratory (MATLAB). LibSVM is a format accepted by the LibSVM software, which numbers each feature for a particular sequence followed by a colon (:).

This is the input file format of SVM.

[label] [index1]:[value1] [index2]:[value2] ...
[label] [index1]:[value1] [index2]:[value2] ...

Before proceeding with SVM-based classification, label the data set, i.e. label the positive data as '1' and negative data as '0'. Labelling can be done by assigning positive data as +1 and negative data as −1.

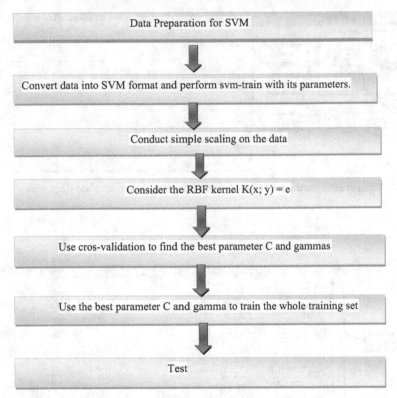

Fig. 3.6 Support vector machine (*SVM*) model building steps in LibSVM

Label: Sometimes referred to as 'class', the class (or set) of your classification, which we usually put as integers. Index: Ordered indexes, which are usually continuous integers. Value: The data for training which are usually lots of real (floating point) numbers.

3.4.2   Waikato Environment for Knowledge Analysis (WeKa)

The WeKa is a popular suite of a large number of feature selection, clustering, classification, association rule mining, regression, etc. [30]. It is best used for data exploration and comparing different ML techniques on the same platform. It has been written in Java and is a freely available software under GNU (General Public License). It contains a collection of tools and algorithms for data analysis. Data preprocessing, clustering, classification, regression, visualization and feature selection can be performed using WeKa. WeKa's main user interface is 'Explorer' (see Fig. 3.7).

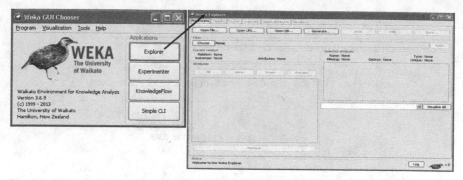

Fig. 3.7 Waikato environment for knowledge analysis (*WeKa*) user interface (Explorer)

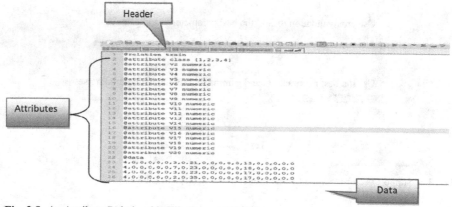

Fig. 3.8 An Attribute Relationship File Format (ARFF)

WeKa uses a specific file format, Attribute Relationship File Format (ARFF).

- It is a text file format to store data in a database and has two sections: header and data section.
- The first line of the header represents the relation name.
- Below header is the list of the attributes (@attribute…) and each attribute is associated with a unique name and a type. It describes the kind of data present in the variable and what values it can have.
- The variables can be numeric, nominal, string and date.
- The header section can also have some comment lines, which is identified with a '%' sign at the beginning and can also describe the database content or give the reader information about the author.

Finally, there is the data itself (@data), each line stores the attribute of a single entry separated by a comma (Fig. 3.8).

WeKa also has its own implementation of Random Forest. It generates correctly classified instances (Features) and incorrectly classified instances along with a confusion matrix.

C-3.3.2.2 Code for building J48 and other classifier models in WeKa

```
import weka.classifiers.Classifier;
import weka.classifiers.Evaluation;
import weka.core.Instances;
import weka.core.OptionHandler;


public class WekaClassifierDemo {

  public WekaClassifierDemo() {
    super();
  }

  public void setClassifier(String name, String[] options) throws Exception {
    m_Classifier = Classifier.forName(name, options);
  }

  public void setFilter(String name, String[] options) throws Exception {
    m_Filter = (Filter) Class.forName(name).newInstance();
    if (m_Filter instanceof OptionHandler)
      ((OptionHandler) m_Filter).setOptions(options);
  }

//SET Training File

  public void setTraining(String name) throws Exception {
    m_TrainingFile = name;
    m_Training     = new Instances(
                        new BufferedReader(new FileReader(m_TrainingFile)));
    m_Training.setClassIndex(m_Training.numAttributes() - 1);
  }

  public void execute() throws Exception {
    // run filter
    m_Filter.setInputFormat(m_Training);
    Instances filtered = Filter.useFilter(m_Training, m_Filter);

    // train classifier on complete file for tree
    m_Classifier.buildClassifier(filtered);

    // 10fold CV with seed=1
    m_Evaluation = new Evaluation(filtered);
    m_Evaluation.crossValidateModel(
        m_Classifier, filtered, 10, m_Training.getRandomNumberGenerator(1));
  }


  public String toString() {
    StringBuffer        result;


    return result.toString();
  }

//E.g., CLASSIFIER weka.classifiers.trees.J48

public static void main(String[] args) throws Exception {
  WekaClassifierDemo        demo;

  String classifier = "";
  String filter = "";
  String dataset = "";
  Vector classifierOptions = new Vector();
  Vector filterOptions = new Vector();
```

```
    int i = 0;
    String current = "";
    boolean newPart = false;

    demo = new WekaDemo();
    demo.setClassifier(
        classifier,
        (String[]) classifierOptions.toArray(new
String[classifierOptions.size()]));
    demo.setFilter(
        filter,
        (String[]) filterOptions.toArray(new String[filterOptions.size()]));
    demo.setTraining(dataset);
    demo.execute();
    System.out.println(demo.toString());
    }
}
```

3.4.2.1   A Tutorial for Building Classification Models Using LibSVM and Weka

In this tutorial we are going to create a binary classification model using the following steps:

1. Create a data set of drugs and nondrugs using available databases of drugs and pharmaceutical leads.
2. Generate descriptors for the compounds using available software (refer to previous chapter).
3. Store the information (molecules along with the descriptors) in an excel sheet or use MATLAB to convert the plain data into spreadsheet format.
4. Convert the output file into CSV format.
5. Convert the CSV file to the LibSVM format.
6. Run the file in the LibSVM (before scaling).
7. Scale the data.
8. Run the scaled file in LibSVM and check the cross-validation accuracies.
9. Create a model file using 'c' and 'g' parameters in LibSVM.
10. Rank the features in WeKa and extract the best features.
11. Convert the CSV file containing the best features to an ARFF file.
12. Run the ARFF file in the WeKa implementation of RF.
13. Check for the accuracy.

Building an SVM model using LibSVM for the Wisconsin Breast Cancer Data Set used for modelling studies [31]:

1. Data preparation for SVM implementation

In data-driven classification/regression applications, it is desirable to collect and utilize maximum data as possible. Data set should contain both positive drug and negative nondrug cases. We need to split the data set into two, one for training and other for testing.

The data set comprising SVM train using the input–output pairs was partitioned in training and test set.

Fig. 3.9 Computing accuracy in LibSVM

2. Convert the data into SVM format and perform svm-train with its parameters, since the SVM algorithm operates on numeric attributes.

So we need to convert the data into the LibSVM format which contains only numerical values. Open the command prompt and give path till the LibSVM folder (Windows).

svm-train—train one or more SVM instance(s) on a given data set to produce a model file. svm-train trains an SVM to learn the data indicated (Fig. 3.9).

Command: svm-train -s 0 -v 10 -q (filename)

3. Conduct simple scaling on the data

The original data ranges maybe too broad or narrow in range, and thus these need to be normalized. The main advantage of scaling is to avoid attributes in greater numeric ranges dominating those in smaller numeric ranges. Another advantage is to avoid numerical difficulties during the calculation. We recommend linearly scaling each attribute, which is linearly scaled between $[-1; +1]$ and $[0; 1]$. This scaling should be done before splitting the data into training and test sets. We have to use the same method to scale both training and testing data.

svm-scale is a tool for scaling input data file.

The syntax of svm-scale is:

svm-scale [options] data_filename.

The output of scaling is the filename.scale file, which is used for creating the model. Bys using the same scaling factors for training and testing sets, we obtain much better accuracy (Fig. 3.10).

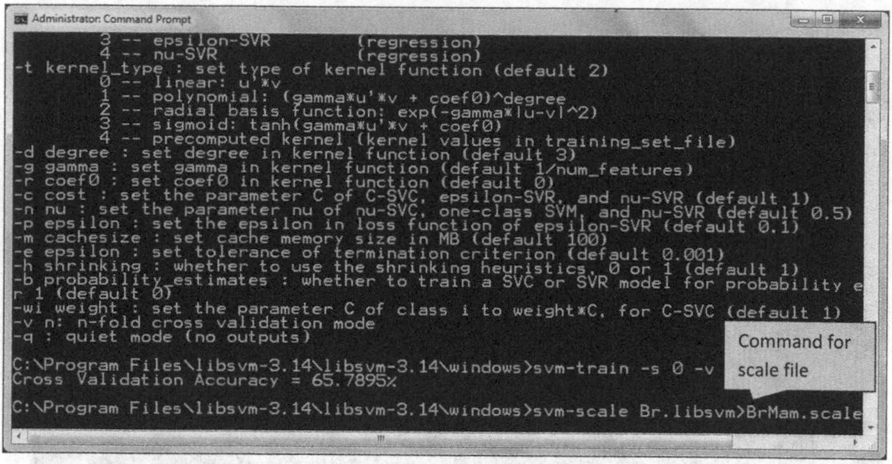

Fig. 3.10 The command to convert the libSVM file to a scale file

4. Running the scale file to obtain the appropriate parameter for best accuracy

After the scale file is created, copy the scale file and paste it to the 'Tools' folder in LibSVM.

Change the path in command prompt, till Tools.

One needs to have Python installed in the tools folder to run the scale file and obtain the appropriate 'c' and 'g' parameters (Figs. 3.11 and 3.12).

Command:

Python grid.py scalefilename.scale

5. Obtaining the accuracy using the best 'c' and 'g' parameters

Before proceeding, change the path in command prompt by coming out of the tools and entering the windows folder.

Syntax: svm-train (c and g parameters) scalefilename

Ranking of Features in WeKa

In order to select the best features or descriptors and improve the model, we should rank the features using information gain as the ranking metric (Fig. 3.13).

Information gain is a measure of the contribution of a particular feature to the model. Ranking using information gain was done using WeKa (Figs. 3.14, 3.15 and 3.16).

For obtaining the best features:

1. Open the Explorer interface in WeKa.
2. Select the Pre-process Tab above and Open an ARFF file and select All in the Attributes section.
3. Go to the Select Attributes Tab.
4. In the Attribute Evaluator Tab, Select Information Gain and Click Start.
5. Select the top-ranked features from the result.

After ranking, the top-ranked features are extracted from the feature set and passed to LibSVM. To obtain the best-ranked features, the ARFF format of the models is

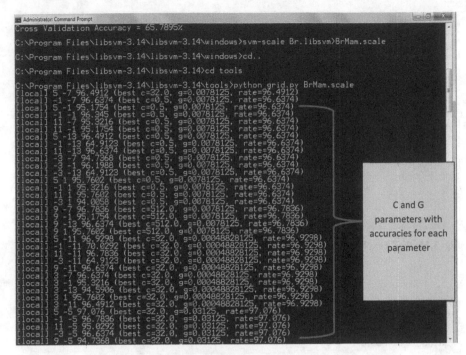

Fig. 3.11 Running the scale file in python to obtain the best 'c' and 'g' parameters

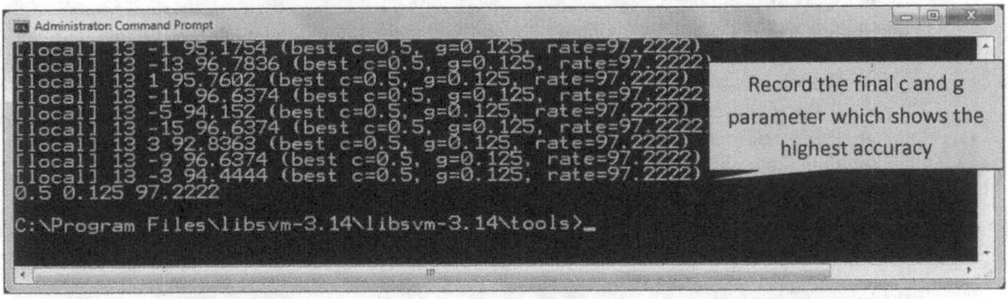

Fig. 3.12 Final optimum of the 'c' and 'g' parameters

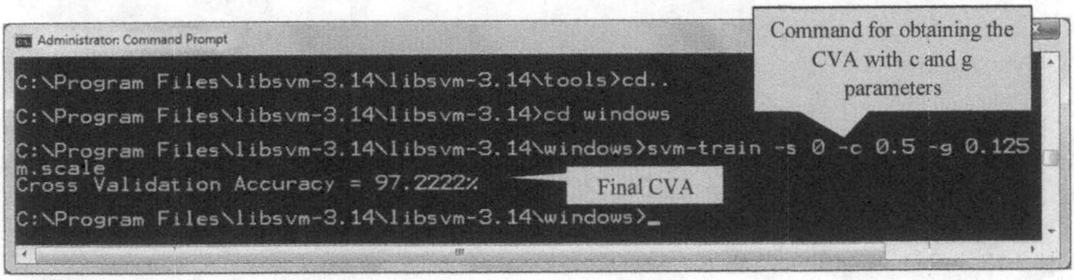

Fig. 3.13 Final cross-validation accuracy

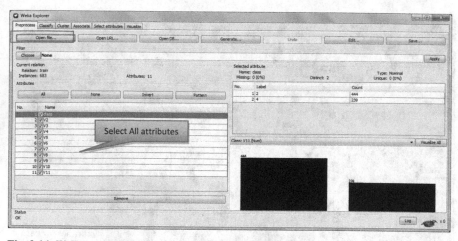

Fig. 3.14 WeKa explorer graphical user interface (GUI)

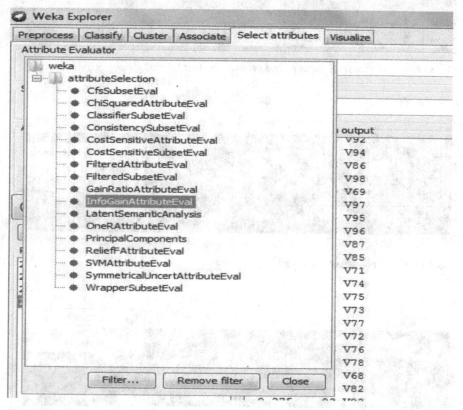

Fig. 3.15 InfoGain attribute Tab

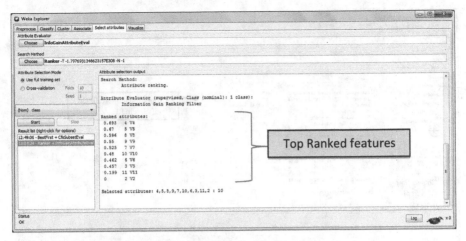

Fig. 3.16 The selected attributes listed are the ranked features

generated by converting the CSV file of the models into ARFF using R-programming language. The ARFF files of the model are first edited to provide the class as first attribute and then opened in WeKa and all the attributes are imported. The selected attributes are then subjected to WeKa Infogain mode, which lists the features in the order of their priority or on the basis of their contribution towards the classification of the model.

Obtaining CVA for Ranked Features The attributes generated are saved and set of topmost attributes or features were selected (top100, 200 and so on). These features were arranged priority wise with the help of coding in MATLAB. Again the MATLAB worksheets of the ranked features are converted to CSV and LibSVM format. The cross-validation accuracies for the ranked features are hence obtained using LibSVM. Ranking of the features helps us to select the best set of features whose contribution towards the classification of the model is the best, which is on the basis of highest cross-validation accuracy.

3.4.2.2 Obtaining Accuracy Using Random Forest

To check the efficiency of our model and the features selected, we can use another classifier, RF, to classify our models. We can use the WeKa implementation of RF.

To classify using RF, the following steps are followed:

1. Open the ARFF file of the model in WeKa.
2. Select all the attributes in the 'Preprocess' window.
3. Select the 'Classify' Tab.
4. Choose 'Trees' from the classifier tab and open 'Random Forest' from the list.
5. Change the RF parameters, Numtrees and NumFeatures, to the required value.
6. Choose Nominal Class (Nom-Class).
7. Start RF (Fig. 3.17).

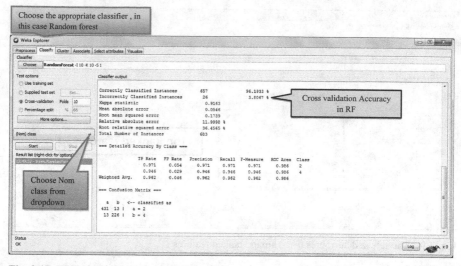

Fig. 3.17 WeKa implementation of RF

		Predicted	
		Negative	Positive
Actual	Negative	a	b
	Positive	c	d

Fig. 3.18 Confusion matrix layout

This starts the classification in RF which takes few seconds or minutes to generate results. The result contains 'Correctly classified instances' which is considered as the accuracy of the model in RF. The result also generates 'Confusion Matrix' which can be utilized to calculate various parameters.

Confusion matrix is a specific table layout which contains information about actual and predicted classifications done by the classification system (Fig. 3.18).

The entries in the confusion matrix have the following meaning:

a   is the number of correct predictions that an instance is negative (True Negative),
b   is the number of incorrect predictions that an instance is positive (False Positive),
c   is the number of incorrect predictions that an instance negative (False Negative), and
d   is the number of correct predictions that an instance is positive (True Positive).

Specificity and Sensitivity are the two statistical measures to detect the performance of a binary classification system:

a. Sensitivity relates to the ability of a test to correctly classify.

$$Sensitivity = d/d + c$$
$$= (True\ Positive/True\ Positive + False\ Negative)$$

b. Specificity relates to the ability of a test to identify negative results.

$$Sensitivity = d/b + d$$
$$= (True\ Positive/False\ Positive + True\ Negative)$$

3.4.3   R Program

R is an open-source tool. It has various packages for building ML models. R is an open source, highly used statistical package with a seamless support of various libraries available on CRAN [32]. This component allows user to input data sets and visualize them after processing for better interpretation and insight.

Jar files needed come with rJava Package, i.e. JRI.jar, JRIEngine.jar, REngine. jar.

C-3.3.3.1 Code to initiate R and compute properties

```
private static void initR() {
        String[] dummyArgs = new String[1];
        dummyArgs[0] = "--vanilla";
        _re = new Rengine(dummyArgs, false, null);
        _re.eval("library(JavaGD)");
        _re.eval("Sys.putenv('JAVAGD_CLASS_NAME'='MyJavaGD2')");
System.out.println("Rengine and JavaGD initialized !!");
        }
    private static void runRCommands(String filename) {
        try {
            System.out.println("PATH : " +
System.getProperty("java.library.path"));
            String[] s = getData(filename);
            for (int i = 0; i < s.length; i++) {
                _re.eval(s[i]);
            }
        } catch (Exception ex) {
        }
    }
```

The single-hidden-layer neural networks are implemented in the package nnet. [33] Tree-structured models for regression, classification and survival analysis, following the ideas in the Classification and Regression Trees (CART) book, are implemented in rpart and tree. The Cubist package fits rule-based models (similar to trees) with linear regression models in the terminal leaves, instance-based corrections and boosting [34]. Two recursive partitioning are algorithms with unbiased variable selection and statistical stopping criterion. Graphical tools for the visualization of trees are available in the package maptree. An approach to deal with the instability problem via extra splits is available in the package TWIX.

Trees for modelling longitudinal data by means of random effects are offered by the packages REEMtree and longRPart [35]. Partitioning of mixture models is performed by recursively partitioned mixture model (RPMM).

Computational infrastructure for representing trees and unified methods for prediction and visualization is implemented in partykit. This infrastructure is used by the package evtree to implement evolutionary learning of globally optimal trees.

The reference implementation of the RF algorithm for regression and classification is available in the package RF. Variable selection through clone selection in SVMs in penalized models (SCAD or L1 penalties) is implemented in the package penalizedSVM. The function svm() from e1071 offers an interface to the LibSVM library and the package kernlab implements a flexible framework for kernel learning (including SVMs, RVMs and other kernel-learning algorithms) [36]. Bayesian additive regression trees (BART), where the final model is defined in terms of the sum over many weak learners (not unlike ensemble methods), are implemented in the package BayesTree. The packages rgp and rgenoud offer optimization routines based on genetic algorithms [37].

3.5   Free Tools for Machine Learning

3.5.1   An Example of SVR-based Machine Learning

The classical multiple regression has a well-known loss function that is quadratic in the prediction errors. However, the loss function employed in SVR is the ε-insensitive loss function. Here, the 'loss' is interpreted as a penalty or error measure. Usage of ε-insensitive loss function has the following implications. If the absolute residual is off-target by ε or less, then there is no loss, that is, no penalty should be imposed. However, if the opposite is true, that is absolute residual is off-target by an amount greater than ε, then a certain amount of loss should be associated with the estimate. This loss rises linearly with the absolute residual above ε.

The SVR algorithm attempts to place a tube around the regression function as shown in Fig. 3.19, wherein the region enclosed by the tube is called as 'ε-insensitive' zone where ε represents the radius of the tube. The diameter of the tube should ideally be the amount of noise in the data. The optimization criterion in SVR penalizes those data points, whose y values lie more than ε distance away from the fitted function (hyperplane).

Tanagra is a free suite of an ML software for research and academic purposes and it is developed by Ricco Rakotomalala at the Lumière University Lyon 2, France [38]. It is basically a free data mining software. Data mining is extracting information from the data set and converting or transforming it into an understandable structure for further use in future. Tanagra proposes several data mining methods from artificial intelligence, exploratory data analysis, statistical learning, ML and database systems. Tanagra supports several standard data mining tasks such as Visualization (includes Correlation Scatter plot, Viewing data set, multiple scatter plot, exporting data set, etc.), Descriptive statistics (includes Univariate continuous statistics, one-way analysis of variance (ANOVA), one-way multivariate analysis of variance (MANOVA), Normality Test, Welch ANOVA, Paired T-test, Paired

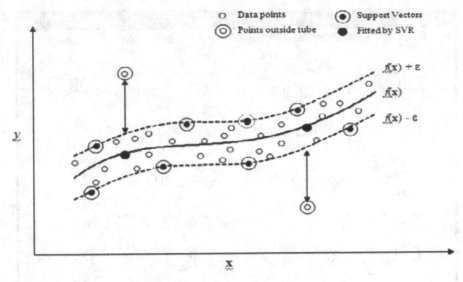

Data points   Support Vectors
Points outside tube   Fitted by SVR

$f(x) + \varepsilon$

$f(x)$

$f(x) - \varepsilon$

Fig. 3.19 A schematic of SVR using ε-sensitive loss function

V-test, Linear Correlation, etc.), Instance selection (includes rule-based selection, sampling, stratified sampling, continuous select examples, discrete select examples, etc.), Feature selection (includes Define status, CFS Filtering, Remove constant, Feature ranking, etc.), Feature construction (includes Trend, radial basis function (RBF), Binary binning, Standardize, etc.), Regression (includes Regression tree, Epsilon SVR, nu-SVR, Multiple Linear Regression, Outlier detection, Regression Assessment, etc.), Factorial analysis (includes Principal Component Analysis, Principal Factor Analysis, Correspondence Analysis, etc.), Clustering (includes K-means, Neighbourhood Graph, VarKmeans, EM-Clustering, etc.), Classification/Spv Learning (includes NB Continous, C-SVC, contingent valuation method (CVM), ID3, C 4.5, Multilayer Perceptron, PLS LDA, etc.), Association rule learning (includes A priori, Spv Assoc rule, Spv Assoc tree, etc.) and Scoring (includes receiver operating characteristic (ROC) Curve, Precision-Recall Curve, Scoring, Lift Curve, etc.). Tanagra is an easy-to-use software for researchers and students and it also provides architecture for them to easily add their own data mining algorithms/methods, and comparing their performances.

For installation just go to Google Search Engine and type 'Tanagra download'. Click on 'SetUp' under 'Reference' Column of the table. Use the software for performing various tasks.

The input file formats or data set formats which are accepted by Tanagra for performing different data mining tasks are .txt, .arff and .xls, and sparse formats include .dat and .data.

After performing the tasks on the data set, the results can be saved in two formats in Tanagra, *.tdm and *.bdm, i.e. text data mining diagram (tdm) and binary data mining diagram (bdm).

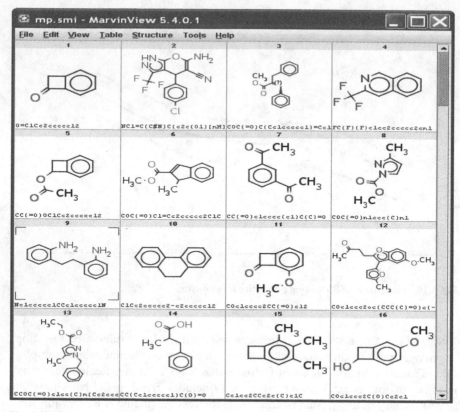

Fig. 3.20 Molecules of the training data set in Marvin view

Here, we will use a melting point data set based on a diverse collection of molecules [39]. It is downloadable from a moltable site [40]. For the present tutorial we will select 100 diverse molecules (Fig. 3.20).

1. Open Tanagra, click on 'File' and select 'New' for choosing the data set of appropriate format and select the checkbox 'Checking Missing Val' and click OK and the downloaded information and data set description will appear on the right window.
2. Click on the 'Data Visualization' palette from the bottom 'Components' window and select 'View Dataset' tab and drag it to the left 'default title' window. Double click on 'View Dataset1' on left window, the whole data set will appear on the right window (Fig. 3.21).
3. Click on 'Instance Selection' palette from the bottom 'Components' window and select 'Continuous Select Examples' tab and drag it to the left 'default title' window. Right click on 'Continuous Select Examples 1' on the left 'default title' window, select 'Parameters' and set Attribute as 'media transfer protocol (MTP)', Operator as '<', Value as '50' and click on 'OK'. Right click on 'Continuous

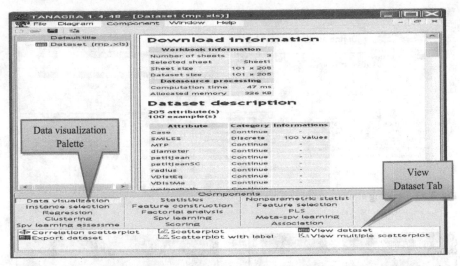

Fig. 3.21 Tanagra showing data set description with components window, below for data visualization

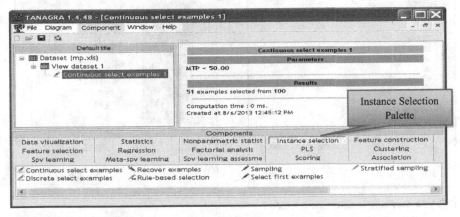

Fig. 3.22 Instance selection process with its result

Select Examples 1' on the left 'default title' window and select 'View' and the results for attribute selection will appear on the right window (Fig. 3.22).

We see that 51 examples are assigned for learning phase, and the other 49 examples will be used for assessment of models.

4. For defining attribute statuses, one can go to 'Feature Selection' palette from the bottom Component window and select 'Define Status' or else can click on the icon below the 'Diagram' tab on the menu bar at the top left of the Tanagra window (Fig. 3.23).

5. Set the attribute 'Case' in 'Target' tab and remaining attribute in 'Input' tab at the right window (except the 'MTP' attribute and 'SMILES') and click on OK.

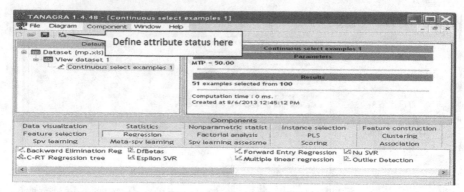

Fig. 3.23 Defining attribute status process

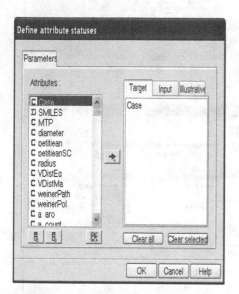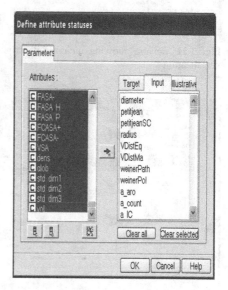

Fig. 3.24 Defining attribute status window

Double click on 'Defined Status 1' on left 'default title' window. The results will appear on the right window (Figs. 3.24 and 3.25).

6. For performing Epsilon SVR, select 'Regression' palette from the bottom 'Components' window and select Epsilon SVR tab and drag it to the left 'default title' window. Double click on 'Epsilon SVR 1' on left 'default title' window and select 'View', the results will appear on the right window showing the following: Epsilon SVR parameters, SVM characteristics, ANOVA and Residual Analysis (Figs. 3.26 and 3.27).

The default kernel is linear. The number of SV is 47 and the Pseudo-R^2 on training sample or selected sample is 0.9905. The regression seems very good.

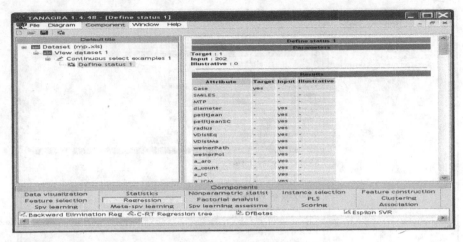

Fig. 3.25 Defining attribute status result for selected examples

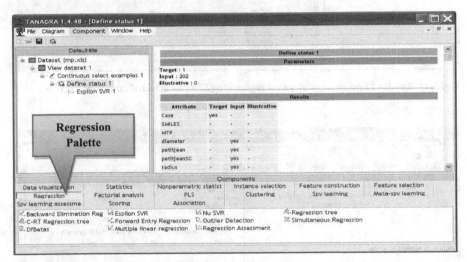

Fig. 3.26 Regression palette from component window showing epsilon support vector regression (*SVR*)

7. For evaluating the unselected samples at Step 3, again set the parameters for defining attribute statuses, for this click on the icon below the diagram tab at the top left of the Tanagra window. Set the attribute 'Case' in 'Target' tab and 'Pred_e_svr_1' in 'Input' tab at the right window of 'Define Attribute Statuses' and click on OK. Double click on 'Define status 2' on the left 'default title' window, the results will appear on the right window (Figs. 3.28 and 3.29).

8. Click on 'Regression' palette from the bottom window and select the 'Regression Assessment' component and drag it to the left 'default title' window.

9. Right click on 'Regression Assessment 1' and click on parameters and set the parameter as 'Unselected' in the dialogue box which appears. Double click on

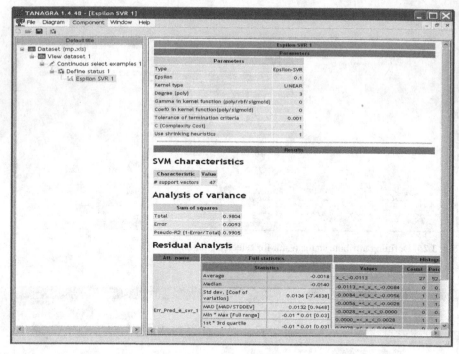

Fig. 3.27 Result of Epsilon support vector regression (*SVR*) for selected examples

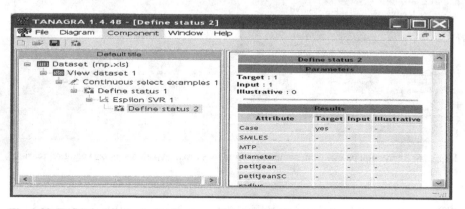

Fig. 3.28 Defined attribute status of test or unselected samples

'Regression Assessment 1' and select view to obtain the results of unselected samples on the right window (Fig. 3.30).

Result Interpretation We obtained Pseudo-$R^2 = 1 - 15.6636/47.0973 = 0.6674$. This is the best result we have obtained on test samples or unselected data set after setting up different parameters and modifying the parameters for Epsilon SVR. The

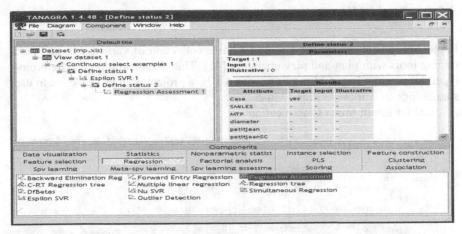

Fig. 3.29 Regression palette showing regression assessment tab from the Component window

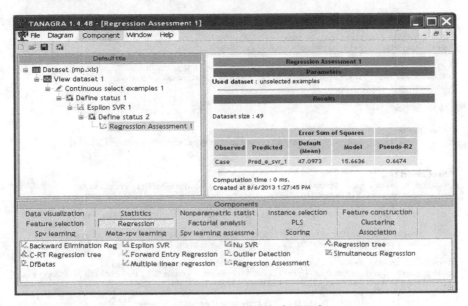

Fig. 3.30 Result after regression assessment for unselected examples

regression results obtained on training samples is quite good as compared to test samples, thus the test set is more dependent on the specificities of the training data set and the results obtained below are the best results obtained after setting up different parameters and using different kernels.

Data set	Pseudo-R^2
Training data set (selected samples)	0.9905
Test set (unselected samples)	0.6674

3.5.2　Rapid Miner

Rapid Miner is an open-source tool and has a collection of various ML and data mining tools with plug and play operators [41]. This demonstration can be prototyped for running Rapid Miner from within a Java project with an objective to use source code to work with Rapid Miner.

Code for Rapid Miner Classification

```java
public static void rapidminer_func(String modelname,String output_filename) throws
OperatorException
    {
    File file = new File("<rapidminer_output_file>");
    if(file.mkdir()){
        System.out.println("Directory is created!");
    }else{
        System.out.println("Failed to create directory!");
    }
    String rapidMinerHome = "<RapidMiner5 Home Folder>";
        System.setProperty("rapidminer.home", rapidMinerHome);
        RapidMiner.setExecutionMode(RapidMiner.ExecutionMode.COMMAND_LINE);
        RapidMiner.init();
```

/* Reading Data */

```java
        try {
        Operator trainingDataReader =
OperatorService.createOperator(RepositorySource.class);

trainingDataReader.setParameter(RepositorySource.PARAMETER_REPOSITORY_ENTRY,
"//NewLocalRepository/resources/com/rapidminer/resources/samples/data/" +
modelname);
```

/* Classifier */

```java
Operator bayesClassifier = OperatorService.createOperator(NaiveBayes.class);
```

/* Save model */

```java
Operator modelWriter = OperatorService.createOperator(ModelWriter.class);
        modelWriter.setParameter("model_file", "<rapidminer_output_dir/>" +
output_filename);
            com.rapidminer.Process process = new com.rapidminer.Process();

        process.getRootOperator().getSubprocess(0).addOperator(trainingDataReader);
process.getRootOperator().getSubprocess(0).addOperator(bayesClassifier);

process.getRootOperator().getSubprocess(0).addOperator(modelWriter);
trainingDataReader.getOutputPorts().getPortByName("output").connectTo(bayesClassifi
er.getInputPorts().getPortByName("training set"));

bayesClassifier.getOutputPorts().getPortByName("model").connectTo(modelWriter.getIn
putPorts().getPortByName("input"));

process.run();

} catch (OperatorCreationException e) {
e.printStackTrace();
        } catch (OperatorException e) {
e.printStackTrace();
        }
            finally
            {
                try
                {
                    System.out.println("Done");
                }catch (Exception ee) {
ee.printStackTrace();
                }
            }
}// end of rapidminer_func
```

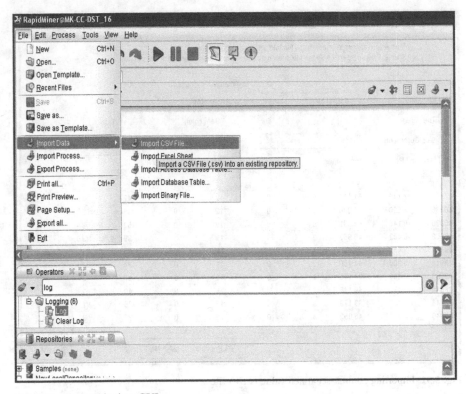

Fig. 3.31 The rapid miner GUI

3.5.2.1 Practice Tutorial for Building Machine Learning Models in Rapid Miner

We will use a dihydrofolate reductase (DHFR) inhibitor data set to build models using various classifiers implemented in rapid miner. The data set is available at a moltable site. The data set consists of 653 training set molecules and 400 test set molecules.

Import the training set file (Fig. 3.31).

Use the import data option to import any type of file. Select the file from the destination folder and click next.

Specify the column separation parameters (Fig. 3.32).

In this step, data types of the attributes are defined. Rapid Miner does the type detection automatically (Fig. 3.33).

In this step attributes can be assigned a special role like an identification (ID) or a label (Fig. 3.34).

Save the file in local repository and click on finish (Fig. 3.35).

Drag the saved file from repository and choose the New building block option from Edit option on the menu bar (Fig. 3.36).

Select the first option to specify the type of validation (Fig. 3.37).

After selecting the validation type, a window appears in which instead of DT other operators can be selected like SVM etc (Fig. 3.38).

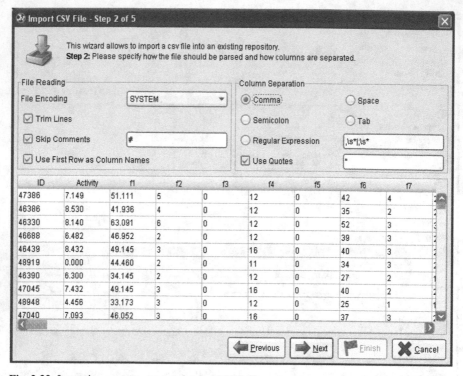

Fig. 3.32 Importing comma-separated value (*CSV*) file

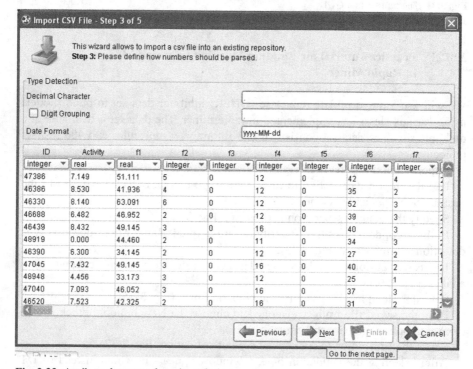

Fig. 3.33 Attribute data-type detection window

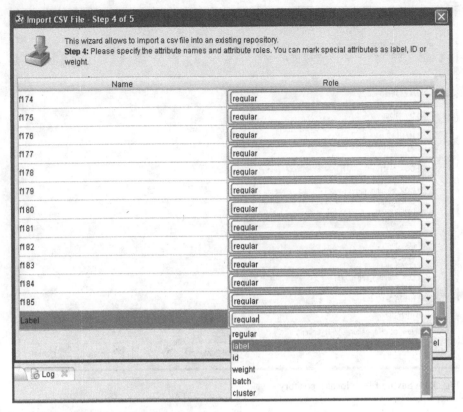

Fig. 3.34 Special role assigned to an attribute

- Use the up arrow button to navigate between different processes.

Select the optimize parameter (grid) operator. Click on the blue window icon at the corner and paste the validation operator inside of it (Fig. 3.39).

Click on the optimize parameter operator and select the parameters to be optimized and specify the value range (Fig. 3.40).

A nested window opens when the blue window icon is clicked. Add a log operator also (Fig. 3.41).

Select the log operator and define the path to store the log file. Edit the log file to select the parameters which are to be optimized (Fig. 3.42).

Click on the run button to start the process (Fig. 3.43).

Results Overview (Fig. 3.44)

- The result gives a set of optimized parameters, performance measure in terms of accuracy, precision, recall and area under the roc curve (AUC).
- A log file is also generated.
- Parameter set (Fig. 3.45).

The results show high accuracy, precision and recall values. The standard equations for calculating these three performance measures are provided below.

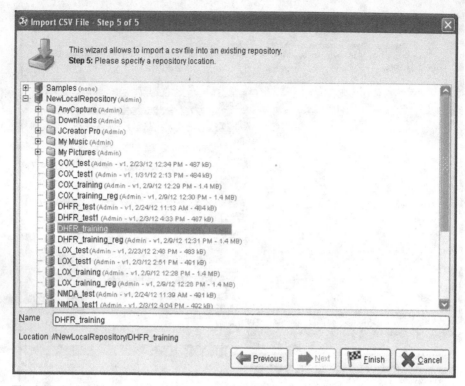

Fig. 3.35 Saving file in local repository

Precision = TP/TP + FP
Recall = TP/TP + FN
Accuracy = (TP + TN)/TP + TN + FP + FN

Apart from these values there are other validation metrics like receiver operating characteristic (ROC) and AUC. An ROC is a two-dimensional (2D) curve that denotes the relation between specificity and sensitivity. AUC is a better classification performance metric as it minimizes the loss of ranking a true negative at least as large as a true positive (Fig. 3.46).

An AUC value >0.6 signifies a good model, anything below this indicates a random prediction. Since we obtained an AUC of 0.9, our model can be considered statistically good.

3.6 Commercial Tools for Building ML Models

3.6.1 Molecular Operating Environment (MOE)

Molecular Operating Environment (MOE) is a comprehensive software system for Life Science [42]. MOE is a combined Applications Environment and Methodol-

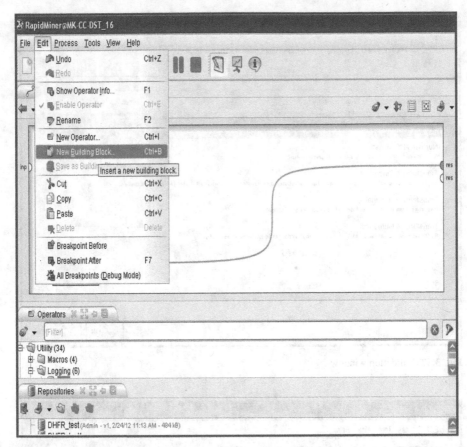

Fig. 3.36 Rapid miner design workspace

ogy Development Platform that integrates visualization, simulation and application development in one package. MOE contains a broad base of scientific applications for general modelling, drug design, homology modelling and library design. It provides a suite of applications for manipulating and analyzing large collections of compounds. It is a fully integrated suite of computational chemistry, molecular modelling and informatics software for life science applications. The suite's applications are written in an embedded programming language, Scientific Vector Language (SVL), and can be easily customized since the SVL source code is provided in the distribution [43]. The Molecular Database is a disk-based spreadsheet central to the manipulation and visualization of large collections of compounds. Compound collections can be 'washed' to remove salts and solvents and to adjust protonation state of acids and bases.

Steps required for QSAR modelling using MOE:

1. Calculating Molecular Descriptors
2. Fitting Experimental Descriptors
3. Cross-Validating Model
4. Performing Graphical Analysis

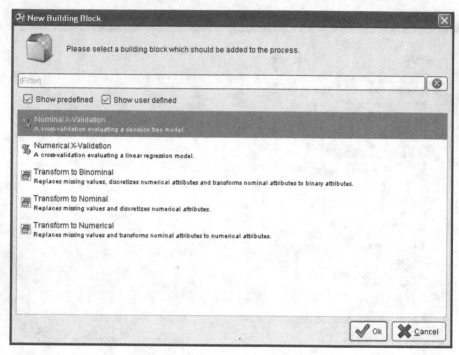

Fig. 3.37 Validation window

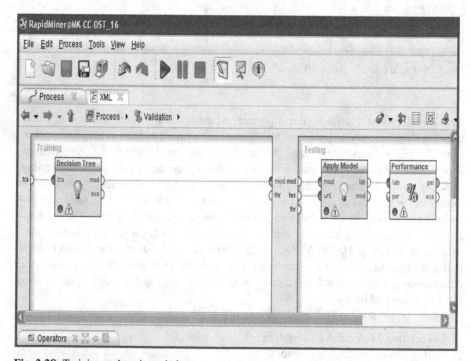

Fig. 3.38 Training and testing window

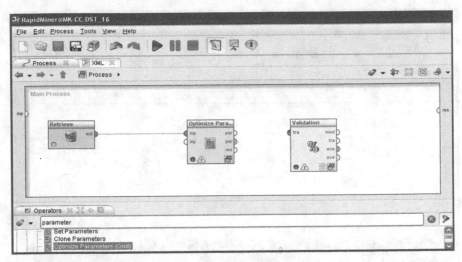

Fig. 3.39 Optimize the parameter operator and the validation operator in the design workspace

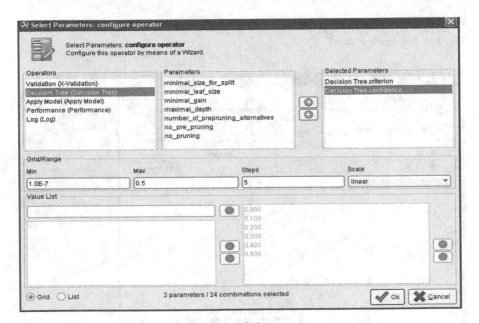

Fig. 3.40 Select parameter: configure operator window

5. Estimating the Predicted activities for the test set
6. Pruning the Descriptors

The QSAR suite of applications in MOE is used to analyze experimental data and build numerical models of the data for prediction and interpretation purposes. Given

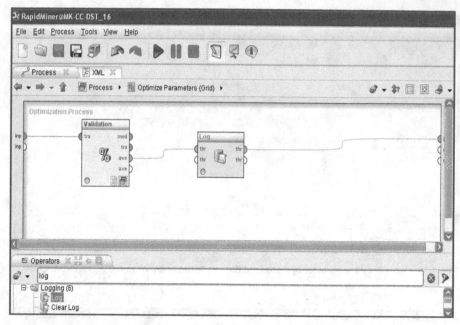

Fig. 3.41 Log operator added

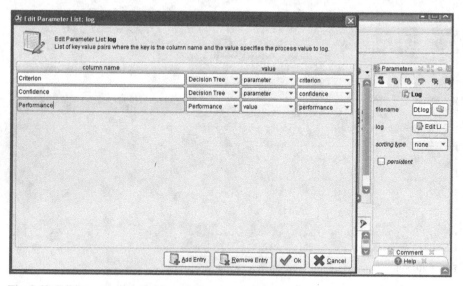

Fig. 3.42 Editing parameter list: log window

a set of molecules whose activity in a particular experiment is known (referred to as a training set or a learning set), a QSAR model correlates these activities with properties inherent to each molecule in the set. These properties are evaluated using molecular descriptors available in MOE (Fig. 3.47).

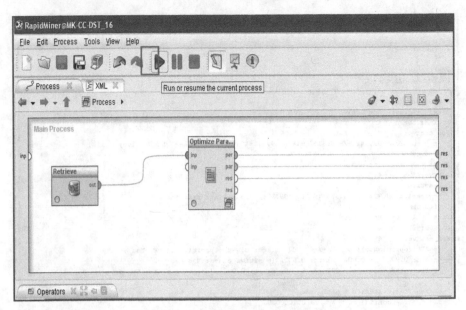

Fig. 3.43 Initializing the grosses after parameter settings

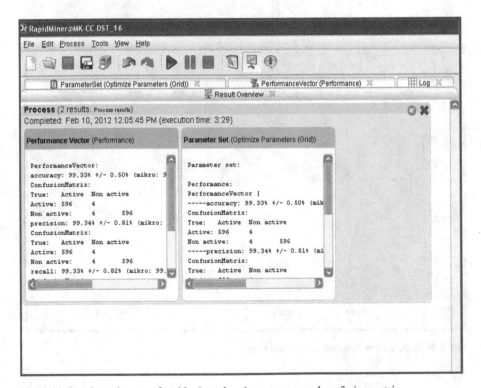

Fig. 3.44 Result workspace of rapid miner showing accuracy and confusion matrix

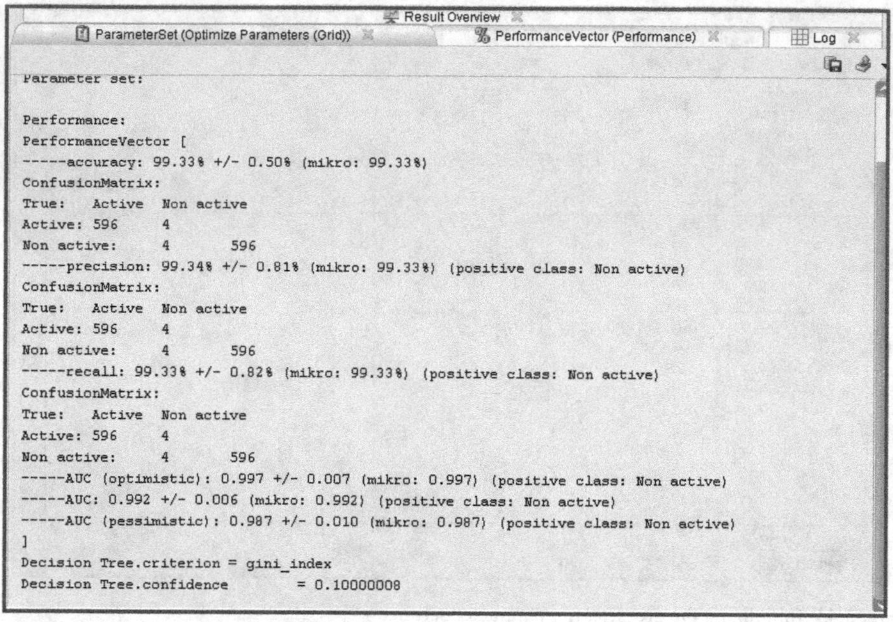

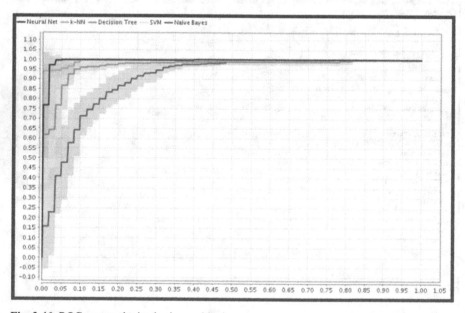

Result Overview
ParameterSet (Optimize Parameters (Grid))
PerformanceVector (Performance)
Log
Parameter set:
Performance:
PerformanceVector [
-----accuracy: 99.33% +/- 0.50% (mikro: 99.33%)
ConfusionMatrix:
True: Active Non active
Active: 596 4
Non active: 4 596
-----precision: 99.34% +/- 0.81% (mikro: 99.33%) (positive class: Non active)
ConfusionMatrix:
True: Active Non active
Active: 596 4
Non active: 4 596
-----recall: 99.33% +/- 0.82% (mikro: 99.33%) (positive class: Non active)
ConfusionMatrix:
True: Active Non active
Active: 596 4
Non active: 4 596
-----AUC (optimistic): 0.997 +/- 0.007 (mikro: 0.997) (positive class: Non active)
-----AUC: 0.992 +/- 0.006 (mikro: 0.992) (positive class: Non active)
-----AUC (pessimistic): 0.987 +/- 0.010 (mikro: 0.987) (positive class: Non active)
]
Decision Tree.criterion = gini_index
Decision Tree.confidence = 0.10000008

Fig. 3.45 Results window in rapid miner

Neural Net
k-NN
Decision Tree
SVM
Naive Bayes

Fig. 3.46 ROC curves obtained using rapid miner

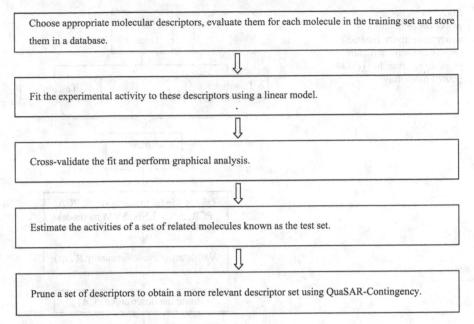

Fig. 3.47 Steps to build a model in MOE

Basic Model Building Steps Performed in MOE:

Structure–activity relationship (SAR) and, more generally, structure–property relationship (SPR) analysis are integral to the rational drug design cycle. Quantitative (QSAR, QSPR) methods assume that biological activity is correlated with chemical structures or properties and that as a consequence activity can be modelled as a function of calculable physiochemical attributes. Such a model for activity prediction could then be used, for instance, to screen candidate lead compounds or to suggest directions for new lead molecules.

The QSAR/QSPR models can be built and applied by following a few steps (Fig. 3.48).

The components of the QuaSAR package are a combination of SVL descriptor modules and SVL programs to operate the fundamental MOE molecular services.

3.6.1.1 A Tutorial for QSAR Model Building of DHFR Inhibitors

We scanned the literature mainly to extract biological activity data for each unique compound in the data set containing total 653 entries. The data was collected in the SDF format and imported into MOE. All the 2D descriptors from MOE were computed for the inhibitors (Fig. 3.49).

1. Prepare the training set.

Fig. 3.48 A general
workflow of quantitative
structure–activity relation-
ship/quantitative structure–
property relationship (*QSAR/
QSPR*) modelling

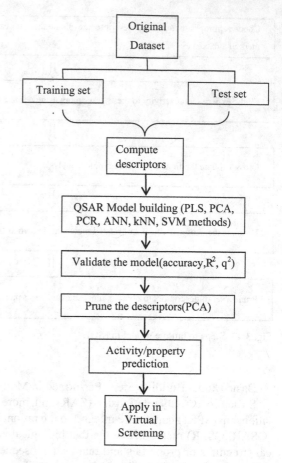

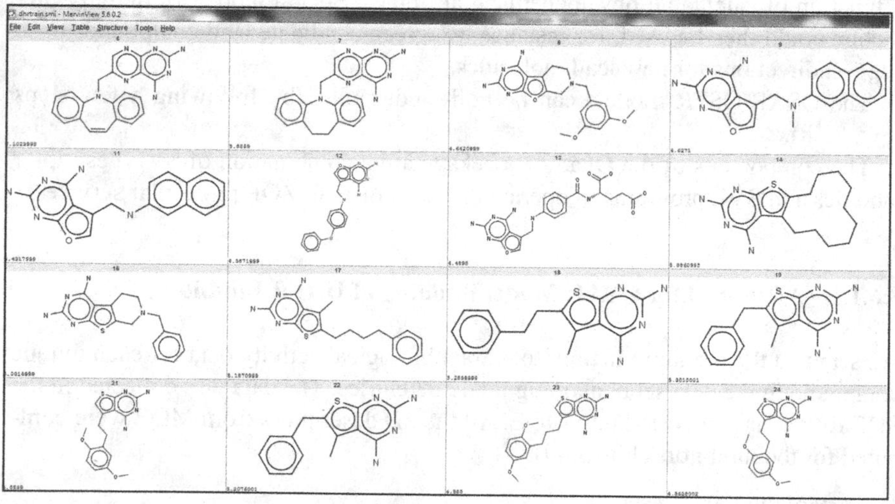

Fig. 3.49 Inhibitors with their IC_{50} and their 2D descriptors

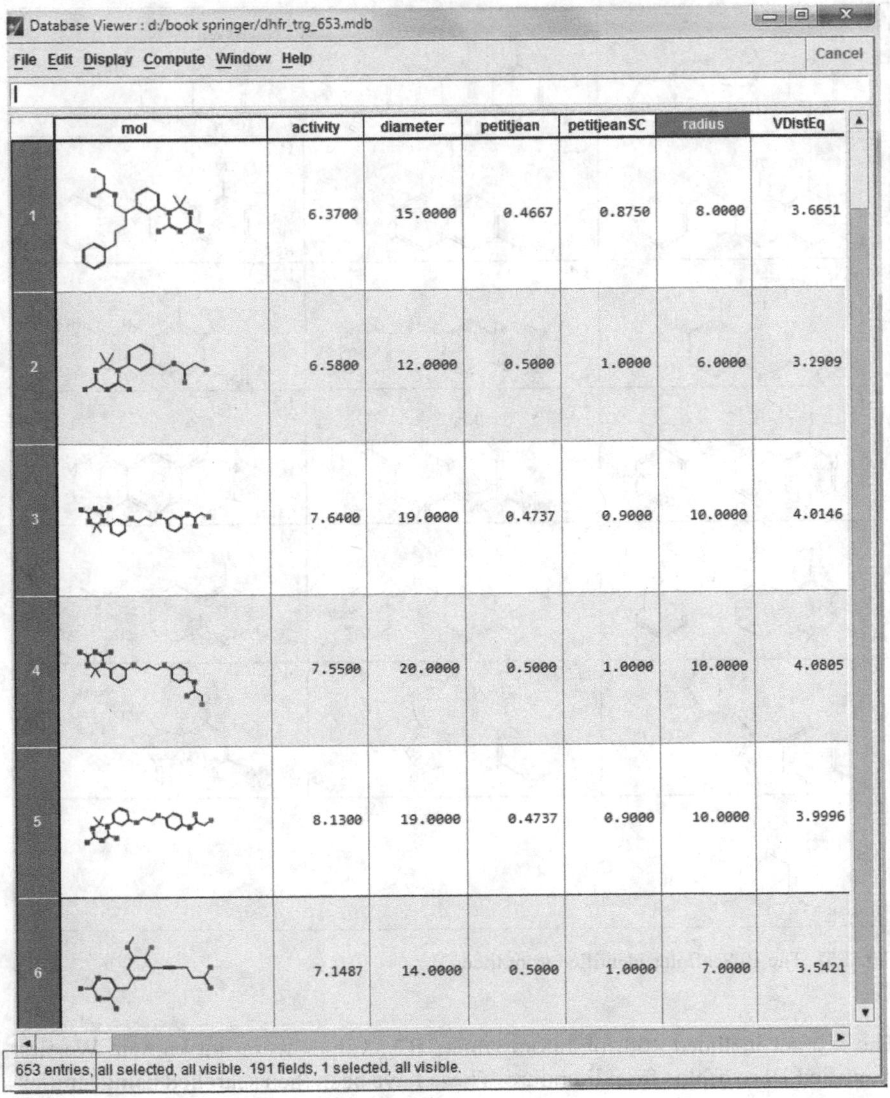

Database Viewer : d:/book springer/dhfr_trg_653.mdb

File Edit Display Compute Window Help

Cancel

	mol	activity	diameter	petitjean	petitjeanSC	radius	VDistEq
1		6.3700	15.0000	0.4667	0.8750	8.0000	3.6651
2		6.5800	12.0000	0.5000	1.0000	6.0000	3.2909
3		7.6400	19.0000	0.4737	0.9000	10.0000	4.0146
4		7.5500	20.0000	0.5000	1.0000	10.0000	4.0805
5		8.1300	19.0000	0.4737	0.9000	10.0000	3.9996
6		7.1487	14.0000	0.5000	1.0000	7.0000	3.5421

653 entries, all selected, all visible. 191 fields, 1 selected, all visible.

Fig. 3.50 Inhibitors with their descriptor values

A number of compounds whose activity is known constitute the training set. The project included 653 inhibitors whose IC_{50} values were known (Fig. 3.50).

1. The Descriptors reported in the literature were computed using MOE for the entire library.
2. Training set values were used to predict and evaluate the model.
3. Prepare the test set.

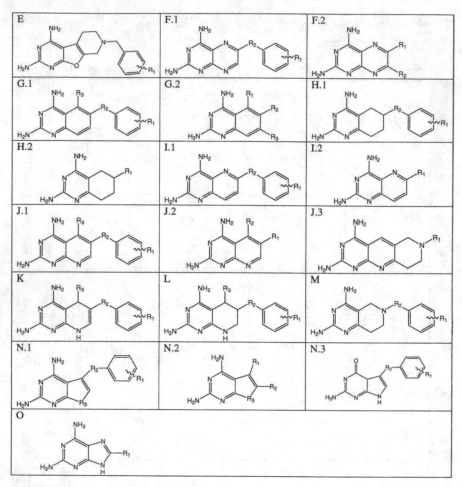

Fig. 3.51 The 19 Scaffolds identified from literature

The test set included 400 inhibitors whose IC$_{50}$ values were not known. We first identified 19 scaffolds from literature. These have been then searched using similarity search methods (Fig. 3.51).

4. After collecting compounds, a MOE-fit file was used to predict the activities of the test set by using the model evaluate option.

The correlation plot between experimental and predicted values is shown in Fig. 3.52.

   Observations: 653
   Descriptors: 329
   Root mean square error (RMSE): 0.49138
   Correlation coefficient (r^2): 0.86270

There are several parameters for validating the model built which parameters include: RMSE, R^2, Q^2 and Leave One Out (LOO) validation method [44].

   RMSE: The RMSE is a frequently used measure of the differences between values predicted by a model or an estimator and the values actually observed. These

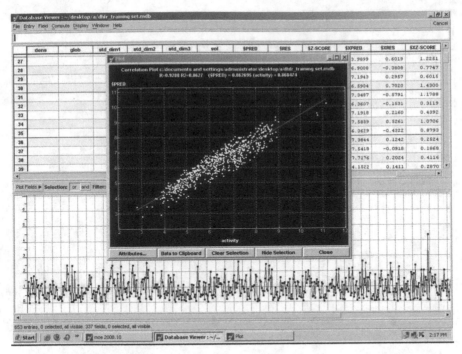

Fig. 3.52 Scatter plots showing predicted versus measured activities, with training set compounds shown using dots

individual differences are called residuals when the calculations are performed over the data sample that was used for estimation, and are called prediction errors when computed out-of-sample.

R^2 indicates how well data points fit a line or a curve. It is a statistics used in the context of statistical models whose main purpose is either the prediction of future outcomes or the testing of hypotheses, on the basis of other related information, i.e. is the proportion of the variance in the dependent variable that is explained by the regression equation (i.e. if $R^2 = 1.0$, then all the actual points lie on the regression line; if $R^2 = 0.0$, then the variance around the regression line is as high as the overall variance of the dependent variable). R^2 is a statistic that will give some information about the goodness of fit of a model. In regression, the R^2 coefficient of determination is a statistical measure of how well the regression line approximates the real data points. An R^2 of 1 indicates that the regression line perfectly fits the data.

In R^2 the same data that is used to build the equation is also used to evaluate it. This can be addressed using Q^2 (sometimes called cross-validated R^2). Here, we make n versions of the equation, each build leaving one of the original known values out (it is thus an example of leave-one-out validation); the Q^2 is then the mean overall variance in using the equation to predict the values left out. Q^2 is always thus less than R^2.

Results The correlation plot obtained after plotting predicted vs. measured activities gave R^2 value of 0.86. R^2 measures the degree of correlation between activity values calculated by model and those measured experimentally. There were very

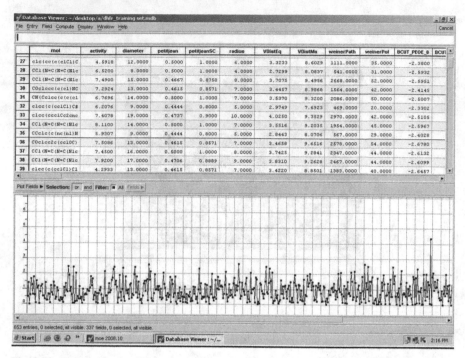

Fig. 3.53 *Z*-score plot for 653 entries is shown. The points with large distances between them are outliers

few outliers. The data range and diversity was very good. The model was validated using Leave One Out (LOO) method (Fig. 3.53).

Model Evaluation: The model was evaluated using a test set of 400 compounds (Fig. 3.54).

3.6.2   IBM SPSS

IBM SPSS is a comprehensive, easy-to-use set of data and predictive analytics tools for business users, analysts and statistical programmers [45]. Its package has a neural network toolbox which includes both Multilayer Perceptron (MLP)-type [46] as well as RBF-type [47] models. Provisions for random number generation (seed) are also provided with this software under the 'Transformations' option. Any data set for neural network modelling purpose has to be partitioned into three partitions:

1. Training
2. Test
3. Validation (or Holdout in SPSS)

The default option in SPSS is to randomly assign cases to these three partitions according to preset portions (e.g. Training 70 %, Test 15 %, Holdout 5 %, etc.) or the data can be manually partitioned with the help of a 'partition variable'. This option can be selected by Analyze > Neural Network > Multilayer Perceptron > Partitions > Use Partitioning Variable (Fig. 3.55).

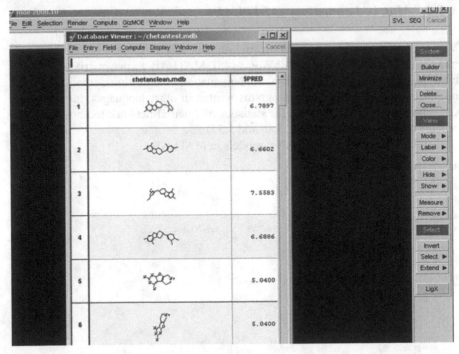

Fig. 3.54 Models showing predicted activity from the test set using QSAR model built in MOE

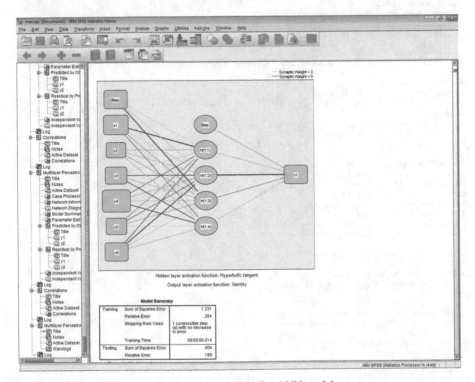

Fig. 3.55 GUI of IBM SPSS showing architecture of an ANN model

3.6.3   Matrix Laboratory (MATLAB)

MATLAB is a numerical computing environment and fourth-generation programming language developed by MathWorks [48]. MATLAB allows matrix manipulations, plotting of functions and data, implementation of algorithms, creation of user interfaces and interfacing with programs written in other languages, including C, C++, Java and Fortran. It provides statistics and neural network toolbox to build reliable predictive models (Figs. 3.56 and 3.57).

C-3.5.3.1 Code for creating ANN models in MATLAB

```
%user specified values
outputs=[];
noi=4;
noo=1;
hidden_neurons = 4;
epochs = 100;
a=xlsread('abc');          % "abc" name of the input data file
tic
inputs=zeros((length(a)),noi);
tf=0;
% ------- load in the data -------
for x=1:noi
    inp=a(:,x);
    inputs(:,x)=inp;
end
train_inp=inputs;
for y=(noi+1):(noi+noo)
    out=a(:,y);
    o=out';
    outputs=[outputs; o];
end
train_out=outputs';
% check same number of patterns in each
if size(train_inp,1) ~= size(train_out,1)
    disp('ERROR: data mismatch')
    return
end
%standardise the data to mean=0 and standard deviation=1
%inputs
mu_inp = mean(train_inp);
sigma_inp = std(train_inp);
train_inp = (train_inp(:,:) - mu_inp(:,1)) / sigma_inp(:,1);
%outputs
mu_out = mean(train_out);
sigma_out = std(train_out);
train_out = (train_out(:,:) - mu_out(:,1)) / sigma_out(:,1);

%read how many patterns
patterns = size(train_inp,1);
%add a bias as an input
bias = ones(patterns,1);
train_inp = [train_inp bias];
%read how many inputs
inputs = size(train_inp,2);
%---------- data loaded ------------

% ---------- set weights -----------------
%set initial random weights
weight_input_hidden = (randn(inputs,hidden_neurons) - 0.5)/10;
weight_hidden_output = (randn(1,hidden_neurons) - 0.5)/10;


%----------------------------------
%--- Learning Starts Here! ---------
%----------------------------------
%do a number of epochs
for iter = 1:epochs
    %get the learning rate from the slider
    for alr=0.1:0.01:1
        alr;
        blr = alr / 10;
```

```
%loop through the patterns, selecting randomly
for j = 1:patterns
    %select a random pattern
    patnum = round((rand * patterns) + 0.5);
    if patnum > patterns
        patnum = patterns;
    elseif patnum < 1
        patnum = 1;
    end
    %set the current pattern
    this_pat = train_inp(patnum,:);
    act = train_out(patnum,1);
    if tf==1
    %calculate the current error for this pattern
    hval = (1/(1+exp(this_pat*weight_input_hidden)))';
    pred = hval'*weight_hidden_output';
    error= pred - act;
    else
    hval = (tanh(this_pat*weight_input_hidden))';
    pred = hval'*weight_hidden_output';
    error= pred - act;
    end
    % adjust weight hidden - output
    delta_HO = error.*blr .*hval;
    weight_hidden_output = weight_hidden_output - delta_HO';

    % adjust the weights input - hidden
    delta_IH= alr.*error.*weight_hidden_output'.*(1-(hval.^2))*this_pat;
    weight_input_hidden = weight_input_hidden - delta_IH';
    end
end
    % --another epoch finished
    %plot overall network error at end of each epoch
    pred = weight_hidden_output*tanh(train_inp*weight_input_hidden)';
    error = pred' - train_out;
    err(iter) =  (sum(error.^2))^0.5;
    figure(1);

    plot(err)
    if (err(iter)^2) < 0.1
        fprintf('converged at epoch: %d\n',iter);
        break
    end
end
    %-----FINISHED---------
    %display actual,predicted & error
    fprintf('state after %d epochs\n',iter);
    weight_inp = weight_input_hidden
    weight_out = weight_hidden_output

    a = (train_out* sigma_out(:,1)) + mu_out(:,1);
    b = (pred'* sigma_out(:,1)) + mu_out(:,1);
    act_pred_err=[a b b-a]
```

3.7 Genetic Programming-Based ML Models

Genetic programming (GP) is an artificial intelligence-based exclusive data-driven formalism [49, 50]. The GP was originally proposed to automatically generating computer codes that execute prespecified tasks. Later, it was extended to perform symbolic regression (SR). Once the data is submitted in the form of pairs of multiple inputs and single output of a model, the GP-based SR searches and optimizes

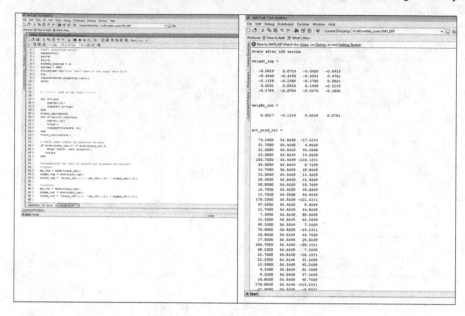

Fig. 3.56 Editor window and the command window in Matrix Laboratory (*MATLAB*)

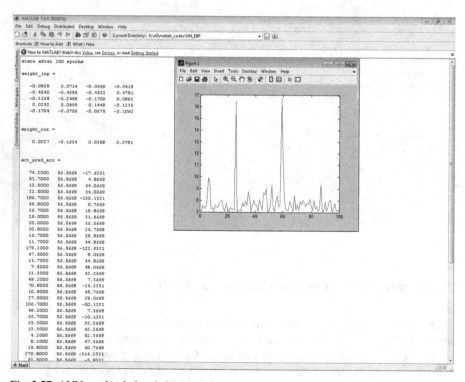

Fig. 3.57 ANN result window in Matrix Laboratory (MATLAB)

both the form (structure) and associated parameters of an appropriate linear/non-linear data-fitting model. The GP does this without making any assumptions about the form of data-fitting function, thereby unravelling the input–output relationships [51]. It may be noted that the basic building block of an MLP or SVR-based model gets fixed depending upon the chosen transfer/basis function. Thus, MLP and SVR models do make certain assumptions pertaining to the data-fitting function. In contrast, the novelty of the GP lies in its ability to secure both the form and parameters of an appropriate linear or nonlinear data-fitting function. The GP has also been found to unravel the natural law that governs the physical phenomena. Other advantages of the SR-based models include providing a human insight, easy interpretation of the models, identification of key variables and ease of deployment [52].

The genetic programming-based SR can be viewed as an extension of the genetic algorithm (GA) [53] wherein the members of the population are not fixed-length binary/real-valued strings encoding candidate solutions to a function maximization/minimization problem, but are mathematical expressions that, when evaluated, represent the candidate solutions to the SR problem [54]. Both GP and GA are based on the Darwinian principles of natural selection and reproduction; however, unlike the former, GAs have been extensively used in the field of drug designing. A number of optimization studies using GAs for QSAR, gene prediction, 3D structure alignment, pharmacophore modelling, combinatorial library generation, docking, etc. have been reported [55]. GAs have been found to significantly improve the prediction values by variable selection in QSAR and also in comparative molecular field analysis [56].

The general form of the model to be secured by the GP-based SR is given as:

$$y = f(X, \alpha) \tag{3.1}$$

where y denotes the model's output (dependent) variable; X refers to an N-dimensional vector of model inputs (independent variables; $X = [x_1, x_2, ..., x_N]^T$); f represents a linear/nonlinear function, and α ($= [\alpha_1, \alpha_2, ..., \alpha_M]^T$) represents a vector of function parameters. Given a multiple input–single output (MISO) example data set, $\{X_i, y_i\}$, $i = 1, 2, ..., K$, consisting of K input–output patterns, the task of the GP-based SR is to obtain an appropriate linear/nonlinear functional form, f, and its parameter vector, α, that best fits the example data.

The implementation of GP-based SR begins by generating a random population of candidate solutions (models/expressions) to the SR problem defined in Eq. 3.1. The expressions are represented in the form of a tree structure. An illustration of a tree structure representing the given expression below:

$$\left(x + \frac{v}{5} \right) * \left(5\sqrt{v} \right) \tag{3.2}$$

is shown in Fig. 3.58a. As can be seen, the tree comprises two types of nodes namely 'operator' (also termed 'function') and 'operand' (terminal) nodes. The first type of nodes represent operations such as addition, subtraction, multiplication, division, exponentiation, logarithm, sine, cosine, etc. while operands denote

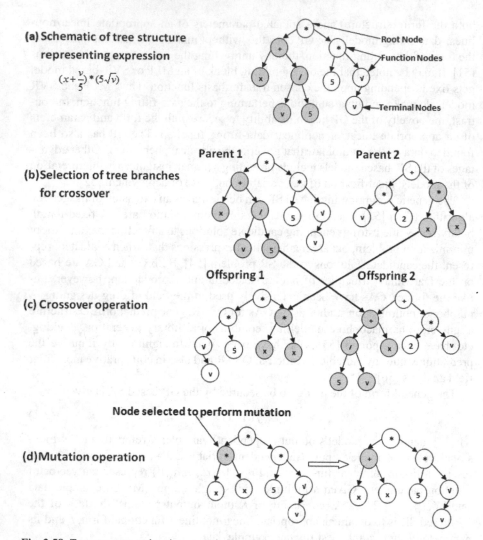

(a) Schematic of tree structure representing expression $(x + \frac{v}{5}) * (5\sqrt{v})$

(b) Selection of tree branches for crossover

(c) Crossover operation

(d) Mutation operation

Fig. 3.58 Tree structure and various genetic implementation operations in GP

the model's input (independent) variables (X) and parameters (α). A single implementation of genetic programming is a competitive search among a diverse population of mathematical expressions, which are coded using functions (operators) and terminals (operands).

Genetic programming procedure iteratively transforms a population of candidate solutions into a new generation of the solutions by employing principles of Darwinian evolution viz. survival of the fittest and genetic propagation of characteristics. Accordingly, GP utilizes analogues of genetic operations such as 'crossover' and 'mutation' occurring in nature. These operations are applied to the candidate solutions selected on the basis of their higher fitness (i.e. ability to better fit the training data). Execution of selection, crossover and mutation steps give rise to a new generation of candidate (offspring) solutions. These steps that bring about transformation in the population of candidate solutions are executed iteratively till convergence is achieved. Prior to implementing the GP procedure, certain preparatory steps need to be executed as given below:

1. Choose a small set of operators (functions) from the large set of available operators that can appear in the candidate solutions. This is necessary to narrow down the solution search space as also avoid long execution times to achieve convergence.
2. Choose an appropriate fitness function for computing the fitness value score of each candidate solution (expression/model) in the population.
3. Choose an error measure, e.g. RMSE, mean absolute percentage error (MAPE), etc., for assessing the output prediction accuracy of the candidate solutions.
4. Partition the available MISO example data set into training and test sets. The test set data should be used to evaluate the generalization performance of the candidate solutions.
5. Choose values of various GP algorithm-specific parameters such as population size, probabilities of crossover and mutation, maximum number of generation over which the GP should evolve, etc.
6. Select an appropriate convergence criterion; the possible criteria are: (1) the GP has evolved over the prespecified maximum number of generations and (2) the fitness value of the best candidate solution (expression/model) no longer increases significantly or remains constant over successive generations (Fig. 3.58).

A generic stepwise procedure for implementing GP-based SR is given below (Fig. 3.59):

1. Create randomly an initial population (Generation = 0) of candidate solutions composed of operators and operands using the tree structure.
2. Repeat.
3. *Fitness computation and ranking:* Evaluate each candidate expression in the population using training input–output data and determine its fitness score using the preselected fitness function; rank the expressions in the order of their decreasing fitness scores.
4. *Selection:* From the ranked population, create a parent pool of candidate solutions with high fitness scores using selection methods such as 'Roulette-wheel selection', 'tournament selection', 'elitist mutation', etc.
5. *Crossover:* Form two offspring candidate solutions (trees) from each randomly selected pair of parent trees from the parent pool. Crossover can be performed multiple ways. For example, in the 'single-point' crossover shown in Fig. 3.59c, a location is selected randomly within the structure of each parent tree. Next, the respective trees are spliced at that location and offspring candidate solutions are created by mutually exchanging and combining the spliced segments of the parent trees.

Fig. 3.59 A flowchart depicting GP-based symbolic regression

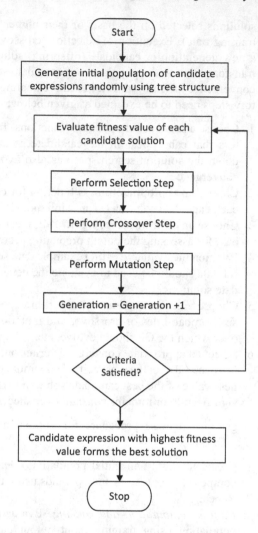

6. *Mutation:* Randomly modify contents of the randomly chosen operator and/or operand node(s) of the offspring trees. Mutation can be conducted two ways: 'node' or 'branch' mutation. In the former, a randomly chosen tree element is replaced by another belonging to the same type. That is, an operator (operand) replaces another operator (operand) see panel d of Fig. 3.59; increment generation index by unity.

7. *Until* convergence condition is fulfilled.

8. *Return* the top ranking, i.e. best candidate expression in the current population (the 'best-so-far' solution) upon convergence as a result of the run.

Genetic programming methods have been applied successfully in the fields of bioprocess monitoring, fermentation models, classification of Raman spectra [57] and optimization of pharmaceutical formulations [58]. Despite its novelty the GP has

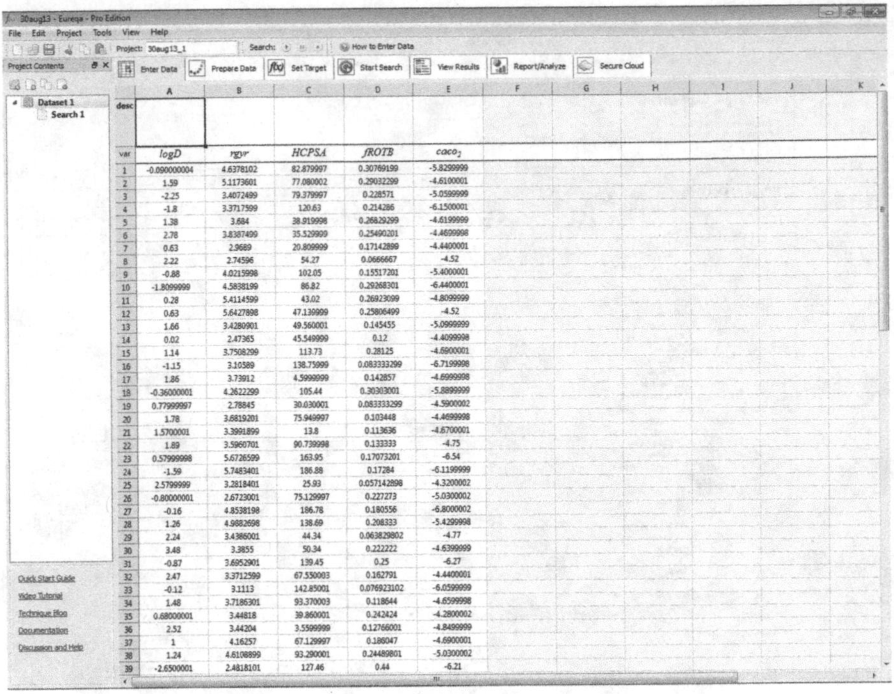

Fig. 3.60 The Eureqa interface with the caco-2 training and test data loaded

var	logD	ηɒɔr	HCPSA	fROTB	$caco_2$
1	-0.090000004	4.6378102	82.879997	0.30769199	-5.8299999
2	1.59	5.1173601	77.080002	0.29032299	-4.6100001
3	-2.25	3.4072499	79.379997	0.228571	-5.0599999
4	-1.8	3.3717599	120.63	0.214286	-6.1500001
5	1.38	3.684	38.819998	0.26829299	-4.6199999
6	2.78	3.8387499	35.529909	0.25490201	-4.4699998
7	0.63	2.9689	20.809999	0.17142899	-4.4400001
8	2.22	2.74596	54.27	0.0666667	-4.52
9	-0.88	4.0215998	102.05	0.15517301	-5.4000001
10	-1.8099999	4.5838199	86.82	0.29268301	-6.4400001
11	0.28	5.4114599	43.02	0.26923099	-4.8099999
12	0.63	5.6427808	47.130999	0.25806499	-4.52
13	1.66	3.4280901	49.560001	0.145455	-5.0999999
14	0.02	2.47365	45.549999	0.12	-4.4099998
15	1.14	3.7508209	113.73	0.28125	-4.6900001
16	-1.15	3.10589	138.75909	0.083333299	-6.7199998
17	1.86	3.73912	4.5990999	0.142857	-4.8999998
18	-0.36000001	4.2622299	105.44	0.30303001	-5.8899999
19	0.77999997	2.78845	30.030001	0.083333299	-4.5900002
20	1.78	3.6819201	75.949997	0.103448	-4.4699998
21	1.5700001	3.3991899	13.8	0.113636	-4.8700001
22	1.89	3.5960701	90.739998	0.133333	-4.75
23	0.57999998	5.6726599	163.95	0.17073201	-6.54
24	-1.59	5.7483401	186.88	0.17284	-6.1199999
25	2.5799999	3.2818401	25.93	0.057142898	-4.3200002
26	-0.80000001	2.6723001	75.129997	0.227273	-5.0300002
27	-0.16	4.8538198	186.76	0.180556	-6.8000002
28	1.26	4.9882698	138.69	0.208333	-5.4209998
29	2.24	3.4386001	44.34	0.063829802	-4.77
30	3.48	3.3855	50.34	0.222222	-4.6399999
31	-0.87	3.6952901	139.45	0.25	-6.27
32	2.47	3.3712599	67.550003	0.162791	-4.4400001
33	-0.12	3.1113	142.85001	0.076923102	-6.0599999
34	1.48	3.7186301	93.370003	0.118644	-4.6599998
35	0.68000001	3.44818	39.860001	0.242424	-4.2800002
36	2.52	3.44204	3.5599999	0.12766001	-4.8499999
37	1	4.16257	67.129997	0.186047	-4.6900001
38	1.24	4.6108899	93.290001	0.24489801	-5.0300002
39	-2.6500001	2.4818101	127.46	0.44	-6.21

not been applied in the drug design field. In this section, we demonstrate the use of GP in Association of Destination Management Executives (ADME) modelling, an important component of drug designing.

3.7.1 A Practical Demonstration of GP-Based Software

There are very few readily available software packages for GP-based SR. There is a commercial software Discipulus which uses automatic induction of binary machine code for predictive modelling [59]. Another GP-based data mining tool is Eureqa Formulize, [60] which is freeware (for limited sue) for generating GP-based models and thereby revealing the input–output relationships hidden in the data. (The software's current limit for free usage is 200 data points and five variables). Here, we illustrate the development of a GP model using Formulize for predicting the caco-2 cell permeability of molecules [61]. The data set used consists of 77 training set molecules and 23 test set molecules; each molecule is represented by four descriptor variables viz. logD, highly charged polar surface area (HCPSA), radius of gyration and fraction of rotatable bonds (fROTB). The GP-based model building process is briefly discussed below, the installation guide, help files and software tutorials can be found at the website of Eureqa Forumulize (Fig. 3.60).

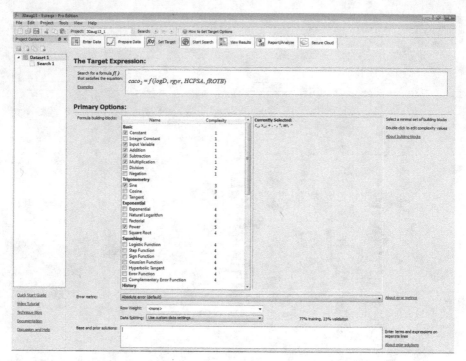

Fig. 3.61 Setting target expression, choosing formula building blocks (Operators), base and prior solution and defining error metric in formulize

The Eureqa formulize homepage appears with a default example set. The spreadsheet-like view is the space provided to enter, edit and inspect the data. The data can be imported in the software as .csv or .txt files. Alternatively, one can copy and paste the training and test data from an excel sheet or from any other source of tabular data or text file. The last column in the data represents the desired model output. The first row defines data labels. A number of data preprocessing options such as smoothing the data, handling missing values, removing outliers, normalizing scale and offset and applying filters, are available. In the variables window, all the variables are specified along with any modification required for better results. Several normalizing options are available in the drop-down menu. One can choose to normalize offset by subtracting the mean, median or interquartile mean or adjust the scale by dividing by the standard deviation, dividing by the interquartile range, or by 10^3, 10^6 or 10^9 (Fig. 3.61).

The software has facility to provide a prior target expression if user wishes to test a specific model as a candidate solution. In the absence of such expressions, the software generates the population of candidate solutions. Primary options provide a list of operators that the software can use to generate model equations. A large set of 54 building blocks (operators) comprising addition, subtraction, sine, cosine, exponential, factorial, Gaussian and if-then-else is available for the stated selection. These building blocks can also be combined in various ways to render the best

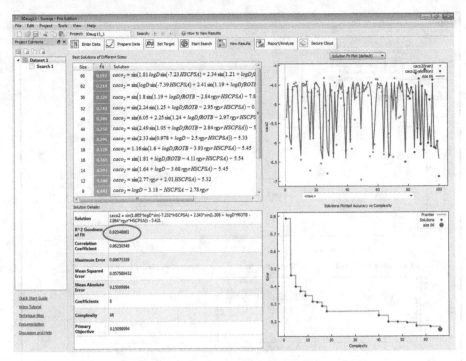

Fig. 3.62 The results summary page showing a correlation coefficient of 0.96

solution. For the case study under consideration the basic four arithmetic (addition, subtraction, multiplication and division) and trigonometric operators were chosen. Owing to the limited number of operators, the solution search space became narrower and more focused when compared with the usage of all possible operators. The operator set used in a GP-implementation typically depends upon the nature of the data-driven modelling problem being solved. If it is a simple data-fitting problem, the four basic mathematic operators will suffice but for a complex task like nonisothermal chemical reaction modelling, advanced operators such as exponentiation need to be employed.

The error metric is a measure of a model's prediction accuracy. The software provides a number of error metrics such as squared error, worst-case error, logarithm error, median error, interquartile absolute error and signed difference for minimization. Additionally, options to maximize the correlation coefficient or the R^2 goodness of fit or experimental hybrid that considers both absolute error and correlation are also available. Data splitting is an important step which divides the data into a training set to generate solutions and a test set to check the accuracy of those solutions (Fig. 3.62).

The search is initialized by clicking on the run button; a log file is created simultaneously to monitor the performance and progress of the ongoing search. The results show the best solutions that have been obtained. The best solutions are

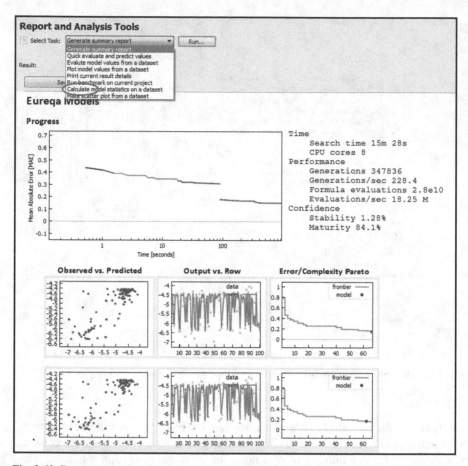

Fig. 3.63 Report generation showing scatter plot of the observed versus model predicted values

determined by two factors: their complexity (size) and their accuracy (fit) on the test (validation) data. The performance metrics for the solution such as correlation coefficient, absolute error and goodness fit are displayed. The best solutions with increasing model complexity can be viewed. Along with other parameters the mean square error is computed for each model. Reports can be generated in html, text or pdf files. The desired report or analysis tool can be changed in the 'select task' drop-down menu. The options available are generate summary report, quick evaluate and predict values, evaluate model values from a data set, plot model values from a data set, print current results details, run benchmark on current project, calculate model statistics on a data set and make scatter plot from a data set (Fig. 3.63).

For the case study under consideration following expression was obtained:

$$y = \sin(1.75x_1 \sin(-7.23x_2) + 2.35\sin(1.21 + x_1x_3 - 2.97x_4x_2)) - 5.42 \quad (3.3)$$

where x_1=logD, x_2=HCPSA, x_3=froTB, x_4=rgyr and y=caco 2 cell permeability.

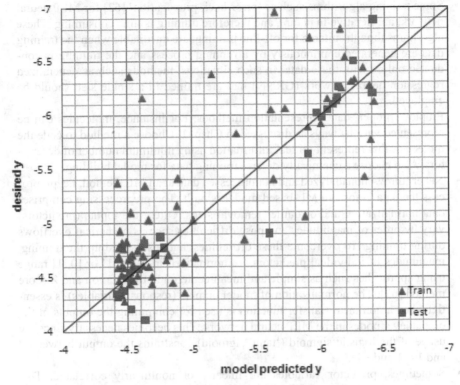

Fig. 3.64 Plot of predicted versus experimental for training and test set molecules

The parity plot of the desired and predicted values of y is shown in Fig. 3.64. The value of 0.96 for the coefficient of correlation between the desired and model-predicted values of y (training and test sets) indicates good prediction accuracy of the model.

3.8 Thumb Rules for Machine Learning-Based Modelling

- Before utilizing the high-end ML-based methods, exclude the possibility that the problem at hand can be solved using conventional statistical and/or algebraic methods. For instance, first explore whether a simple multivariable linear regression is yielding good results before using an ANN or SVR.
- Use as much example data as possible for training the ML model since a model trained on a large data set is likely to possess better prediction and generalization ability. Note that data adequacy depends on various factors including the dimensionality of the system being modelled.
- ANNs and SVR can handle qualitative inputs and/or outputs provided these are appropriately represented in numeric quantities.

- Some training algorithms such as 'error-back-propagation' [62] for MLP neural network, are iterative in nature, and therefore training is time consuming. These algorithms though suitable for off-line training, are unsuitable when the training data are generated continuously by a running process and the training is conducted online with these data. In such situations, methods such as generalized regression neural network (GRNN) that are trained in a single step should be employed.

- Employ 'proper' data representation methods. For instance, molecules can be represented various ways as discussed in Chap. 1. Choose a method to code the molecules that represent critical information using minimum number of descriptors. Also choose a chemically diverse training data for model building.

- Never 'throw' nonanalyzed and nonprocessed data at an ML method, i.e. preprocess the data before an ML-based model is built. The preprocessing comprises data normalization and/or outlier removal steps. Often, the inputs (predictors) vary by order of magnitudes and pose difficulties such as numerical overflows during training. To overcome these difficulties, and also speed up the training, magnitudes of individual predictors are normalized in $[-1, +1]$ or $[0, 1]$ range using approaches such as simple and mean-centered normalization and Z-score method [63]. The normalization of model outputs (response variables) is essential when a nonlinear transfer function is used for computing the outputs of the output layer nodes in the MLP neural network. This becomes necessary since the usage of the logistic sigmoid ('tanh' sigmoid) constrains the output between 0 and 1 (-1 and $+1$).

- Sometimes, predictor variables are linearly or nonlinearly correlated. This unnecessarily increases the dimensionality of the input space thereby enhancing the computational load in training the model. The issue of linearly correlated inputs can be addressed using principal component analysis (PCA) [64] which transforms correlated inputs into a new set of linearly uncorrelated inputs. Using PCA, it becomes possible to use fewer uncorrelated transformed inputs that capture maximum amount of variance in the original data. This feature can be used to effect reduction in the dimensionality of the input space thereby reducing the computational load in ML-based modelling. There also exist techniques such as kernel PCA to perform nonlinear PCA to transform nonlinearly correlated inputs and thereby effect dimensionality reduction of the model's input space employ 'proper' data representation methods [65].

- Avoid overtraining of an ML-based model: use 'test' set, which is different from the training set for assessing the generalization ability of the network. Also, ensure that the training set data are well-distributed and the test set is a true representative of the training set.

- An MLP neural network is capable of performing multiple input–multiple output (MIMO) nonlinear mapping using a single neural network architecture. However, avoid mapping multiple functions using a single MIMO MLP neural network. The reason being in an MIMO–MLP model, the same weights between the input and hidden layer nodes as also between multiple hidden layer nodes appear in the computation of all the outputs, which limits the flexibility of model training. Accordingly, it is desirable to develop a separate MISO–MLP model for each output.

- Develop parsimonious models with low complexity (i.e. with fewer parameters and terms in the model) since such models tend to possess better generalization ability than their more complex counterparts. In ANNs, this can be achieved by using only one or two hidden layers and as few neurons as possible in them. While building an SVR model, complexity can be reduced by using as small as possible the number of SVs. In the GP, a model consisting of a small number of terms and parameters is selected while ensuring a good prediction and generalization performance by that model.
- No single paradigm of the various ML-based modelling methods, such as ANNs, SVR and GP, is capable of consistent out-performance in every modelling task. It is therefore at most important to utilize and compare the performance of all the ML methods for a particular modelling task to arrive at the best possible model. Within a class of methods such as ANNs, there exist multiple architectures (e.g. MLP and RBF networks) for performing nonlinear function approximation and supervised classification tasks. Accordingly, all such alternatives within a class of ML methods also need to be tested.
- Use validation parameters like ROC and AUC for reporting results of virtual screening experiments.

3.9 Do it Yourself (DIY)

- Build a neural network-based binary classification model for antibacterial and antiviral class of compounds using any of the free machine learning tools.
- Using WeKa program build a SVM model for the Wisconsin breast cancer data set.

3.10 Questions

1. What are the known supervised and unsupervised methods in machine learning?
2. How machine learning methods can be used in drug discovery studies?
3. Enumerate the steps involved in building a QSAR/QSPR model.
4. Briefly explain how genetic algorithms and genetic programming-based models can be applied in drug design efforts.
5. Define machine learning.
6. What are the various parameters which need to be assessed from the molecular structures?
7. Enlist the various machine learning methods.
8. Which is the most widely used computer aided filter?
9. What are the drawbacks of ANN, associated with prediction method?

10. Explain in detail how SVM is different from RF?
11. Explain the kernel trick in SVM.
12. Enlist and explain the programs that involve the SVM's open-source tool LibSVM.
13. What is the purpose of ranking the features in a particular data set? Explain the methods used for ranking the features.
14. Explain in detail the file format used in LibSVM and WeKa.
15. What is the purpose of scaling a data and how is that carried out in LibSVM?
16. How can the best c and g parameters be extracted for a particular data set?
17. What is Information Gain in WeKa?
18. Explain the various components of WeKa implementation of RF.
19. What is confusion matrix?
20. State the difference between GP and GAs.

References

1. Breiman L (2001) Statistical modeling: the two cultures. Stat Sci 16(3):199–231
2. Murphy RF (2011) An active role for machine learning in drug development. Natl Chem Biol 7:327–330. doi:10.1038/nchembio.576
3. Gramatica P (2007) Principles of QSAR models validation: internal and external. QSAR Comb Sci 26:694–701
4. Tropsha A, Gramatica P, Gombar V (2003) The importance of being earnest: validation is the absolute essential for successful application and interpretation of QSPR models. QSAR Comb Sci 22:69–77
5. Devillers J (2004) Prediction of mammalian toxicity of organophosphorus pesticides from QSTR modeling. SAR QSAR Environ Res 15:501–510
6. Okey RW, Stensel DH (1993) A QSBR development procedure for aromatic xenobiotic degradation by unacclimated bacteria. Water Environ Res 65(6):772–780
7. Sahigara F, Mansouri K, Ballabio D et al (2012) Comparison of different approaches to define the applicability domain of QSAR models. Molecules (Basel, Switzerland) 17:4791–4810
8. Cao DS, Liang YZ, Xu QS et al (2010) A new strategy of outlier detection for QSAR/QSPR. J Comput Chem 31:592–602
9. Clarke B, Fokoue E, Zhang HH (2009) Principles and theory for data mining and machine learning. J Am Stat Assoc 106(493):379–380
10. Michie D, Spiegelhalter DJ, Taylor CC, Campbell J (1995) Machine learning, neural and statistical classification. Overseas press, New York
11. Kotsiantis SB (2007) Supervised machine learning: a review of classification techniques. Informatica 31:249–268
12. Handfield LF, Chong YT, Simmons J, Andrews BJ, Moses AM (2013) Unsupervised clustering of subcellular protein expression patterns in high-throughput microscopy images reveals protein complexes and functional relationships between proteins. PLoS Comput Biol 9(6):e1003085. doi:10.1371/journal.pcbi.1003085
13. Maetschke SR, Madhamshettiwar PB, Davis MJ, Ragan MA (2013) Supervised, semi-supervised and unsupervised inference of gene regulatory networks. Brief Bioinforma. doi:10.1093/bib/bbt034
14. Sun Y, Peng Y, Chen Y, Shukla AJ (2003) Application of artificial neural networks in the design of controlled release drug delivery systems. Adv Drug Deliv Rev 55(9):1201–1215

15. Kisi O, Guven A (2010) Evapotranspiration modeling using linear genetic programming technique. J Irrig Drain Eng 136(10):715–723
16. Kirew DB, Chretien JR, Bernard P, Ros F (1998) Application of Kohonen neural networks in classification of biologically active compounds. SAR QSAR Envssss Res 8:93–107
17. Klon AE (2009) Bayesian modeling in virtual high throughput screening. Comb Chem High Throughput Screen 12:469–483
18. Olivas R (2007) Decision trees: a primer for decision-making professionals
19. Statnikov A, Wang L, Aliferis CF (2008) A comprehensive comparison of random forests and support vector machines for microarray-based cancer classification. BMC bioinforma 9:319
20. Svetnik V, Liaw A, Tong C (2003) Random forest: a classification and regression tool for compound classification and QSAR modeling. J Chem Inf Comput Sci 43:1947–1958
21. Breiman L (2001) Random forests. Mach Learn 45:5–32
22. Cortes C, Vapnik V (1995) Support-vector networks. Mach Learn 20:273
23. Scholkopf B, Smola AJ (2002) Learning with kernels: support vector machines, regularization, optimization, and beyond. MIT Press, p 626
24. Burges CJC (1998) A tutorial on support vector machines for pattern recognition. Data Min Knowl Discov 2(2):121–167
25. Hofmann T, Scholkopf B, Smola AJ (2008) Kernel methods in machine learning. Ann Stat 36(3):1171–1220
26. Nalbantov G, Groenen PJF, Bioch JC (2005) Support vector regression basics 13(1):1–19
27. Chang CC, Lin CJ (2011) LIBSVM: a library for support vector machines. ACM Trans Intell Syst Technol 2(27):1–27
28. http://www.csie.ntu.edu.tw/~cjlin/libsvm/infogain weka
29. Pyka M, Balz A, Jansen A et al (2012) A WEKA interface for fMRI data. Neuroinformatics 10:409–413. doi:10.1007/s12021-012-9144-3
30. http://www.cs.waikato.ac.nz/ml/weka/
31. http://archive.ics.uci.edu/ml/datasets.html
32. http://www.r-project.org/
33. http://ftp.iitm.ac.in/cran/
34. Kuhn M, Weston S, Keefer C, Coulter N (2013) C code for Cubist by Ross Quinlan. Packaged: 2013-01-31
35. Sela RJ, Simonoff JS (2011) RE-EM trees: a data mining approach for longitudinal and clustered data. Mach Learn 86:169–207. doi:10.1007/s10994-011-5258-3
36. http://cran.r-project.org/web/packages/kernlab/vignettes/kernlab.pdf
37. Ouyang Z, Clyde MA, Wolpert RL (2008) Bayesian kernel regression and classification, bayesian model selection and objective methods. Gainesville, NC
38. http://eric.univ-lyon2.fr/~ricco/tanagra/en/tanagra.html
39. Karthikeyan M, Glen RC (2005) General melting point prediction based on a diverse compound data set and artificial neural networks. J Chem Inf Mod 45:581–590
40. http://moltable.ncl.res.in/web/guest
41. http://rapid-i.com/content/view/181/
42. Molecular Operating Environment (MOE) (2012) Chemical Computing Group Inc., 1010 Montreal, QC, Canada, H3A 2R7, 2012
43. http://www.chemcomp.com/journal/svl.htm
44. http://i571.wikispaces.com/Quantitative+Structure-Activity+Relationships+%28QSAR%29+and+Predictive+Models
45. http://www-01.ibm.com/software/analytics/spss/
46. Rosenblatt F (1962) Principles of neurodynamics: perceptrons and the theory of brain mechanisms. Spartan Books, Michigan
47. Park J, Sandberg IW (1991) Universal approximation using radial-basis-function networks. Neural Comput 3:246–257
48. http://www.mathworks.in/products/matlab/
49. Koza JR (1990) Genetic programming: a paradigm for genetically breeding populations of computer programs to solve problems. Stanford University, Stanford

50. Tsoulos IG, Gavrilis D, Dermatas E (2006) GDF: a tool for function estimation through grammatical evolution. Comput Phys Commun 174(7):555–559
51. Poli R, Langdon WB, McPhee NF (2008) A field guide to genetic programming (With contributions by Koza JR). Lulu enterprises. http://lulu.com, http://www.gp-field-guide.org.uk
52. Kotanchek M (2006) Symbolic regression via genetic programming for nonlinear data modeling. In: Abstracts, 38th central regional meeting of the American Chemical Society, Frankenmuth, MI, United States, 16–20 May 2006, CRM–160
53. Goldberg DE (1989) Genetic algorithms in search optimization and machine learning. Pearson Education, Boston
54. Koza JR, Poli R (2003) A genetic programming tutorial. In: Burke E (ed) Introductory tutorials in optimization, search and decision support. http://www.genetic-programming.com/jkpdf/burke2003tutorial.pdf
55. Gasteiger J (2001) Data mining in drug design. In: Hoeltje H-D, Sippl W (eds) Rational approaches to drug design: proceedings of the 13th European symposium on quantitative structure-activity relationships, Duesseldorf, Germany, pp 459-474, Aug. 27–Sept. 1 2000
56. Terfloth L, Gasteiger J (2001) Neural networks and genetic algorithms in drug design. Drug Discov Today 6(15):102–108
57. Hennessy K, Madden MG, Conroy J, Ryder AG (2005) An improved genetic programming technique for the classification of Raman spectra. Knowl-Based Syst 18:217–224
58 Barmpalexis P, Kachrimanis K, Tsakonas A, Georgarakis E (2011) Symbolic regression via genetic programming in the optimization of a controlled release pharmaceutical formulation. Chemom Intell Lab Syst 107:75–82
59. http://www.rmltech.com/
60. http://www.nutonian.com/
61. Hou TJ, Zhang W, Xia K, Qiao XB, Xu XJ (2004) ADME evaluation in drug discovery. 5. correlation of caco-2 permeation with simple molecular properties. J Chem Inf Comput Sci 44:1585–1600
62. Rumelhart DE, Hinton GE, Williams RJ (1986) Learning representations by back-propagating errors. Nature 323:533–536
63. Tambe SS, Kulkarni BD, Deshpande PB (1996) Elements of artificial neural networks with selected applications in chemical engineering, and chemical & biological sciences. Simulation & Advanced Controls, Louisville
64. Geladi P, Kowalski BR (1986) Partial least squares regression (PLS): a tutorial. Analytica Chimica Acta 85:1–17
65. Scholkopf B, Smola A, Klaus-Robert Muller KR (1998) Nonlinear component analysis as a Kernel Eigen value Problem. Neural Comput 10(5):1299–1319

Chapter 4
Docking and Pharmacophore Modelling for Virtual Screening

Abstract Protein and ligand molecules as two separate entities appear and behave differently, but what happens when they come together and interact with each other is one of the interesting facts in modern molecular biology and molecular recognition. This interaction can be well explained with the concept of docking which in a simple way can be described as the study of how a molecule can bind to another molecule to result in a stable entity. The two binding molecules can be either a protein and a ligand or a protein and a protein. Irrespective of which two molecules are interacting, a docking process invariably includes two steps—conformational search through various algorithms and scoring or ranking. Even though prolific research has been carried out in this field, yet it is still a topic of current interest as there is a scope for improvement to rationalize binding interactions with biological function using docking program. This chapter focuses on how to set up and perform docking runs using freeware and commercial software. Most of the known docking protocols like induced fit docking, protein–protein docking, and pharmacophore-based docking have been discussed. The use of pharmacophore queries as filters in virtual screening is also demonstrated using suitable examples.

Keywords Docking · Conformation · Structure-based drug design · Pharmacophore

4.1 Introduction

Structure-based drug design approaches generally employ docking and pharmacophore modelling techniques. Even though computationally intensive compared to the latter, docking is now routinely used by biologists, pharmacists, and medicinal chemists alike. Since protein interactions among themselves and with other molecular components drive the cellular machinery, docking studies play an important role in understanding cellular biology. It is the basis of rational drug design. There are three main objectives of docking—predict the correct conformation (pose) of the ligand, provide the binding affinity between a ligand and a protein, and apply it as an efficient filter for virtual screening [1]. Docking results help in improving bind-

M. Karthikeyan, R. Vyas, *Practical Chemoinformatics,*
DOI 10.1007/978-81-322-1780-0_4, © Springer India 2014

ing affinities by suggesting changes in the molecular structure as the key binding regions are identified. Ranking of compounds based on the docking score helps in the identification of lead compounds in drug design. Additionally, docking can also be used to predict the potential target of orphan compounds. This process is usually known as reverse docking. A compound of known bioactivity is docked against different protein targets and the protein hits are obtained. These hits are then chosen for further experimental validation. A web-based tool known as Target Fishing Dock (TarFisDock) is available for this purpose where a compound given as input will be docked against the proteins present in potential drug target database (PDTD) and protein hits are given as output [2].

There are a number of excellent reviews on the theory of docking and its limitations. In simple terms, docking denotes placing potential binders into the hydrophobic pockets of the tertiary structure of a protein and score their complementarity in three-dimensional (3D) space. Binders are generally small organic molecules, although metal ions, cofactors, and water molecules are also often present in the crystal structure of the protein. Docking programs predict the binders using shape, surface, or chemistry complementarity features with the receptor. The predicted binders are given a score which reflects the strength of the binding, by employing any of the scoring schemes, viz. empirical scoring, force field scoring, knowledge, or consensus-based scoring [3]. The poses of the binders is predicted by employing any of the search techniques like systematic search, molecular dynamics simulations, annealing, genetic algorithms, incremental construction, and rotamer libraries [4].

The main challenge in protein–ligand docking is the enormous number of degrees of freedom (translational, rotational, conformational). Covering such a huge search space is computationally demanding. Other difficulties include taking take of protein flexibility and conformational changes induced upon binding. Despite its drawbacks and challenges, docking has found tremendous use as a filter in virtual screening in drug design context. Hence, it is important that we should be able to set up and perform a docking experiment to generate the binding score of a library of compounds and prioritize them.

To perform the docking and other concepts in a computer, both commercial and open source programs are available. Every software has its own approach but the basic concept or method is the same in all of them (Fig. 4.1).

It is very well known that to perform docking we must have a known 3D protein structure (a crystallized protein will always give good results than a predicted one) and one or more bound ligands.

4.2 A Practice Tutorial: Docking Using a Commercial Tool

The docking run is performed by the Grid-based Ligand Docking with Energetics (GLIDE) module of Schrödinger [5]. This software has many modules which can perform a wide variety of tasks. As an example, we have taken an MTB protein 1G3U (Protein Data Bank identification, PDB ID) and three first-line antitubercular

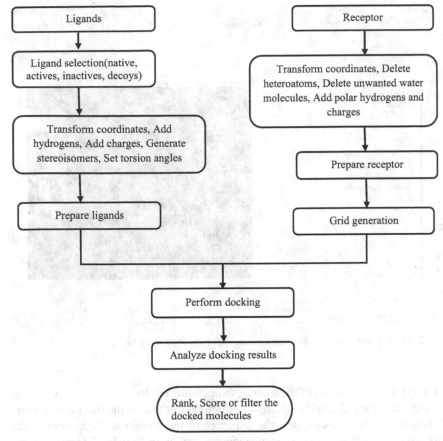

Fig. 4.1 A general docking protocol

drugs ethambutol [6], isoniazid [7], and pyrazinamide [8]. Let us perform docking in a stepwise manner using GLIDE. The downloaded 3D protein structure must be processed or prepared so that it can be used in further steps. This is always required because a typical PDB structure can be multimeric and also consists of heavy atoms, cofactors, and metal ions, and it may have problems like its terminal amide group could be misaligned because X-ray crystallography [9] usually cannot distinguish between oxygen and NH_2. Also, bond orders are not assigned and ionization and tautomeric states are not generated which when used as such for docking may produce inaccurate results. These are important because GLIDE uses all atom force fields for energy evaluations which require properly assigned bond orders and ionization states. The protein preparation wizard in Schrödinger takes care of all these factors while processing the given protein structure. A brief outline of all the steps is given here; for a detailed account, readers are advised to refer to manuals available at the Schrödinger site. The first step essentially consists of opening a new project and creating a directory (Fig. 4.2).

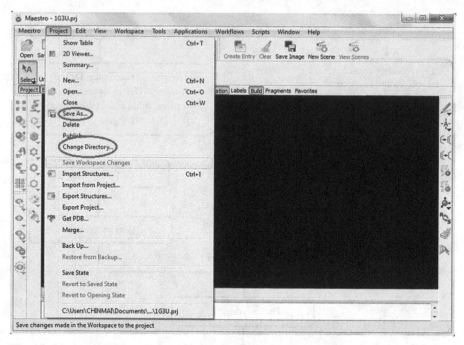

Fig. 4.2 Maestro interface and dropdown menu of project from main menus

Export the downloaded PDB structure (in.pdb format) from the respective folder into the workspace through the "import structures" option in the drop-down menu of "project" menu. We can view the sequence of the protein at the bottom of the workspace after clicking the "sequence viewer" option in "window" menu.

In "workflows," click the protein preparation wizard [10] option in its drop-down table (Fig. 4.3).

In this window, we can find different options which can be chosen according to the requirement like assigning bond orders, adding hydrogens as per valency, creating disulphide bonds, filling side chains and missing loops, creating zero-order bonds, and removing water molecules from a respective area which can be given by user in Å (default is 5 Å). For our study, we will use the default parameters. When we click the "preprocess" option, running of the job appears below the protein preparation window (Figs. 4.4 and 4.5).

Now the next step is "review and modify"; on clicking this option, we can see the details of the protein-like number of chains, list of water molecules, heteroatoms, etc. We can view each water molecule and delete the unnecessary ones and also inspect the ionization states of the heteroatoms and choose the correct one (Fig. 4.6).

In our structure, we will remove all water molecules and heteroatoms except the substrate for simplicity. The final step "refine" includes H-bond assignment (optimize) and restrained minimization (minimize) (Fig. 4.7).

The H-bond assignment step automatically optimizes and reorients the hydroxyl positions along with the flip positions of Asn, His, and Gln amino acids. The

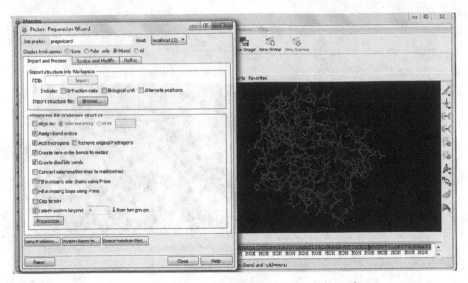

Fig. 4.3 Protein preparation wizard panel along with 1G3U protein in workspace

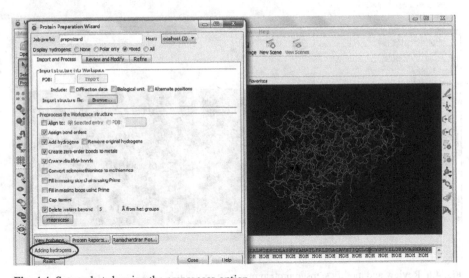

Fig. 4.4 Screenshot showing the preprocess option

option "restrained minimization" adjusts the atom coordinates by using the force field OPLS-2005 (user can define the force field). We can minimize only hydrogens and we can define the root-mean-square deviation (RMSD). The job can be monitored through "monitor jobs" present in the drop-down menu of "applications" in the main menu bar (Fig. 4.8).

After completion of the job, the output file is generated in .mae format. We can also save the output structure in .pdb format through the export option in the project table or workspace (Fig. 4.9).

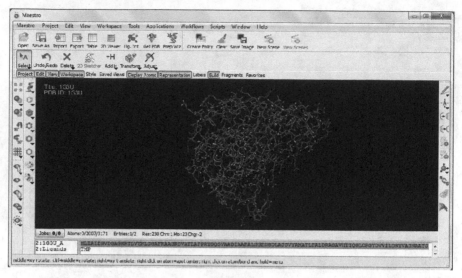

Fig. 4.5 Changes in the input protein after preprocessing

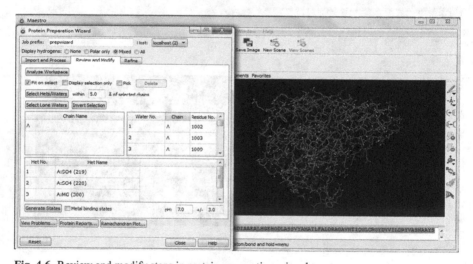

Fig. 4.6 Review and modify steps in protein preparation wizard

The receptor is now ready for docking. The second important step is preparing ligands in order to have the low-energy 3D structures for the study. For this step, LigPrep [11] module is used from Schrödinger. As mentioned earlier, the three ligands are ethambutol, isoniazid, and pyrazinamide. We can download the structures from PubChem [12] directly in .sdf or .mol2 format and import into the workspace. To initiate LigPrep, go to the "applications" in the main menu and choose "LigPrep" which opens the window (Fig. 4.10).

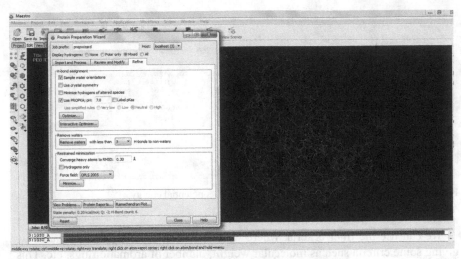

Fig. 4.7 Refine step in protein preparation wizard

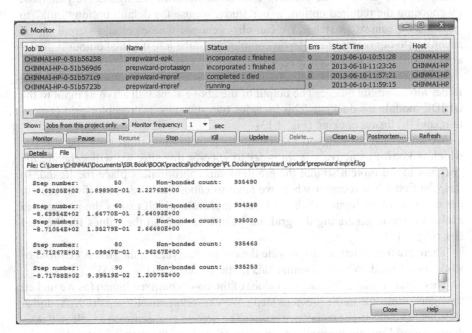

Fig. 4.8 A monitor window

In our example, we will choose the option "from Project Table (selected entries)." The option "file" below it is also for input. If we already have the ligand file, we can directly browse it from the respective folder (it accepts different formats like .mae, .sdf, .smi, and .csv) (Fig. 4.11).

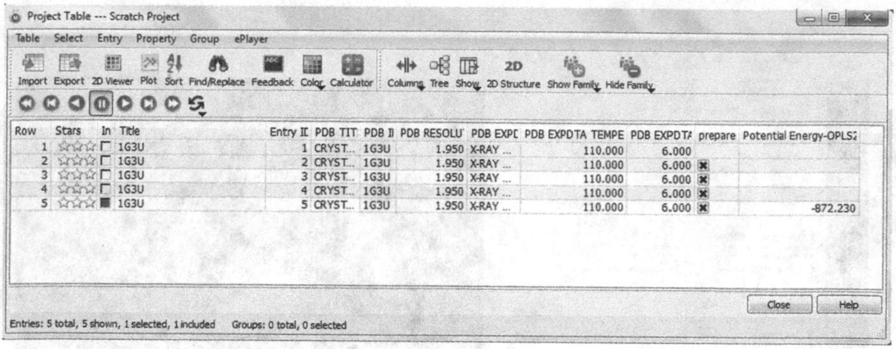

Fig. 4.9 Project table with final structure along with its potential energy

We can use the filter criteria, where we can selectively choose the ligands by giving some criteria such as molecular weight, number of aromatic rings, etc. This will be useful when we have a huge set of compounds. The force field can also be chosen (OPLS_2005 is the default). Ionization states of the ligand are generated by choosing the required option. In our study, we use the default option "Epik" to generate the ionization states. We can also generate tautomers and stereoisomers according to our requirement. Upon completion of the job, the final output is generated in .maegz or .sdf format (Fig. 4.12).

After completion of the job, the output file is generated automatically in the folder *job name-out.maegz*. The output of the above steps will serve as input to the GLIDE module for docking. This includes two substeps: (a) receptor grid generation, and (b) ligand docking (Fig. 4.13).

We know that for a ligand to bind to the protein there must be a specific site which is known as active site. In order to specify this site, we will generate the grid box in the protein so that the program can appropriately place the ligand. In this, the first tab is receptor where we have to choose only receptor if our protein has co-crystallized ligand. Pick the ligand molecule which makes the program to exclude it while generating the grid. In the site tab, assign the values for the grid box (Fig. 4.14).

There are three methods to provide the coordinate values; the first is centroid of workspace ligand. We can visualize this in the workspace (Fig. 4.15).

This will automatically take the values of the co-crystallized ligand (as we picked the ligand molecule in the previous step). We can automatically see the X, Y, and Z coordinate values in the respective boxes. The second one, centroid of selected residues, is useful mainly for predicted models. If we have the details of the active site residues, we can give them here. The third is where we can give the coordinate values directly. This can be obtained from the literature or from the .pdb file (Fig. 4.16).

After this step, we can see the magenta coloured grid box around the ligand specifiying the active site where our molecule will go and bind in the workspace (Fig. 4.17).

There are a few other options in this grid generation step which are worth a mention. They are constraints, rotatable groups, and excluded volumes tabs. These are

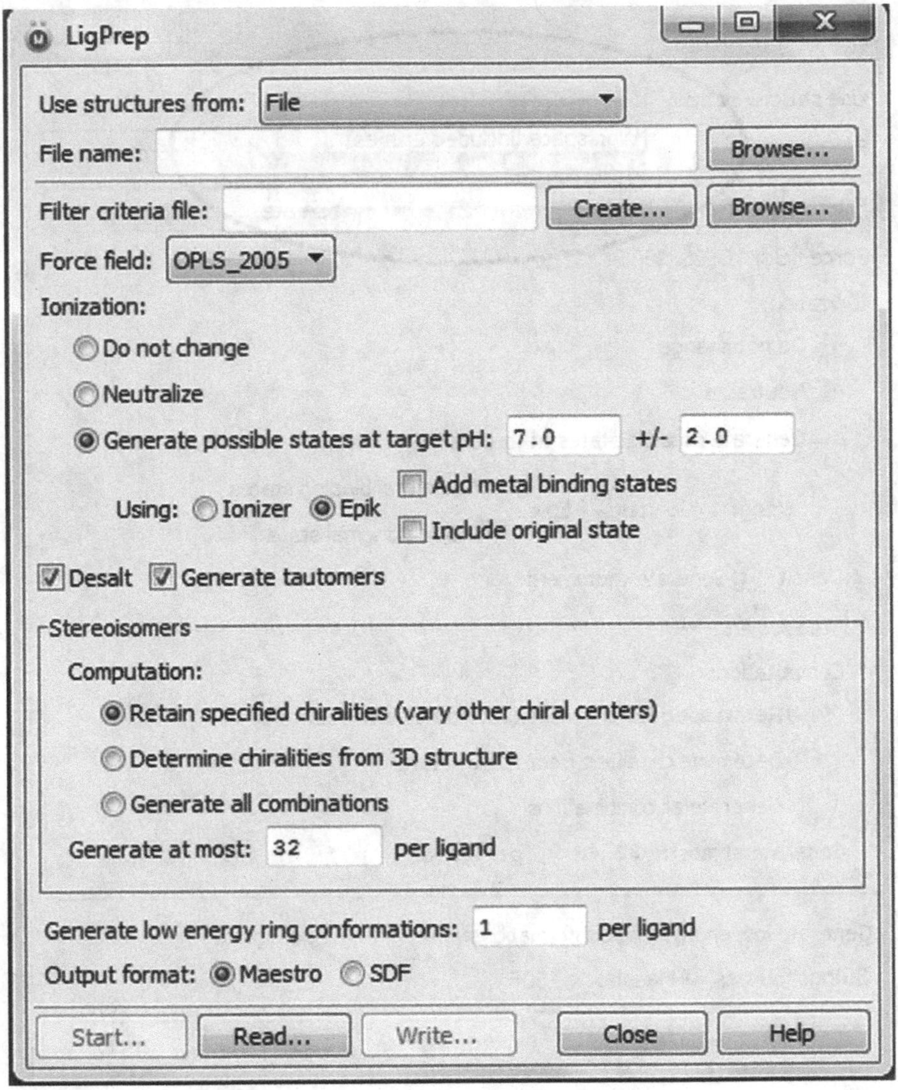

Fig. 4.10 Ligprep window

not used in our present example, but knowing about them will be useful. Through the literature, if we know that any specific interactions are important, we can set them as constraints through this tab which will screen the ligands based on these criteria. They can be positional constraints, metal constraints, and hydrophobic constraints. The rotatable groups allow the hydroxyl and thiol groups (serine, tyrosine, threonine, and cysteine) to be flexible during docking if we know that rendering flexibility provides better binding of ligand. If we want our ligand to not get bound at other sites (other than active site), we can pick those residues under excluded volumes so that those regions can be excluded from docking. After choosing the

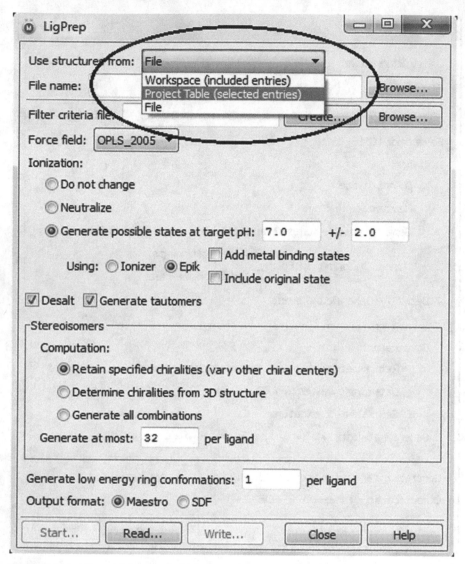

Fig. 4.11 Input ligand dialog box in LigPrep

required options, click on start which takes a few seconds and the result is generated in .maegz and.zip formats which is used in ligand docking step. Now, the second and final step in docking is the ligand docking (Fig. 4.18).

In this, first we have to give the grid file as input. We can browse it from the respective folder to the input box (Fig. 4.19).

Next is setting the docking parameters. First is precision, where we can find three options high-throughput virtual screening (HTVS), standard precision (SP), and extra precision (XP). HTVS rapidly screens very large number of compounds and cannot score in place. SP is the default step where we can screen a large unknown

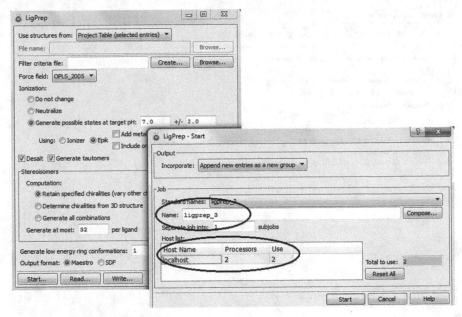

Fig. 4.12 Job submission in ligprep

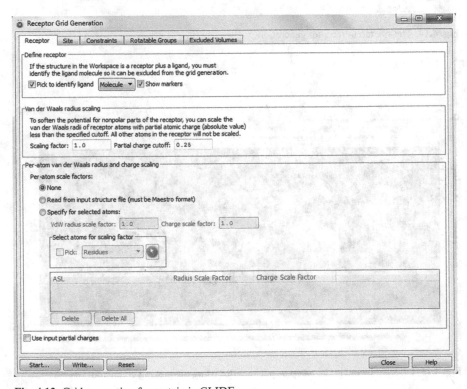

Fig. 4.13 Grid generation for protein in GLIDE

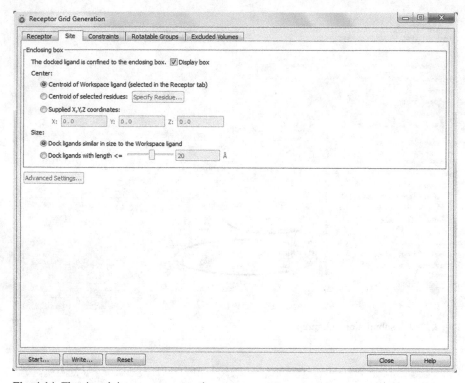

Fig. 4.14 The site tab in receptor generation

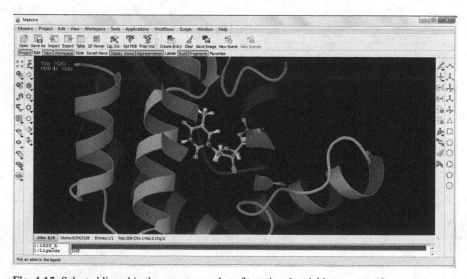

Fig. 4.15 Selected ligand in the receptor pocket after using the picking command

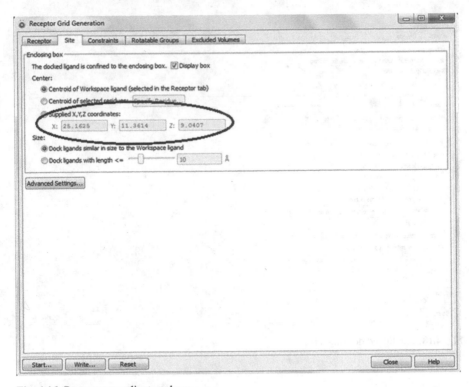

Fig. 4.16 Receptor coordinate values

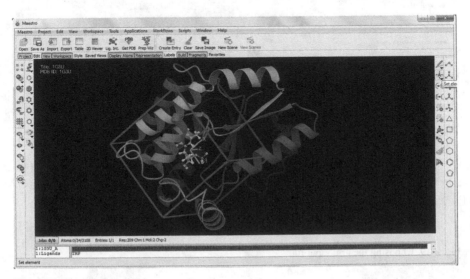

Fig. 4.17 The receptor grid box

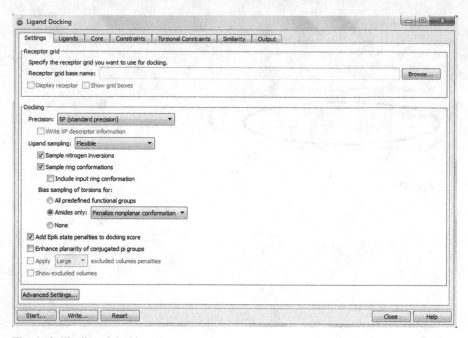

Fig. 4.18 The ligand docking tab

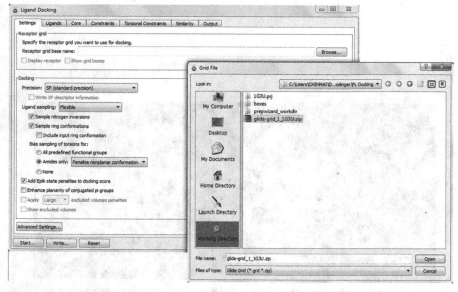

Fig. 4.19 Choosing the grid input file

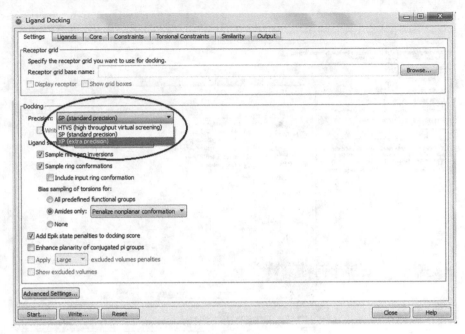

Fig. 4.20 The docking precision parameters

dataset. XP is a more powerful step which takes more time and gives more accurate results. Generally, it is recommended to first perform SP, then choose 10–30% of the best results, and finally run XP to get better results (Fig. 4.20).

Ligand sampling provides us options to choose flexible (ligand conformations are used) or rigid docking (ligand is considered as rigid and no conformations are used) and flexible is the default one. The other default parameters are chosen and the next is ligands tab, in which the ligand file obtained after LigPrep is given as input and other scaling factors are taken as such (Fig. 4.21).

The core tab allows those ligands to dock that match the core of the reference ligand and excludes others. This is explained as ligand-based constraint. The constraints tab next to the core tab is relative to the constraints set during the grid generation step, if any. The constraints that are set in that step are displayed here, and to use them during docking, we have to select them here. The torsional constraints tab provides an option to constrain the torsional degrees of freedom in the ligand. The final and most important tab is the output tab where we can set the output parameters. The number of poses per docking and the number of poses per ligand can be fixed according to our requirements. It also performs post-docking minimization, calculate per residue interaction energy, and RMSD for input ligand. Once we define these parameters, the job begins, which may take a few minutes to hours based on the number of ligands and the parameters we gave (Fig. 4.22).

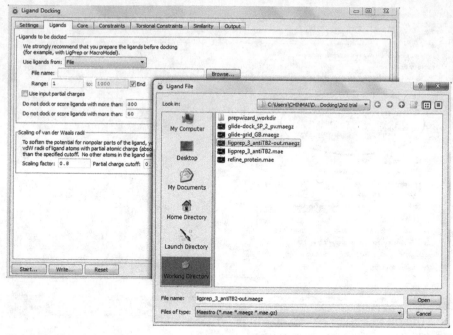

Fig. 4.21 Ligands input for the docking run in GLIDE

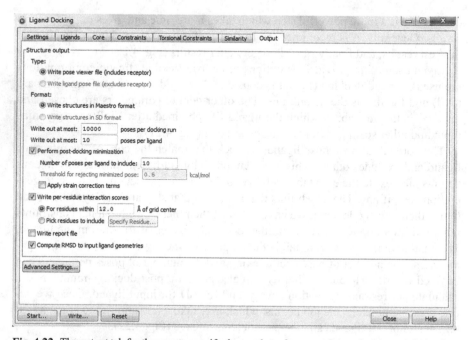

Fig. 4.22 The output tab for the user to specify the number of poses and post-docking minimization

Row	Title	docking score	glide gscore	glide emod	res:168 vd	res:168 cou	res:168 hbo	res:168 dst	res:168 Err	res:120 vd	res:120 cou	res:120 hbo	res:120 dst	res:120 Err	res:105 vd	res:105 c...
17	1G3U															
18	3767	-6.536	-6.536	-41.561	-0.014	0.100	0.000	8.888	0.086	-0.046	0.011	0.000	6.263	-0.037	-0.016	0.00
19	1046	-6.448	-6.448	-37.872	-0.014	1.085	0.000	9.982	1.071	-0.082	0.066	0.000	5.505	-0.016	-0.065	0.03
20	3767	-6.405	-6.405	-38.008	-0.021	0.476	0.000	8.736	0.454	-0.014	-0.026	0.000	8.563	-0.040	-0.010	-0.01
21	3767	-6.261	-6.261	-36.003	-0.021	-0.879	0.000	8.107	-0.900	-0.024	0.002	0.000	7.497	-0.022	-0.014	0.00
22	3767	-6.169	-6.169	-37.698	-0.019	-6.392	0.000	8.419	-0.410	-0.026	0.031	0.000	7.240	0.006	-0.012	0.02
23	3767	-6.117	-6.117	-37.321	-0.019	0.497	0.000	8.485	0.478	-0.022	-0.036	0.000	7.532	-0.058	-0.012	-0.03
24	3767	-6.109	-6.109	-38.451	-0.012	0.174	0.000	9.002	0.161	-0.048	-0.003	0.000	6.139	-0.051	-0.015	-0.04
25	1046	-6.090	-6.090	-37.409	-0.008	0.597	0.000	10.270	0.589	-0.063	0.076	0.000	6.138	0.012	-0.026	0.01
26	1046	-5.824	-5.824	-35.782	-0.009	-0.347	0.000	10.298	-0.356	-0.045	0.098	0.000	6.164	0.052	-0.014	0.04
27	1046	-5.794	-5.794	-34.869	-0.010	-0.556	0.000	9.653	-0.566	-0.045	0.072	0.000	6.186	0.027	-0.014	0.02
28	3767	-5.692	-5.692	-33.848	-0.020	0.164	0.000	8.661	0.144	-0.025	-0.013	0.000	7.079	-0.038	-0.015	-0.02
29	3767	-5.679	-5.679	-37.066	-0.013	-0.578	0.000	8.193	-0.591	-0.046	0.045	0.000	6.334	-0.002	-0.013	0.04
30	1046	-5.542	-5.542	-35.283	-0.009	-0.060	0.000	9.280	-0.069	-0.042	-0.039	0.000	6.998	-0.081	-0.016	-0.04
31	1046	-5.417	-5.417	-37.161	-0.009	-0.156	0.000	9.298	-0.165	-0.055	0.084	0.000	6.213	0.029	-0.016	0.07
32	3767	-5.408	-5.408	-33.309	-0.011	-0.566	0.000	9.044	-0.577	-0.047	0.030	0.000	6.281	-0.016	-0.012	-0.06
33	1046	-5.278	-5.278	-33.226	-0.017	-0.938	0.000	8.786	-0.955	-0.021	0.013	0.000	7.285	-0.067	-0.014	0.00
34	1046	-5.230	-5.230	-34.320	-0.009	-0.081	0.000	9.403	-0.090	-0.042	-0.050	0.000	7.428	-0.092	-0.015	-0.05
35	14052	-4.273	-4.374	-36.946	-0.022	-26.931	0.000	8.357	-26.954	-0.039	0.456	0.000	6.438	0.416	-0.021	0.63
36	14052	-4.139	-4.241	-33.414	-0.023	-25.986	0.000	8.370	-26.009	-0.053	0.560	0.000	5.722	0.515	-0.015	0.81
37	1046	-4.064	-4.064	-29.276	-0.003	0.498	0.000	13.565	0.495	-0.001	-0.011	0.000	14.614	-0.013	-0.001	-0.00
38	14052	-4.018	-4.120	-37.394	-0.024	-26.011	0.000	8.152	-26.035	-0.048	0.552	0.000	6.034	0.504	-0.014	0.78
39	14052	-3.990	-4.092	-33.533	-0.021	-26.845	0.000	8.405	-26.866	-0.030	0.465	0.000	7.003	0.435	-0.017	0.64
40	14052	-3.395	-4.488	-40.592	-0.023	-13.783	0.000	7.907	-13.805	-0.028	0.269	0.000	6.767	0.240	-0.018	0.33
41	14052	-3.338	-4.431	-38.547	-0.023	-13.704	0.000	7.784	-13.727	-0.021	0.225	0.000	7.145	0.204	-0.015	0.31
42	14052	-2.962	-3.064	-32.828	-0.025	-25.310	0.000	8.082	-25.335	-0.050	0.602	0.000	5.937	0.632	-0.017	0.96
43	14052	-2.659	-2.761	-29.164	-0.016	-25.617	0.000	9.069	-25.633	-0.054	0.571	0.000	6.029	0.517	-0.013	0.82
44	14052	-2.534	-3.627	-34.836	-0.020	-14.053	0.000	8.394	-14.073	-0.027	0.208	0.000	7.175	0.181	-0.015	0.26
45	14052	-2.432	-3.525	-40.428	-0.022	-13.988	0.000	8.494	-14.010	-0.045	0.254	0.000	6.596	0.209	-0.013	0.38
46	14052	-2.424	-3.517	-37.051	-0.019	-12.843	0.000	9.198	-12.862	-0.029	0.206	0.000	6.780	0.177	-0.019	0.34
47	14052	-2.135	-3.229	-30.114	-0.024	-14.149	0.000	8.070	-14.173	-0.024	0.180	0.000	6.816	0.156	-0.017	0.28
48	14052	-1.984	-2.086	-33.211	-0.024	-27.601	0.000	8.638	-27.624	-0.049	0.510	0.000	6.069	0.461	-0.028	0.70

Entries: 55 total, 55 shown, 39 selected, 2 included   Groups: 2 total, 1 selected

Fig. 4.23 The docking results viewed in the project table

The results of SP docking are shown in the project table. The table includes docking score, GLIDE score, and per residue interaction score for residues within 12 Å of grid (as given in the output parameters).

The output is saved automatically in the destination folder in pv.maegz format.

This can be imported into the project table and workspace at any time using either of the options.

Applications → Glide → View Poses → Import Glide Results Or Project Table → Entry → View Poses Setup.

We can visualize the docking of ligand into the protein in the workspace. For our example, among the three drugs ethambutol, isoniazid, and pyrazinamide, the order of binding to the protein is isoniazid>pyrazinamide>ethambutol (-6.536, -5.542, and -4.273) based on the GLIDE score. The hydrogen bond contacts can be seen in the respective figures (Figs. 4.23, 4.24, 4.25, 4.26 and 4.27).

After SP docking, we can always perform XP docking where the steps are similar to SP docking except choosing the precision parameter as XP and choosing to write the descriptor file in the settings tab of ligand docking. The results will be generated in pv.maegz and .xpdes formats which can be seen through XP visualizer.

4.3 Docking Using Open Source Software

Autodock [13] is one of the most cited docking software. It has two main programs, AutoDock and AutoGrid, which perform docking of the ligand to a set of grids describing the target protein and pre-calculation of grids. Autodock Vina [14] is the improved version which can be performed in a batch mode and is also known to be

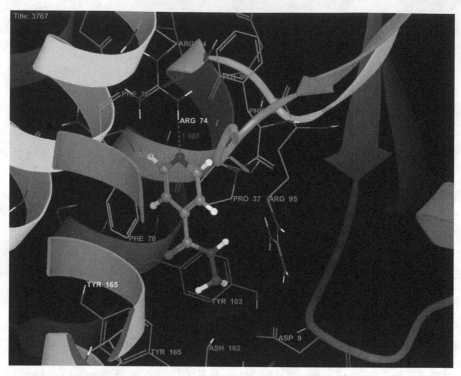

Fig. 4.24 The isoniazid molecule interacting with the protein (the dotted line indicates hydrogen bond)

more accurate than Autodock. Let us see the working steps of both the tools in the following sections.

4.3.1 Autodock Steps

1. Preparing the Grid Parameter File (.gpf)
   i.   Grid→Macromolecule→receptorH.pdb→Open
   ii.   This opens the .pdb structure of receptor and converts it into .pdbqt inside the same path, name it as receptor.pdbqt, and save
   iii. Grid→Set Atom types→Directly
   iv.   This will open a new window where we need to specify the atoms in ligand; for generalizing the study, we can specify the atoms in window in Fig. 4.28 which can be used.
         Accept A C H Cl Br I F S P HD N NA OA SA
   v.   Grid→Grid box
   vi.   This command opens a new window where we need to specify the 3D space for docking which is called GRID. To set the grid, we need to know the

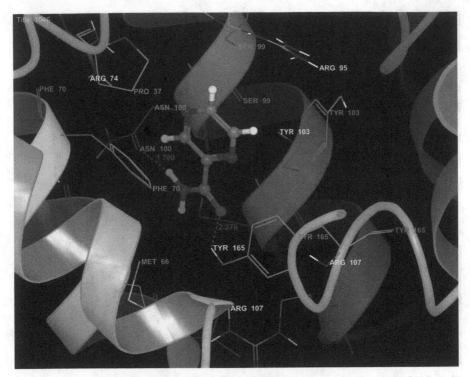

Fig. 4.25 Pyrazinamide molecule interacting with the protein

amino acids of receptor which commonly interact with known ligands. By taking the average of XYZ coordinates of interacting amino acids, we will enter the values in this window (Fig. 4.29).

To obtain interacting amino acids,

vii. Open PDBsum [15] and type the PDB code (of receptor having ligand in the cavity) in the search bar, e.g., 2ZD1 (Fig. 4.30).

A new window is displayed. Click on ligand name, a new window with ligand and interacting amino acid with receptor is opened, note down the names. You can save the ligPlot in .pdf format (Figs. 4.31 and 4.32).

To obtain XYZ co-ordinates, open receptor.pdb with WordPad and search for co-ordinate values of the heteroatom present in the receptor. Then copy the three coordinate columns of the heteroatom in to the excel sheet. Now calculate the average of each column and enter the respective values in the X, Y, and Z grid boxes.

For human immunodeficiency virus (HIV)-1 reverse transcriptase (1HMV), the grid is calculated as follows (Fig. 4.33):

i. The number of points in X, Y, and Z coordinates are set to 60 in most of the studies.
ii. In grid options window, Center→Pick an atom
    Will pick a center and adjust the grid accordingly.

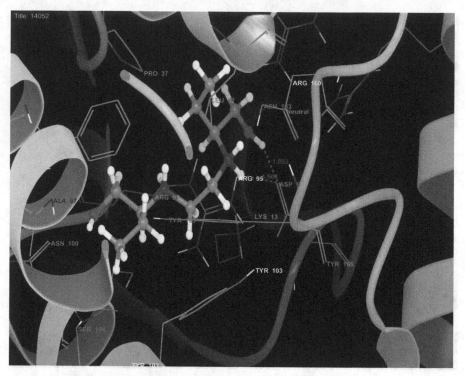

Fig. 4.26 Ethambutol molecule interacting with the protein

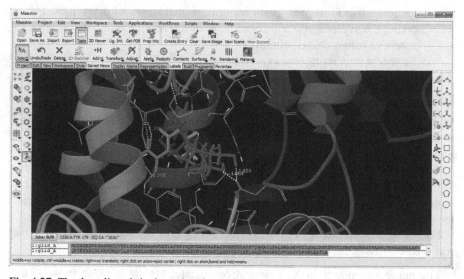

Fig. 4.27 The three ligands in the active site pocket

Fig. 4.28 Atom specifications for any ligand

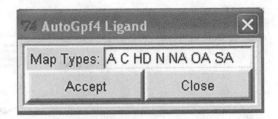

Fig. 4.29 Grid specifications for the receptor

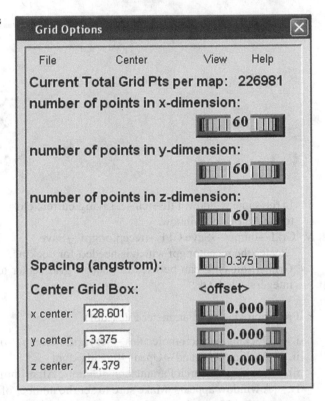

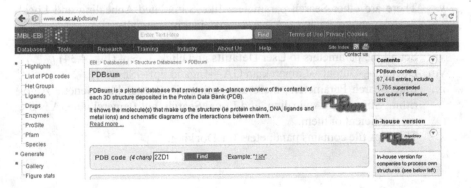

Fig. 4.30 Interface of PDBsum

Fig. 4.31 Receptor–ligand
interaction information
obtained from PDBsum

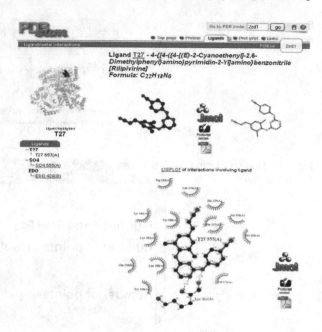

iii. In the same window, File→close, saving current closes the window and returns
to the AutoDock window.
iv. Grid→output→Save GPF→receptor.gpf→Save
Creates the receptor.gpf which is needed for docking.
v. Grid→Edit GPF can be used to change grid center and coordinate values to
integers. Click Ok.

2 Preparing Docking Parameter File (.dpf)

i.   Docking→Macromolecules→set rigid filename→receptor.pdbqt
ii.   Docking→Ligand→Open→ligand.pdbqt
iii. Docking→Search Parameters→Genetic Algorithm
iv.   A window appears. Make sure to set the number of genetic algorithm (GA)
      runs to 10 and the population size to 150. Click Accept.
v.   There are other Search parameters like Stimulated Annealing and Local
      Search parameters, but GA is most efficient of them all.
vi.   Docking→Docking Parameters
vii. Set all the parameters to User Defaults and click Accept (Fig. 4.34)
viii. Docking→Output→Lamarckian GA→Ligand.dpf→Save
ix.   Like Search Parameters, there are options in Output like Genetic Algo-
      rithm, Stimulated Annealing, and Local Search, but Lamarckian GA is the
      most efficient of them.
x.   This .dpf file contains parameters for Docking.

Fig. 4.32 PDBSum page

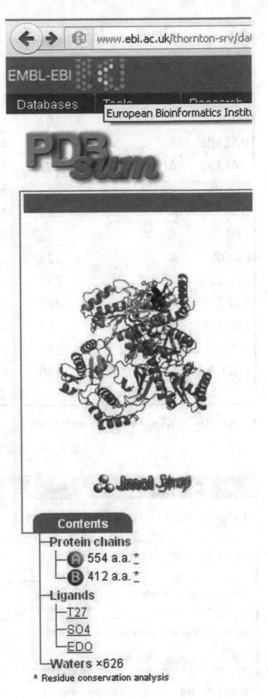

AA	Chain	No.	X	Y	Z
PRO95	A	95	30.793	99.896	209.815
LEU100	A	100	37.074	105.029	212.735
LYS101	A	101	38.913	107.742	214.735
LYS103	A	103	44.137	105.739	217.654
VAL106	A	106	45.526	99.268	217.909
VAL179	A	179	37.802	100.769	222.454
TYR181	A	181	34.702	97.036	218.016
GLN138	A	182	33.428	93.67	217.075
TYR188	A	188	39.454	95.697	216.756
PRO225	A	225	50.407	98.298	210.456
PHE227	A	227	45.648	95.36	210.149
TRP229	A	229	39.764	94.512	210.369
LEU234	A	234	44.101	99.866	208.926
HIS235	A	235	47.353	101.867	208.606
PRO236	A	236	47.719	104.689	211.256
			44.05864	107.1027	229.0651

Fig. 4.33 The X, Y, and Z coordinates of the receptor

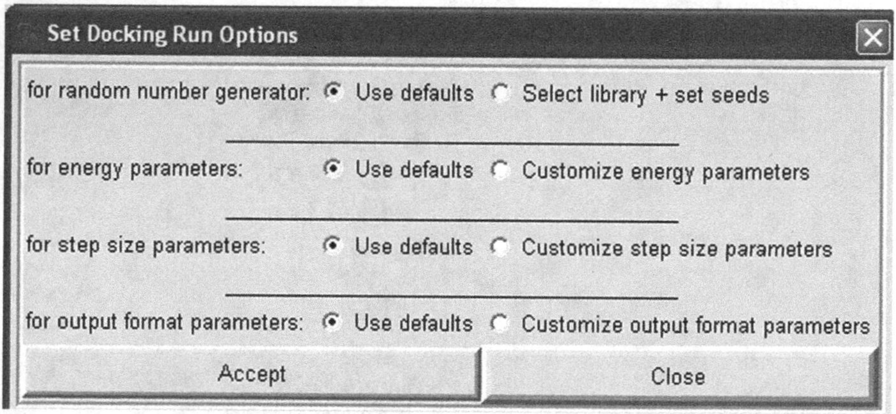

Fig. 4.34 Setting options for a docking run

Fig. 4.35 Command for running the grid file

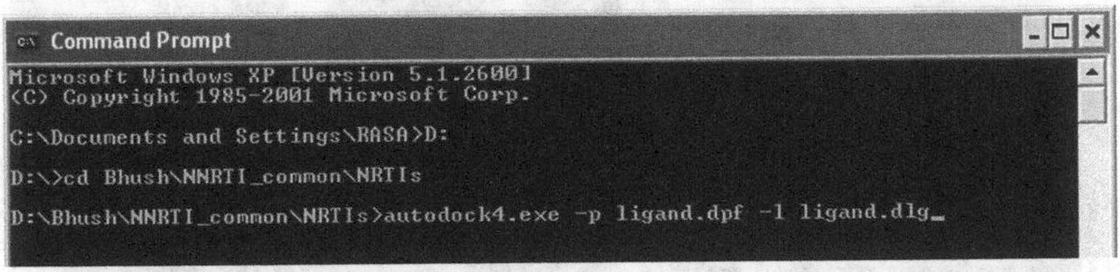

Fig. 4.36 Command for running the dock file

Docking

i. All the required files (receptor.pdbqt, ligand.pdbqt, receptor.gpf, ligand.dpf) should be stored in one folder.
ii. Open command prompt and change directory to the above-mentioned folder.
iii. First, run the grid file by command.
   Autogrid4 –p receptor.gpf –l receptor.glg (Fig. 4.35)
iv. Then run Docking file by command.
   Autodock4 –p ligand.dpf –l ligand.dlg (Fig. 4.36)
v. This will create the ligand.glg file which can be visualized using AutoDock.

Active site identification:

3. Analyzing docking results (ligand.dlg)

   i. Open AutoDock and click
      Analyze→Dockings→Open→ligand.dlg
   ii. For better visualization
       Analyze→Macromolecules→Open→receptor.pdbqt
   iii. To visualize best-interacting conformation
        Analyze→Conformations→Play
   iv. The above command opens a new window, enter the exact number of conformation and press enter to visualize it.
   v. To know the best conformation, open ligand.dlg with WordPad and search the RMSD table. The conformation with least-binding energy is ranked first which is most of the times the best-fitting conformation.
   vi. Use Commands like build current and write complex to obtain the PDB file of the best-docked conformation (Fig. 4.37).

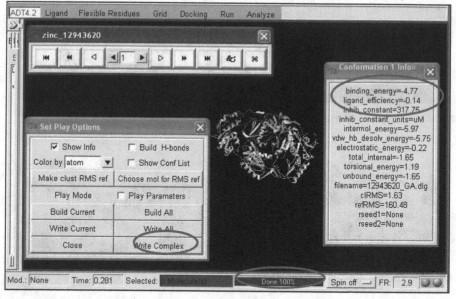

Fig. 4.37 Analyzing the docking results.

4.3.2   Docking Using AutoDock Vina

Let us try one example using the Autodock Vina program. It has an improved searching method and allows the use of multi-core setups. Autodock Vina calculates the grids internally and instantly which is done by autodock and autogrid in AutoDock4. Also in Vina, there is no need to prepare the grid (.gpf) and docking parameter (.dpf) files. There is availability of Autodock tools to visualize the results.

For our example, we will be using the following software:

1. Schrödinger for saving the ligand and protein in .pdb format.
2. Multiple granularity locking (MGL) tools for generating the coordinate files (required by Vina).
3. PyMOL for veiwing the results.

First, download the protein structure from PDB. Then separate the protein and ligand and save them separately in .pdb format. Save the ligand of interest, which we want to dock with our protein in .pdb format. This was done using Schrödinger, where we imported the pdb structure 1G3U to the workspace, separated the ligand, and saved it in .pdb format. Then remove the water molecules using the delete option from the menu and save the protein structure without ligand in .pdb format. Now these files can be used in MGL tools.

These files are converted into .pdbqt format (as they are needed by Autodock Vina) through MGL tools. Click on the autodock icon on your desktop to open the workspace. Go to File menu on the top and then to Read molecule and from there browse the protein .pdb file into the workspace (Fig. 4.38).

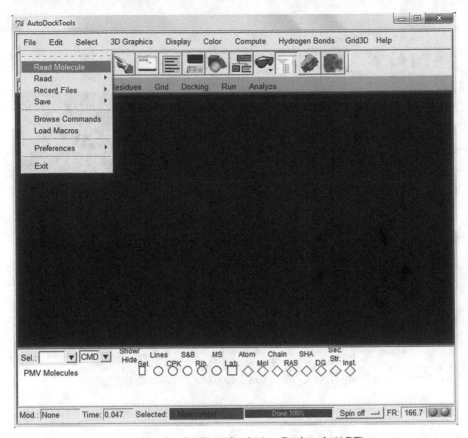

Fig. 4.38 The menu bar and read molecule option in AutoDock tools (ADT)

Then we have to add polar hydrogens to that protein structure as most of the .pdb files do not contain hydrogens. This is done through edit menu where we have to choose "polar only" option (Figs. 4.39 and 4.40).

Then to save it in. pdbqt format, go to Grid Macromloecule Choose. This opens a window where our protein name is displayed; select it and save it in .pdbqt format in the respective folder (Fig. 4.41).

The next step is to generate grid, go to Grid Grid box (Fig. 4.42).

Choose the dimensions and spacing. Usually, the spacing is taken as 1 Å. Enter the X, Y, and Z coordinate values of the co-crystallized ligand (active site) which can be obtained from the text file of the pdb structure. These values are also needed while preparing the config file for Vina (discussed later).

Similarly for getting .pdbqt format of ligand, go to ligand input open (Fig. 4.43).

Browse the .pdb file of the respective ligand into the workspace. Autodock tools automatically add the polar hydrogens to the structure when we open the file. Here, we took isoniazid as the ligand which is the first line drug for tuberculosis (TB). We can visualize and change the rotatable bonds of the ligand structure by going to Ligand torsion tree choose torsions. Rotatable bonds are represented in green

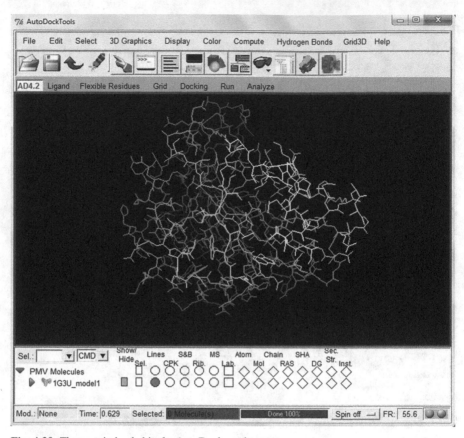

Fig. 4.39 The protein loaded in the AutoDock workspace

color in the structure. Now save the structure by ligand output save as .pdbqt in the respective folder (Figs. 4.44, 4.45 and 4.46).

After preparing the .pdbqt files for target and ligand, we can use either Autodock or Autodock Vina.

We have to open the command prompt, from the start menu go to all programs, then to command prompt. Before running Vina in the command prompt, we have to create configuration file in a text document which includes the receptor name, ligand name, dimensions, and coordinates of the grid (active site) which we gave in the AutoDock tools (mentioned above). This is named as conf.txt file (Fig. 4.47).

Now change the directory to the respective folder where the input files are saved using the change directory (cd) command. Now run the following command.

vina –config conf.txt –out out.txt –log log.txt

or

vina –config conf.txt –receptor receptorname.pdbqt –ligand ligandname.pdbqt –out out.txt –log log.txt and press the enter button to view the results in the command window (Fig. 4.48).

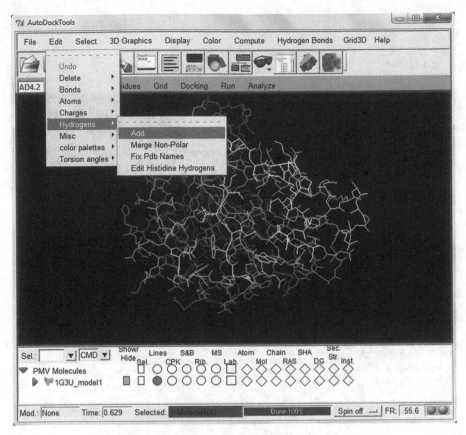

Fig. 4.40 The addition of polar hydrogens to the protein molecule

It gives the protein-binding affinity in kilocalorie per mole (kcal/mol) for every conformation and also the RMSD. Vina automatically detects the processors and displays it. The log files and output files can be seen in the folder. The outfile (out.pdbqt) can be visualized through PyMOL [16] or AutoDock tools (Figs. 4.49 and 4.50).

In PyMOL, we can visualize all the conformations by clicking on the arrow button or the play button.

4.4 Other Docking Algorithms

Ludi Ludi [17] is a product of Accelrys (Insight II) which can be used in both structure-based drug design (protein structure is known) and ligand-based drug design (protein structure is not known). Based on this, it runs in two modes: receptor mode and active analog mode. This mainly follows the fragment approach where initially

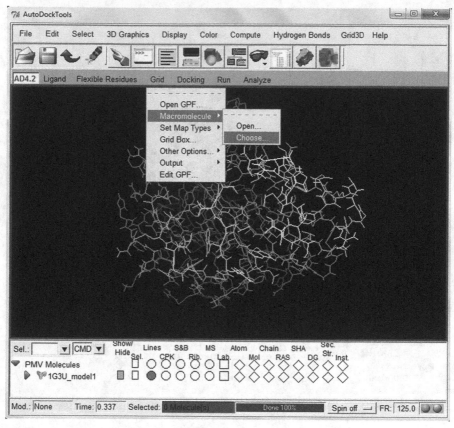

Fig. 4.41 The protein structure is saved as .pdbqt

small fragments are allowed to make hydrogen bond and hydrophobic interactions within the active site, and then these fragments are linked by spacer fragments to get a whole new compound.

FlexX FlexX [18], one of the most-cited method, is also a commercial tool for protein–ligand docking. Here, the protein is rigid and both flexible and rigid ligands can be docked. It follows the robust incremental construction algorithm. The ligand is decomposed into pieces and then flexibly built up in the active site, using a variety of placement strategies. The docking approach is like incremental construction where the ligand is placed incrementally in the active site rather than the whole ligand placed at one time [19].

4.4.1   Induced Fit Docking

The docking programs discussed or mentioned above follow the usual docking method, where the protein is rigid and the ligand is flexible to generate conforma-

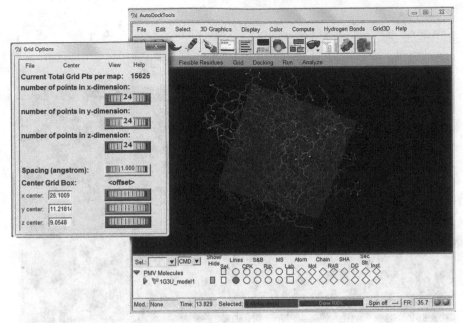

Fig. 4.42 The grid box and its dimensions in ADT

tions and each conformation binds to that protein. But factually, proteins in our body when bound by a ligand undergo changes in their structure which, in turn, leads to adjustments in the binding site in order to generate a better binding mode. This is referred to as an induced fit docking [20], one of the emerging area of research. Induced fit docking has two advantages over the general docking procedures as, first, it simulates the flexible protein as present in the body leading to more accuracy in results. Second, it helps us in retaining even those molecules which are poor binders in one conformation of protein but may be good in another. For this, Schrödinger has a module named induced fit docking [21], which can be picked from workflows present in the main menu.

4.4.2   Flexible Protein Docking

As discussed above, side-chain movement or protein flexibility plays an important role in a docking process. Usually, uncertainty in side-chain placement or loop modelling arises in protein structures predicted through homology modelling. Only a few docking tools consider these, like FlexE. This is a tool which considers the protein structure variations or flexibility to dock flexible ligands [22] (Fig. 4.51).

It has options like Receptor, Ligands, GLIDE Docking, Prime Refinement, and GLIDE Redocking. This process, however, takes high computation time for good results.

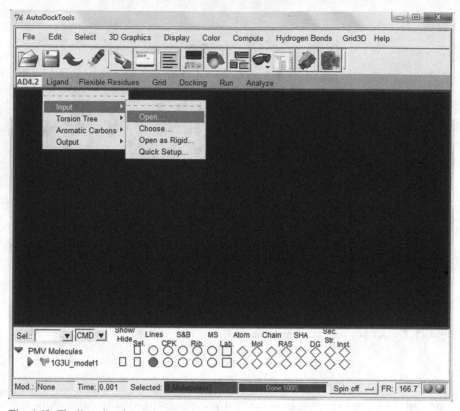

Fig. 4.43 The ligand molecule imported into workspace

4.4.3 Blind Docking

The importance of recognizing the active site for a given 3D structure of a protein cannot be overstated. Most of the docking tools use the already known functional site (active site) details to perform docking [23]. But literature reports prove that docking can also be used to locate the binding site. This is termed as blind docking, i.e., the docking algorithm is unable to see the binding site but still can find it where the protein–ligand interaction information can be known without the knowledge of the specific binding site. The search space includes limited region of protein if the active site is known, but in blind docking the search space includes the whole protein surface which requires more computational time. Therefore less feature points representing a rigid conformation of the ligand to be docked are considered. [24].

4.4.4 Cross Docking

In all the methods discussed above, the docking procedure involved docking of ligands to the native conformation of protein termed as self-docking. There is another

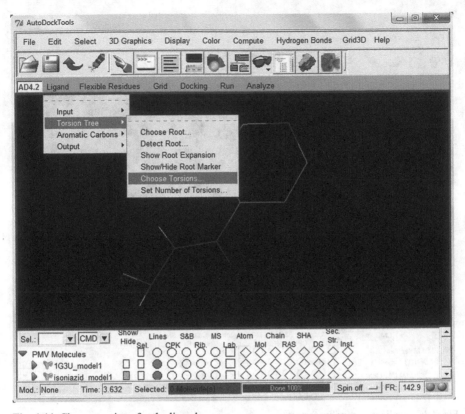

Fig. 4.44 Choose torsions for the ligand

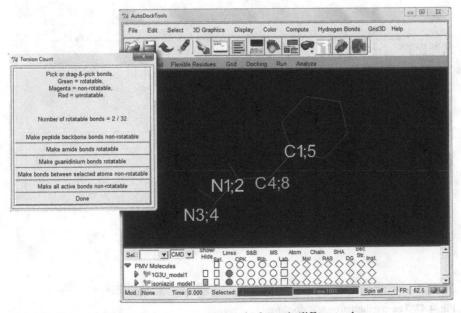

Fig. 4.45 The rotatable bonds and unrotatable bonds shown in different colors

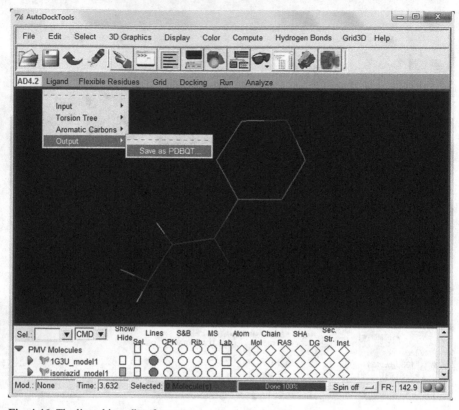

Fig. 4.46 The ligand in .pdbqt format

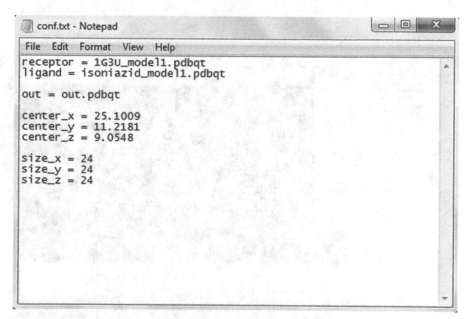

Fig. 4.47 The configuration text document

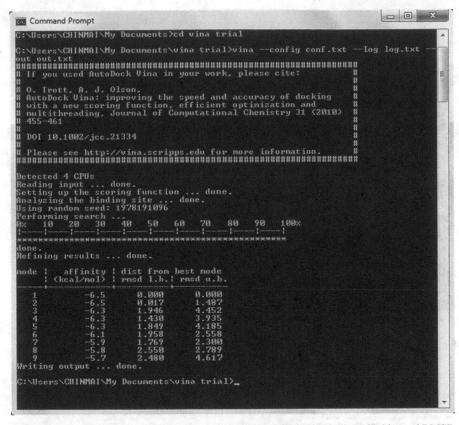

Fig. 4.48 The docking results file in the command prompt showing binding affinities and RMSD values

method in which the ligands are docked into the protein targets whose structural determination was performed using different ligands. This is helpful in identifying ligands active against different set of proteins [25, 26]. The different algorithms that have been used in benchmarking studies are CDocker [27], Fred [28], Rocs [29], etc. In Schrödinger, the GLIDE module can be used for cross docking. Using the script file xglide-gui.py, ligand and protein preparation can be automated for cross-docking ligands from complexes and analysis of results.

4.4.5 Docking and Site-Directed Mutagenesis

An important application of docking is in site-directed mutagenesis, wherein one can make specific changes in amino acid residues in the active site of a protein [30]. This generates a different set of modified proteins which can be docked to compare binding affinities with the original wild protein. Thus, the key amino acids impacting the biological activity of a protein can be identified. In silico studies coupled

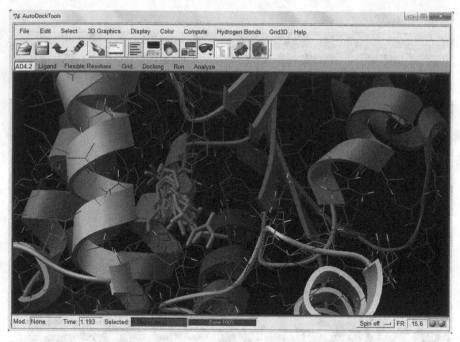

Fig. 4.49 The docked conformations of isoniazid in the active site in Autodock

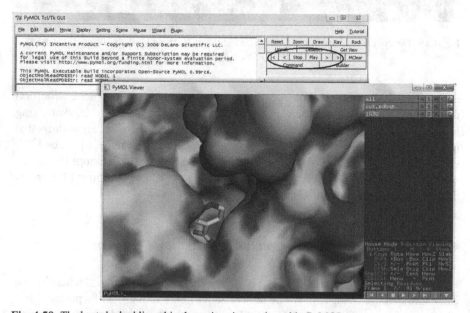

Fig. 4.50 The best-docked ligand in the active site as viewed in PyMOL

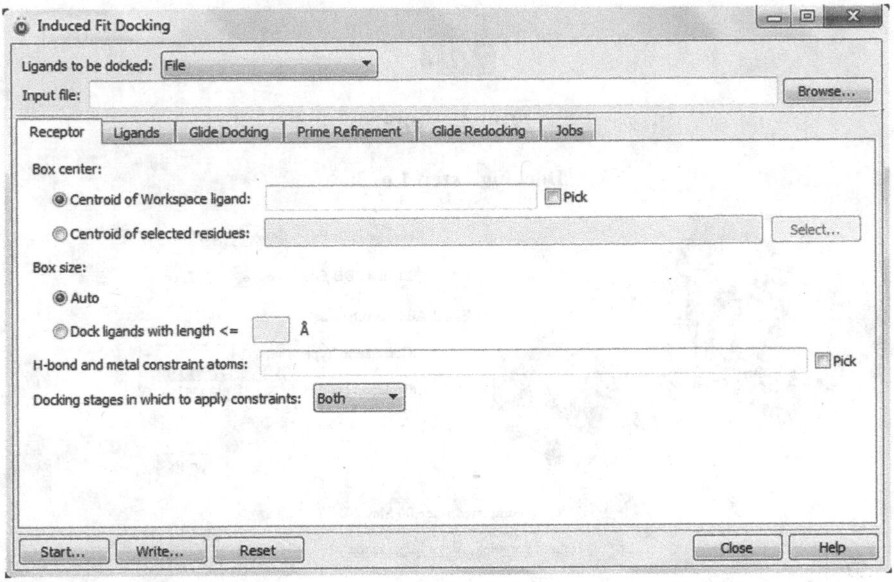

Fig. 4.51 The induced fit docking window in Schrödinger

with experimental data enable protein screening, active-site structure elucidation, mapping binding modes, etc. [31].

4.5   Protein–Protein Docking

Protein–protein interactions are the mediators of several functions and are considered as vital for many biological processes, fundamental to understand cellular organization, signal transduction, metabolic control, gene regulation, etc. [32]. These protein–protein interactions have a major part in diseases like prion diseases, where host protein is converted to pathogenic protein through interactions. The protein–protein interactions are at the core of the entire interatomic system of any living cell to express any biological function. Designing small molecule inhibitors against protein–protein interaction targets is gaining importance these days [33]. Pharmaceutically targeting these protein–protein interactions have shown to have greater significance in diseases like cancer and HIV [34]. Before designing the inhibitors, it is essential to have a good knowledge regarding the actual interactions occurring between two proteins and this can be accomplished by protein–protein docking studies.

While genome-wide proteomics studies provide an increasing list of interacting proteins, only a small fraction of the potential complexes are amenable to direct experimental analysis. Thus, it is important to utilize protein–protein docking methods that can explain the details of specific interactions at the atomic level. Furthermore,

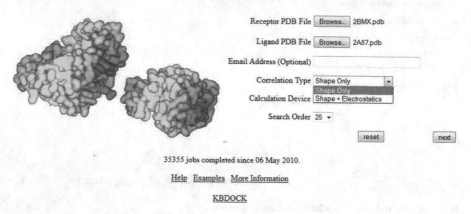

Fig. 4.52 The Hex server interface with the two proteins given as input

the precise understanding of protein–protein interactions for disease-implicated targets is ever more critical for the rational design of biologic-based therapies [35]. Hence, understanding these inter-molecular interactions through protein–protein docking may provide us with novel therapeutic molecules. Many databases are available with the information regarding the protein–protein interactions like Database of Interacting Protein (DIP) [36], STRING [37], BioGRID [38], etc. Generally, the molecular structure of these protein complexes is experimentally studied through techniques like principal component analysis (PCA), yeast two hybrid system, X-ray crystallography, and nuclear magnetic resonance (NMR) along with the emerging theoretical method of protein–protein docking. To demonstrate this approach, we selected two proteins which are known to interact from STRING database [39], viz. thioredoxin reductase (2A87) and alkyl hydroperoxidase (2BMX), and submitted them in few protein–protein docking servers discussed below.

Hex Hex [40] is a free software used to study the docking modes of proteins and ligands, protein pairs, and DNA molecules. It is an interactive molecular graphics program which can be downloaded or performed online. It uses the spherical polar Fourier (SPF) correlations for the docking calculations (Fig. 4.52).

Docking is performed in two steps in this server. In the first step, we will upload the two proteins and choose the calculation type. In the second step, we have to give the details of origin residues and interface residues for both receptor and ligand, along with the required number of results. Now the job is submitted and the results are shown in the same page or sent to the given mail address.

Fig. 4.53 The ZDOCK server interface

ZDOCK This is also a freely available automatic protein docking server [41]. It uses the fast Fourier transform (FFT) algorithm to search the binding modes of the protein [42]. Its evaluations are based on the shape complementarity, desolvation energy, and electrostatic parameters (Fig. 4.53).

As mentioned above, the two proteins were submitted to the server along with the e-mail address where the results are sent (Fig. 4.54).

The next step is to select any specific blocking residues which can be taken from the literature. For our example, we chose none. After completion of the job, output link is mailed which includes output file; pdb files of two proteins, where one is considered as receptor and the other as ligand; and tar file of top five predictions.

GRAMM-X

This is also another free sever for protein–protein docking which follows the FFT algorithm [43] (Fig. 4.55).

Other servers are also available for this purpose like Rosettadock [44], clusPro [45], etc.

MEGADOCK It is known that druggability of protein–protein interactions is an emerging area of research and studying a protein–protein interaction network requires huge computation. MEGADOCK [46] is a recently reported protein–protein docking engine which is shown to be efficient on a large number of protein pairs. It is a high-throughput and ultra-fast pixels per inch (PPI)-predicting system with hybrid parallelization technique which makes it work on parallel supercomputing systems. It follows the rigid body docking considering the tertiary structural data of the proteins.

Piper Piper [47] is a state-of-the-art protein–protein docking program based on a multistaged approach and advanced numerical methods that reliably generate accurate structures of protein–protein complexes. Piper program has been used for protein–protein complexes prediction in previous CAPRI experiments [48].

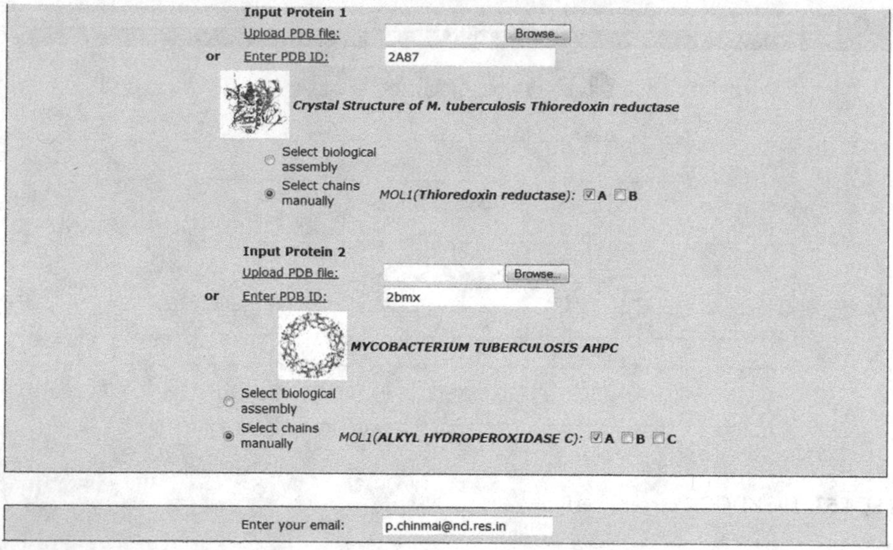

Fig. 4.54 The proteins submitted for docking studies in ZDOCK

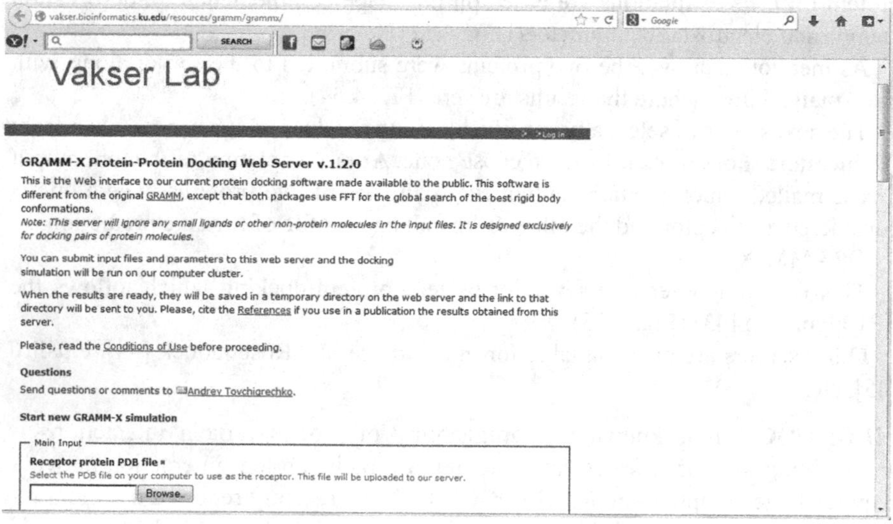

Fig. 4.55 The GRAMM-X server interface

4.6   Pharmacophore

This is a familiar word in the field of lead molecule design or drug discovery. It is defined by International Union of Pure and Applied Chemistry (IUPAC) as "an ensemble of steric and electronic features that is necessary to ensure the optimal supra molecular interactions with a specific biological target and to trigger (or block) its biological response." Although docking is the best method to understand the protein–ligand interactions and we can screen huge number of small molecules based on the score, yet the use of pharmacophore concept before docking and screening can lead to better compounds [49, 50]. A pharmacophore in a usual way is a feature of the compound responsible for a specific biological activity. So every compound which is known to be active will have some features accountable to its activity. Hence, knowledge of these features can be used as a filter to screen unknown dataset of molecules. This can be of different types like ligand based where a set of active and inactive ligands are analyzed and information about the receptor is not used. It can also be complex based where a protein–ligand complex is analyzed and, finally, target based where only the structural data of receptor is used [51].

The typical pharmacophores are listed below:

1. Hydrogen bond acceptor (A)
2. Hydrogen bond donor (D)
3. Hydrophobic group (H)
4. Aromatic ring (R)
5. Positively charged group (P)
6. Negatively charged group (N)

There are commercial software like Schrödinger, MOE, Discovery Studio, etc. to generate the pharmacophore query or model and use it as a filter to screen compounds and for building a quantitative structure–activity relationship (QSAR) model. Catalyst is a commercial software by Accelrys [52]. It creates a hypothesis in terms of chemical features that are important to bind at active sites which can be used for further screening of databases. HipHop, a part of catalyst performs the feature-based alignment of given set of compounds without considering activity. Hypogen is the algorithm which generates the activity-based pharmacophore model.

4.6.1 Pharmacophore Modelling in SCHRÖDINGER

The module which can be used for pharmacophore generation in this suite is Phase [53]. Any number of ligands with their activity value (half maximal inhibitory concentration, IC_{50}) can be used for the generation of common pharmacophore, but always a set of compounds that are studied under similar experimental results will provide good results. For example, we have chosen few compounds which are known inhibitors of dihydroorotate dehydrogenase enzyme in *Plasmodium falciparum* [54]. This set of compounds contains active and inactive compounds (based on the IC_{50} values). Draw the molecules either in maestro workspace or in any tool like ChemSketch [55] or ChemDraw [56] and save them in .mol format. Now import these molecules into the project table (Fig. 4.56).

When we import the structures into the project table, they can be visualized in the workspace. In the project table, choose and add the biological activity (IC_{50}) values of the molecules obtained from the literature. Usually for the purpose of easy calculations, we convert the micromolar values into molar using formula through the project table calculator (Fig. 4.57).

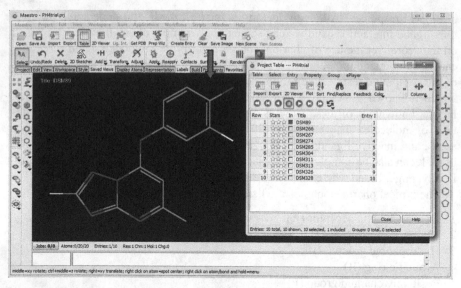

Fig. 4.56 The molecule viewed in project table and workspace

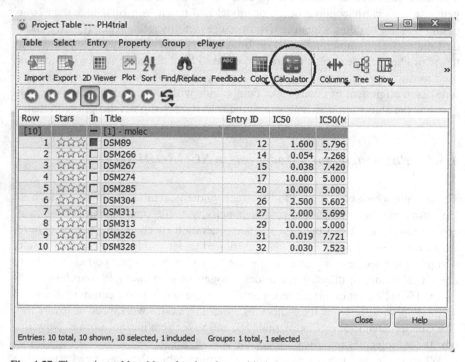

Fig. 4.57 The project table with molecules along with their IC_{50} values

Fig. 4.58 The common pharmacophore hypotheses table

Phase provides a wide range of options like simplified pharmacophore modelling and screening (where we find pharmacophore features for a single ligand, edit them, manage them, and screen against a database of molecules), creating pharmacophore hypotheses manually, managing the hypotheses, generating a database, shape screening, etc. For our study, we develop pharmacophore hypotheses where we can find a series of steps to follow along with building a QSAR model (optional).

Go to main menu Applications → Phase → Develop common pharmacophore hypotheses.

It opens a window with further options (Fig. 4.58).

The first step here is to prepare ligands. Import the molecules from either file or run or project table. In our example, we use "From Project" option which will show a table with the molecules and an option to choose the property. Choose the property (converted IC_{50} values) IC_{50} (M) (Fig. 4.59).

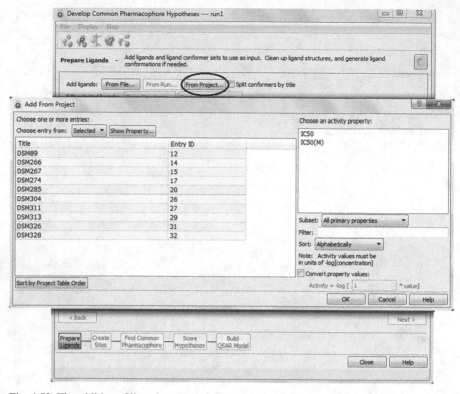

Fig. 4.59 The addition of ligands to start pharmacophore modelling

This will show all the ligands with IC_{50} values and default "active" in the pharm set column (Fig. 4.60).

Now click on the activity thresholds button on the right down corner and define the values above 7 as active and below 5 as inactive. This will change the pharm set parameters accordingly as active and inactive. And the ligands with activity between 7 and 5 are considered as moderately active and the pharm set parameter will be empty (Fig. 4.61).

The clean structures option will convert the two-dimensional (2D) structures to 3D, remove counter ions and non-compliant structures, generate stereoisomers, and perform energy minimization. Default parameters were used (retained specified chiralities and original states) in this example (Fig. 4.62).

The generate conformers button generates all possible conformations using confgen method and OPLS_2005 force field (default parameters). Here, rapid sampling method is used to generate the core conformations first and then the peripheral conformations are sampled one by one. We also have the option to choose continuum solvation methods for water namely distance-dependent dielectric (default) and GB/SA. The redundant conformers are eliminated by using cut-off RMSD of 1 Å (Fig. 4.63).

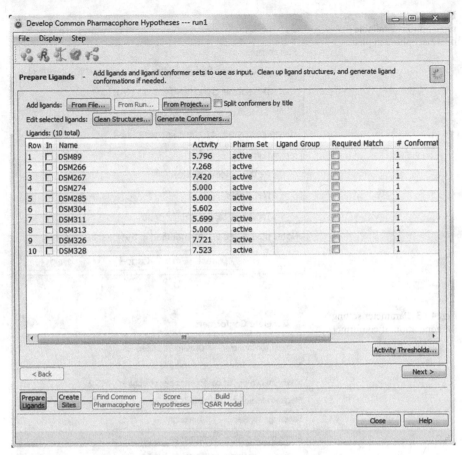

Fig. 4.60 The ligands with activity and the default "active" pharm set

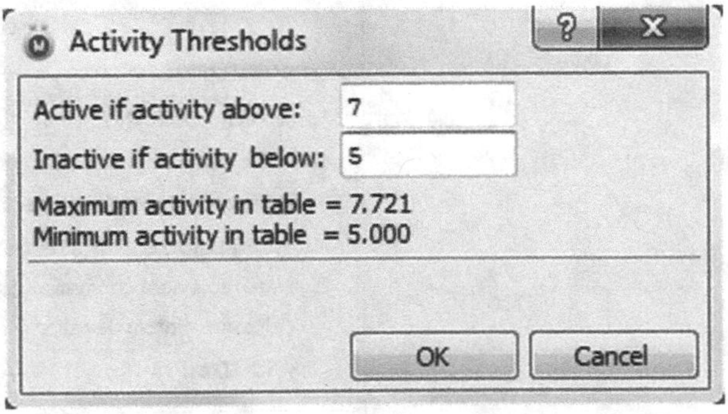

Fig. 4.61 The activity thresholds

Fig. 4.62 The clean structures options available in pharmacophore in phase

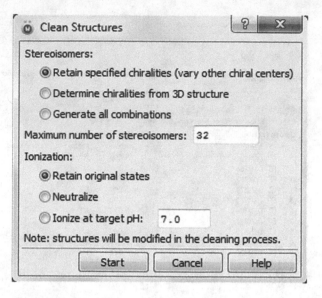

Fig. 4.63 Parameter setting for generation of conformers

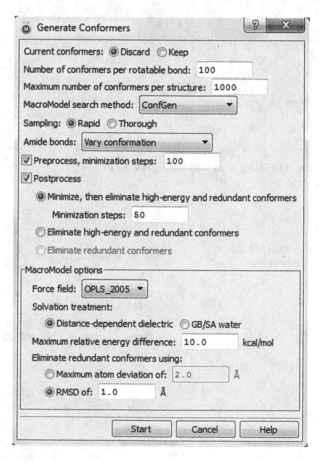

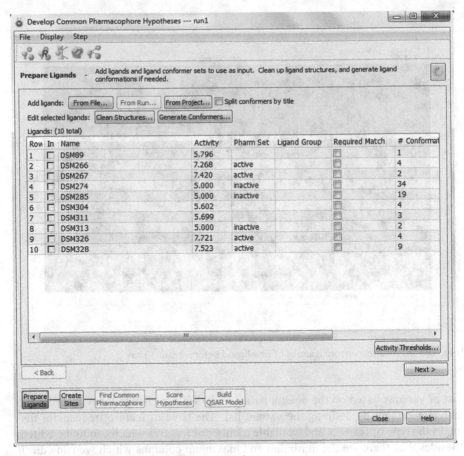

Fig. 4.64 The number of conformations generated

We can see the number of conformations generated for each ligand in the conformations column (Fig. 4.64).

The next step is to create sites. In this step, for every conformation of the ligand, the sites of each feature are found among the pharmacophore features present (inbuilt features mentioned above). There is an option of editing the features whose detailed description is always available in the manual (http://www.schrödinger.com/supportdocs/18/13/). We can visualize the features of every ligand before creating sites (Figs. 4.65 and 4.66).

We can see the features displayed in the ligands box after completion of the job. In the next step, we "Find Common Pharmacophores" through which we can find the common pharmacophores for selected variant list (Fig. 4.67).

The common pharmacophore hypothesis is the description of how the ligand binds to the receptor. And Phase follows a tree-based partitioning technique to find the common pharmacophores. Immediately after entering this step, we can see the

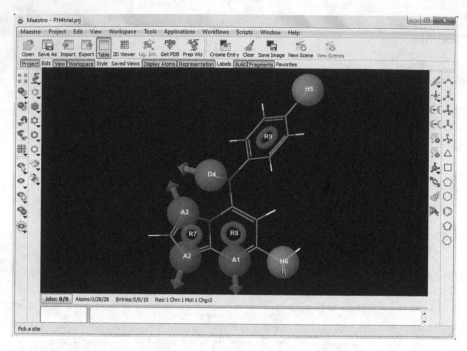

Fig. 4.65 All pharmacophore features of first ligand before creating sites

list of variants based on the default parameters. We can change them according to our requirement by defining the variant list. The table of feature frequencies displays the type of features and available number of features which cannot be edited. Besides this, there are the minimum and maximum columns which we can edit. It means we can define the number of features that should be present in a variant.

We can also define the minimum and maximum number of sites for a variant. It ranges from three to seven. Care must be taken while defining the number of site points because less number of sites may not contain all the features and more sites may result in no pharmacophores. Usually, default value 5 is considered. The must-match-at-least box will display the total number of actives present in our data by default. We can reduce the number in order to widen the search (less number of actives will give more number of variant lists). The completion of job displays the maximum number of hypotheses found for each variant (Fig. 4.68).

The next step is to score the hypotheses where we can choose the best hypothesis based on the score analysis for further screening of database of molecules or to build a QSAR model out of that hypothesis. In this step, initially we have to calculate the score for actives by clicking on the score actives button by using the default vector and site filtering options (Fig. 4.69).

Clicking on the start button runs the job and on completion gives the survival values. Next, click on the score inactives (use default parameters). It will give the inactive values along with the survival inactive values. If we want to re-analyze

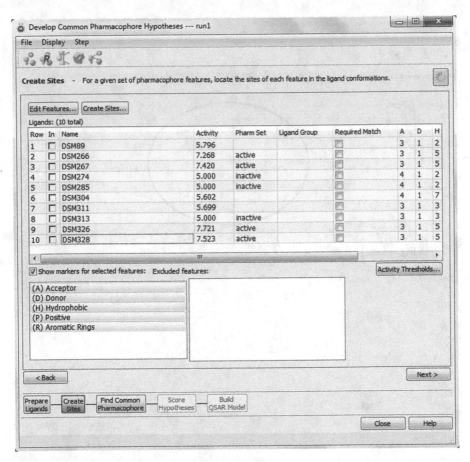

Fig. 4.66 The pharmacophore features for each ligand

the hypotheses, we can rescore it by clicking on the rescore button which will give the post hoc values according to the changes we have made in the parameters (Fig. 4.70).

After generating all the scores, we can examine them individually by clicking on the box in "In column." When we click on the box of "In column," it shows the alignment details for that single hypothesis with the fitness value. The fitness value shows how good the conformation of ligand matches the hypotheses. The perfectly matched will show a value of 3. The selection of the best hypotheses from the list requires keen analysis and visualization. Usually, the one with the highest survival inactive score is considered as good hypothesis, but it is not compulsory to choose that for further study (Figs. 4.71 and 4.72).

The fitness value usually differentiates the active and inactive ligands. All active ligands will have a fitness value nearing 3. And the inactive ligands will have less

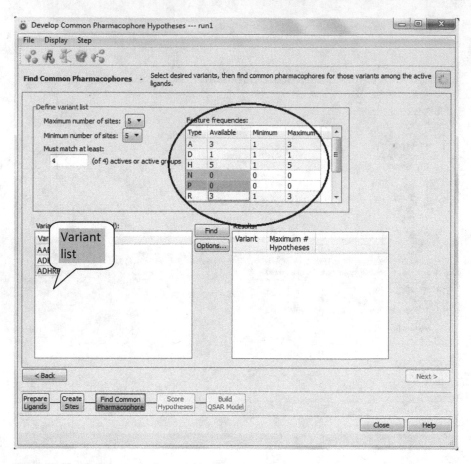

Fig. 4.67 The find pharmacophore step in phase

fitness score. As a default, the inactive ligands cannot be aligned on the hypothesis (Fig. 4.73).

But we have the alignments options button from where we can align and examine the inactive ligands (Fig. 4.74).

Check the box "align non-model" ligands which will enable the inactive ligands and by selecting them, we can superimpose them on the hypothesis (Fig. 4.75).

Now the final use of the selected hypothesis can be in developing a QSAR model which is the next step provided by Phase, and it can also be used directly as a filter in searching a 3D database to find similar matches as that of the hypothesis. This may result in novel compounds which can be further validated through docking (Fig. 4.76).

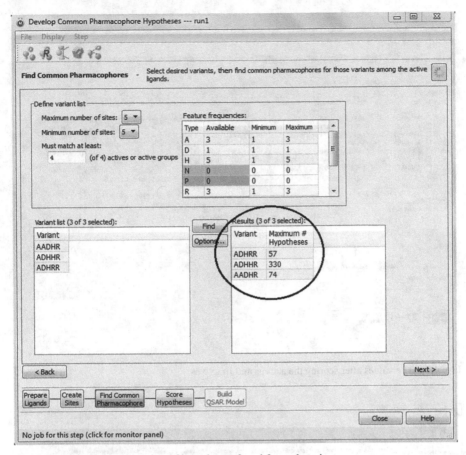

Fig. 4.68 The maximum number of hypotheses found for each variant

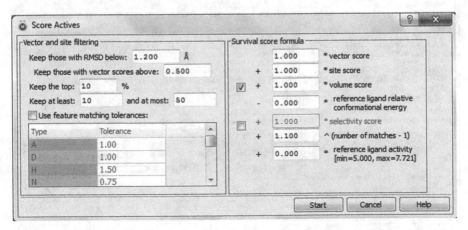

Fig. 4.69 The default values for scoring the actives

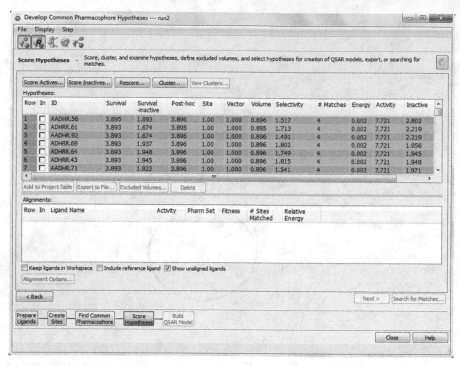

Fig. 4.70 The values after scoring the actives and inactives

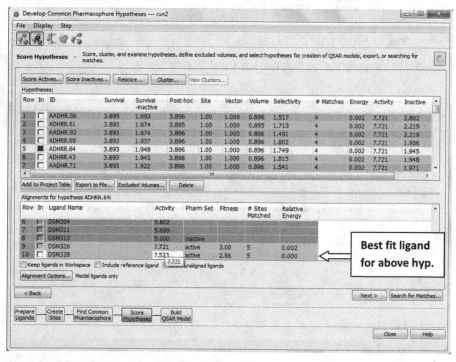

Fig. 4.71 A hypothesis and its alignment details

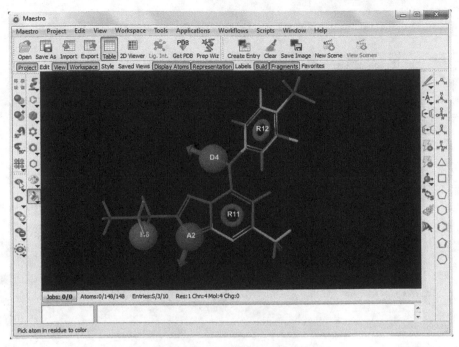

Fig. 4.72 All superimposed active ligands of the selected hypothesis

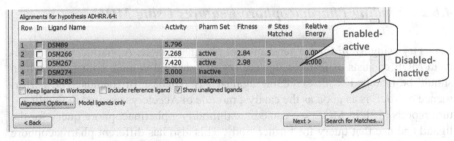

Fig. 4.73 The disabled inactive ligands

Fig. 4.74 The alignment
options window

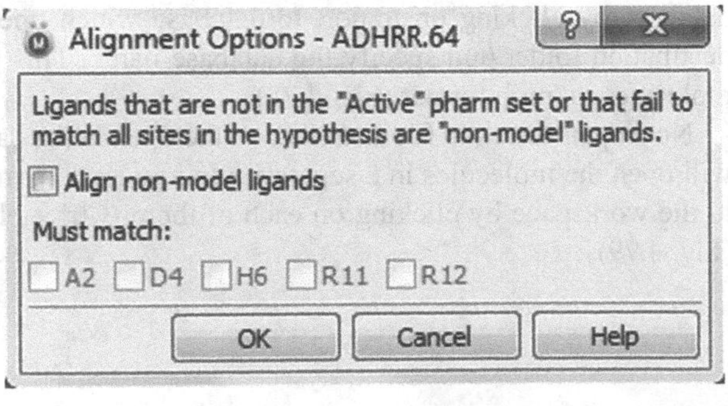

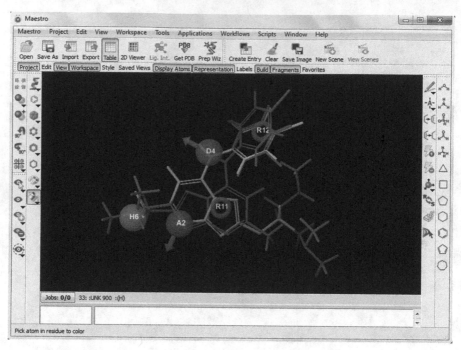

Fig. 4.75 The superimposed inactive ligands on the selected hypothesis

4.6.2 Finding Pharmacophore Features Using MOE

Let us try using MOE for the generation of pharmacophore. A pharmacophore query in MOE consists of features, feature constraints, and volume constraints which should be matched while screening against any database of ligands. The performance of MOE is as good as the catalyst module of Accelerys as cited in few literature reports [57]. In MOE, we can edit and modify a pharmacophore query for the ligand and use that query for further study. This also has different pharmacophore features defined in it which can be edited manually by the user.

Let us see an example. Go to the main menu File Open (Fig. 4.77).

Browse the selected molecules which are in .sdf format and import them to a database by clicking on import to database which opens a window. Browse the destination folder and specify the database name. This will create the database of molecules in .mdb format (Fig. 4.78).

Now open the. mdb file and click on the open in the database viewer button. This will open the molecules in a separate window and the molecules can be visualized in the workspace by clicking on each of them (Ctrl + click for multiple selection) (Fig. 4.79).

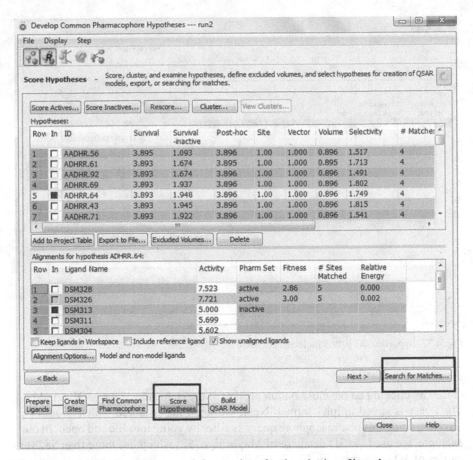

Fig. 4.76 Build QSAR model or search for matches after the selection of hypotheses

Now double click the first molecule in the database viewer table which will display the ligand in the workspace. Go to main menu compute pharmacophore query editor (Fig. 4.80).

Clicking on the query editor will automatically generate the pharmacophore features along with the opening of the editor window (Fig. 4.81).

Clicking on the "info" button opens a window where we can choose the specific pharmacophore features as required for the study by selecting and deselecting the boxes. Each ligand can be visualized for analyzing its features (Figs. 4.82 and 4.83).

In this example, the pharmacophore query was generated using the most active ligand. Features were created by selecting the respective feature in the workspace ligand and then clicking on the feature button to create the pharmacophore (Fig. 4.84).

This query was saved in the respective folder by clicking on the save option. This query will be used to search any database for the hits. Now all windows can be

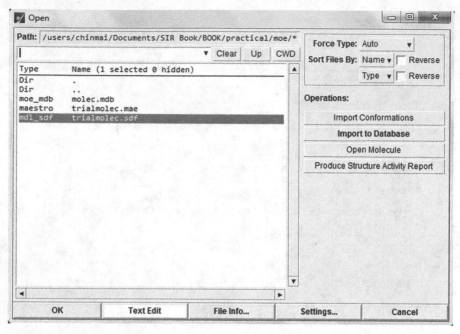

Fig. 4.77 Importing the ligand molecules in MOE

closed by clicking on the close button present at the right side of the MOE window. The query is saved in .ph4 format. Next, we open the database which should be screened using this pharmacophore query as filter by going into file and open. In our example, we used the 3D database of Maybridge [58] which has more than 58,000 entries of 3D conformations (Fig. 4.85).

Now in the database viewer window, go to compute and then to pharmacophore search option (Fig. 4.86).

It opens a window as shown in Fig. 4.87. This will have the input file by default. The query is then entered through browsing it from its respective folder. Finally, set the output path and click on the search button to start the job.

The search process takes few moments depending on the number of entries in the database. After completion of the search, it shows the number of hits found (Fig. 4.88).

Clicking on the report button present beside the search button will show the following window where we can see the details of the search process (Fig. 4.89).

Here, the numbers of hits are more. But a pharmacophore query can always be modified to focus the search and narrow the hits. Here, we tried to modify the query as shown in Fig. 4.90 which led us to three final hits (Fig. 4.91).

Pharmacophore always helps in choosing better hits as it contains the features that are known to be important for their activity. This pharmacophore query can be

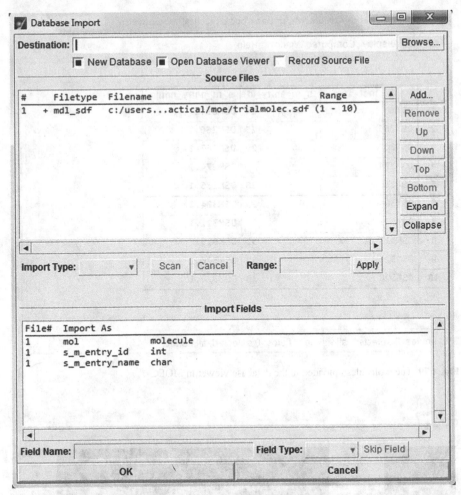

Fig. 4.78 The creation of database of molecules for PH4 query generation

used either before docking to screen a database (as explained above) or while docking to screen a set of ligand conformations.

For this, the prepared protein structure is imported into the MOE window.

Then go to compute → simulations → dock. This will open the window as shown in Fig. 4.92.

After setting the receptor and the site options, the pharmacophore option is set by browsing the pharmacophore query (PH4 file) and the ligand option is set for .mdb file which contains the set of ligands. After choosing the other parameters, click on run which performs the docking of ligands using the pharmacophore query we gave as the filter (Fig. 4.93).

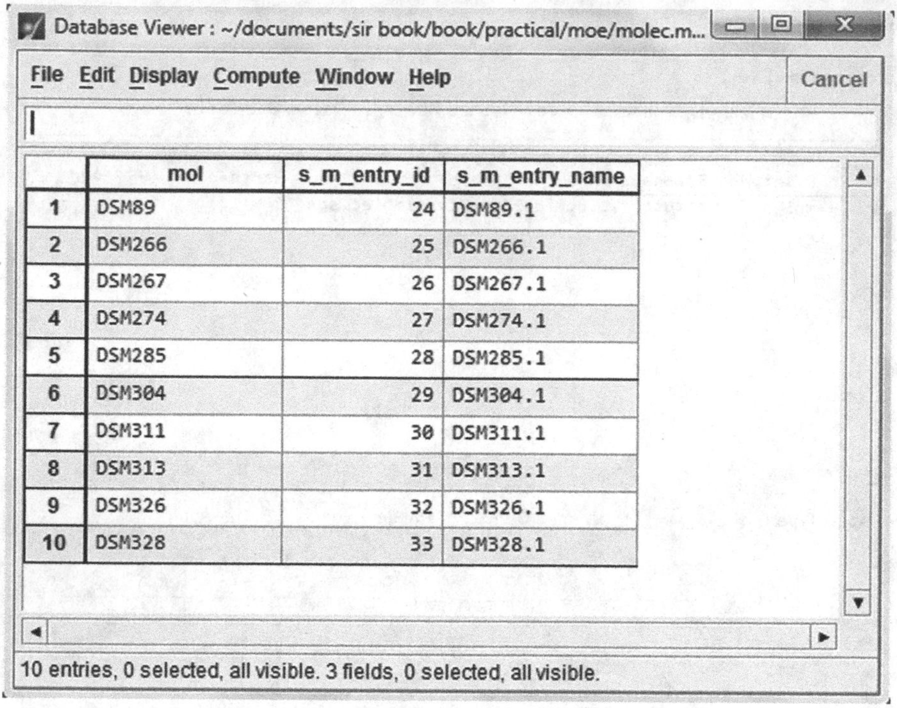

Database Viewer : ~/documents/sir book/book/practical/moe/molec.m...
File Edit Display Compute Window Help
Cancel
mol
s_m_entry_id
s_m_entry_name
1 DSM89 24 DSM89.1
2 DSM266 25 DSM266.1
3 DSM267 26 DSM267.1
4 DSM274 27 DSM274.1
5 DSM285 28 DSM285.1
6 DSM304 29 DSM304.1
7 DSM311 30 DSM311.1
8 DSM313 31 DSM313.1
9 DSM326 32 DSM326.1
10 DSM328 33 DSM328.1
10 entries, 0 selected, all visible. 3 fields, 0 selected, all visible.

Fig. 4.79 The molecules uploaded in the database viewer in MOE

moe 2010.10
File Edit Selection Render Compute GizMOE Window Help
SVL SEQ Cancel
Potential Energy
Protonate 3D...
Partial Charges...
Energy Minimize...
Site Finder...
Surfaces and Maps...
Ligand Interactions...
Protomers...
2D Molecules...
Pharmacophore
Fragments
Conformations
Biopolymer
QuaSAR
Simulations
Priority
Query Editor...
Elucidation...
Search...
System
Open
LigX
Constrain
Close
Center
SiteView
Hydrogens
Hide
Show
Ligand
Surface
Measure
Builder
Minimize
Select
Extend
Delete
R S L
MMFF94x PBC
Atoms Ribbon Contacts Fog ZClip

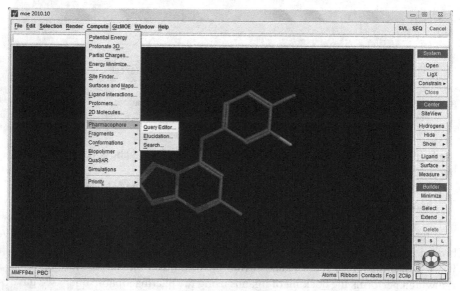

Fig. 4.80 The ligand and the query editor option in MOE

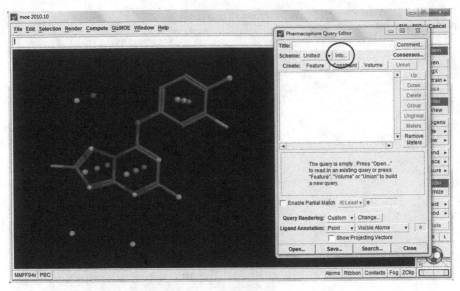

Fig. 4.81 The ligand with (default) pharmacophore features and editor window

4.7 Open Source Tools for Pharmacophore Generation

PharmaGist [59, 60] is a freely available web server for detecting pharmacophore from a group of ligands known to bind to a particular target. The output is given as a list of pharmacophores (Fig. 4.94).

One can download structures from any database and open in Marvin view program and convert .sdf file format to .mol2 format. Then click on save and upload the saved .mol2 file in PharmaGist. We can set the number of output Pharmacophores (2, 5, 10, 20) and manage advanced options to set weightage for aromatic ring, charge, hydrogen bond, and hydrophobic. If not specified, default values (0.3 for hydrophobic and 1 Å for rest) will be considered. User can also set a pivot molecule which will be considered as basic framework with which other structures are aligned and screened for similarity. Results are submitted to the specified e-mail address. As an example, we chose few non-nucleoside reverse transcriptase (NNRT) inhibitors which act against the reverse transcriptase enzyme.

Pharmacophore generation from known ligands using PharmaGist (Figs. 4.95 and 4.96):

Click on Jmol to obtain the Pharmacophore models which can be saved as .pdb (Fig. 4.97).

The downloaded file can be opened in WordPad. It displays the XYZ coordinates of the pharmacophore groups (Fig. 4.98).

Open ZINCPharmer (http://zincpharmer.csb.pitt.edu/) where the XYZ coordinates of Pharmacophore groups can be used to screen ZINCDatabase. ZINCPharmer is the free and open source pharmacophore search software which can identify

Fig. 4.82 The pharmacophore features list

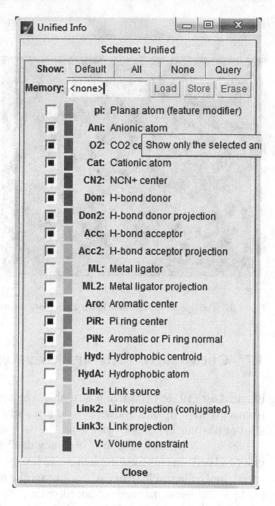

pharmacophores by itself or we can import the pharmacophore definitions from MOE or LigandScout (Fig. 4.99).

Use "Load feature" to load pharmacophore structure of PharmaGist (Fig. 4.100).

The given pharmacophore query is used to screen the molecules present in the ZINC database and gives the hits which can be downloaded. These screened compounds can be further validated through docking against the respective enzyme (Fig. 4.101).

4.8 Rules of Thumb for Structure-Based Drug Design

• Study the structural details in the context of biochemical pathways, recognize role of solvent, and cofactors in the binding process while performing docking studies.

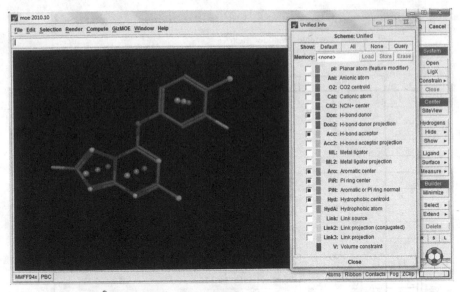

Fig. 4.83 The selected features for the first ligand

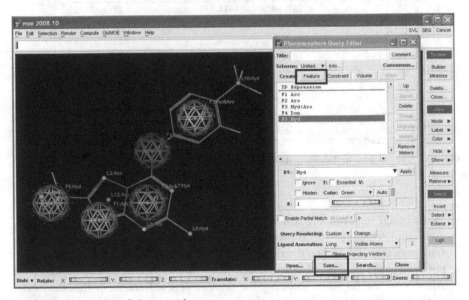

Fig. 4.84 The creation of pharmacophore query

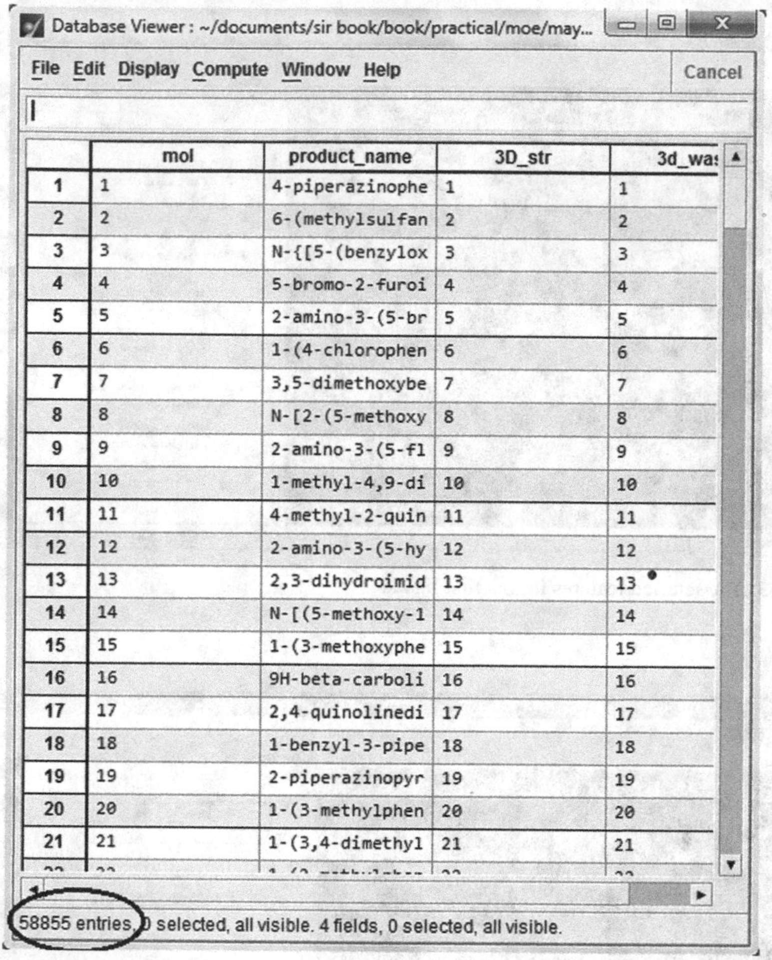

Fig. 4.85 The database molecules in database viewer

- In the initial stages of any structure-based drug design project, always choose a well-validated target with known inhibitors and a good X-ray structure of resolution more than 2 Å and an R value of 0.2. Poorly resolved crystal structures may not have easily distinguishable isoelectronic groups in molecules.
- Always be careful while selecting the docked poses. Remember that the best-docked conformation need not be the closest one to the native bioactive conformation.
- The active site coordinates have to be supplied properly. In case of multi-domain proteins, extra caution needs to be exercised to define the active site.
- In the drug discovery scenario, it is important that receptor and ligand interaction is not viewed via the rigid lock and key mechanism. Protein flexibility although

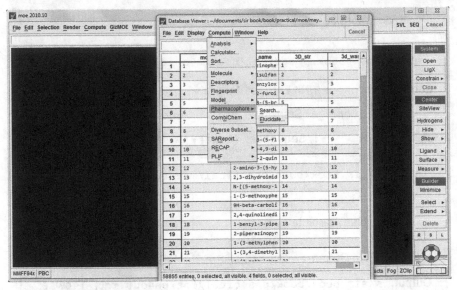

Fig. 4.86 The pharmacophore search option in database viewer

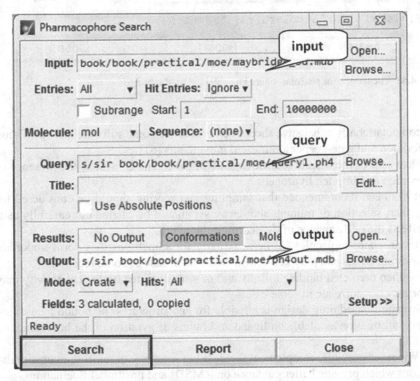

Fig. 4.87 The pharmacophore search window

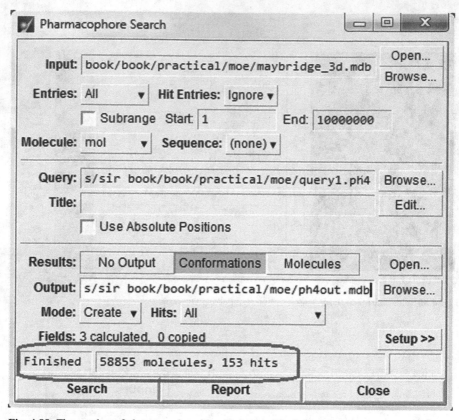

Fig. 4.88 The number of pharmacophore hits obtained in Maybridge

computationally exhaustive should be considered; some software have provision for loop and gate-keeper amino acid movements [61].

- Take care of the correct ionization state of the ligand as well as the presence of explicit and implicit hydrogens.
- It has been recommended that sampling and scoring should be considered together; selection of training and decoy set should be carried out carefully as it affects the performance of the scoring function [62].
- Retain water molecules in the active site for sometimes they form important bridging bonds and play key role in catalysis in case of some receptors.
- In silico predicted binding affinity and in vitro obtained biological activity may or may not correlate in some cases.
- Structure-based drug design is possible for apo structures (no bound ligand) using information available on ligand and radius of gyration of the holo structure [63].
- Whenever possible, complement the docking results with multi-domain simulations which provide better guidance on RMSDs and positional fluctuations.

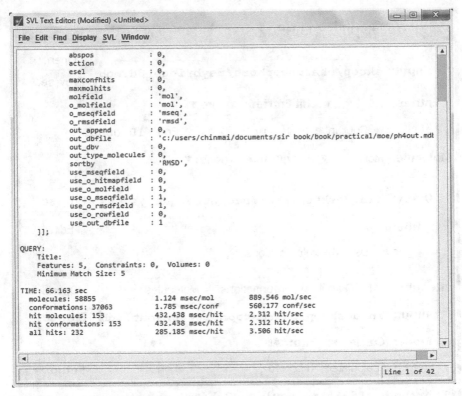

Fig. 4.89 The report of the search process

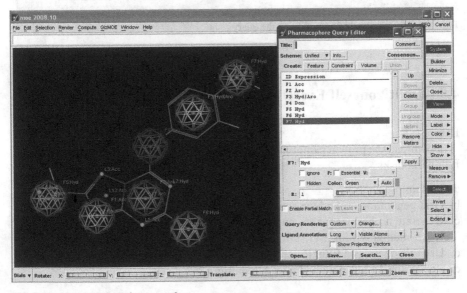

Fig. 4.90 The modified pharmacophore query

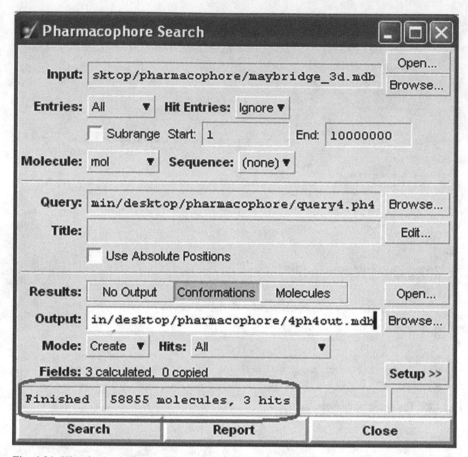

Fig. 4.91 Hits obtained after modifying the query

4.9 Do it Yourself Exercises

Exercise 1

Download the protein panthothenate synthetase (PDB ID 3IUB) and perform the following tasks:

a. Protein preparation
b. Grid generation around the active site
c. Generate a pharmacophore query using the co-crystallized ligand (5-methoxy-N-[(5-methylpyridin-2-yl)sulfonyl]-1H-indole-2-carboxamide)
d. Use this query as a filter to screen a public database like ZINC
e. Report top 15 hits obtained after screening.

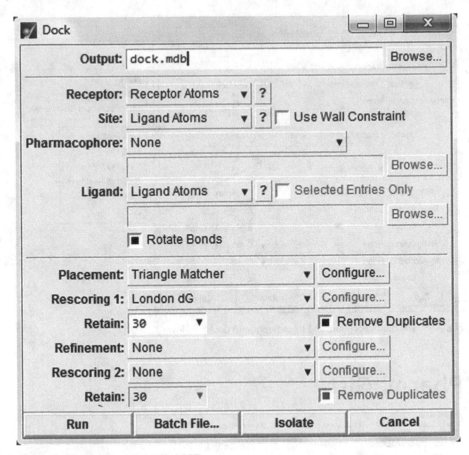

Fig. 4.92 The docking window in MOE

4.10   Questions

1. Define docking. What are the steps involved in it?
2. What is rigid docking and flexible docking?
3. List a few online tools that are used for docking.
4. How is Autodock program different from Autodock Vina?
5. How is protein preparation helpful before docking?
6. What is the LigPrep application used for?
7. Distinguish between protein–ligand docking and protein–protein docking.
8. Induced fit docking considers the flexibility of the protein. Discuss.
9. Define pharmacophore. Enlist some pharmacophore features.
10. How does a pharmacophore hypothesis helps in better screening of ligand databases?

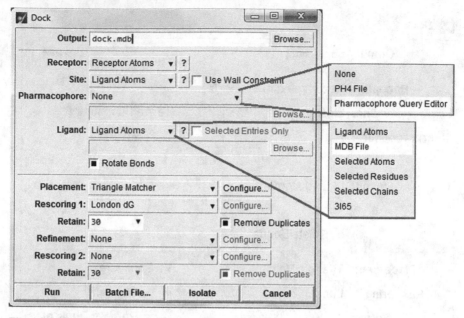

Fig. 4.93 The pharmacophore and ligand option in dock window

PharmaGist

Webserver
[About] [WebServer] [Download] [FAQ] [Help / Getting Started] Contact: ppdock@tau.ac.il

Upload Input Molecules in Mol2 Format   (e.g. input examples) [   ] [Browse...]
No. of Output Pharmacophores: [5 ▾]
E-Mail Address: [   ]
[Submit Query] [Clear]

Advanced Options:
[Show] [Hide]

Set a key-molecule: [None ▾]
Min no. of features in pharmacophore [3 ▾]

Feature Weighting
Aromatic ring: [3.0]   Charge (anion/cation): [1.0]   Hydrogen bond (donor/acceptor): [1.5]   Hydrophobic: [0.3]

User Defined Feature
Feature assignment file: [   ] [Browse...]
Feature weight: [100.0]

[Submit Query] [Clear]

If you use this webserver, please cite:
1. Inbar Y, Schneidman-Duhovny D, Dror O, Nussinov R, Wolfson HJ. Deterministic Pharmacophore Detection via Multiple Flexible Alignment of Drug-Like Molecules. In Proc. of RECOMB 2007, vol. 3692 of Lecture Notes in Computer Science, pp. 423-434. Springer Verlag.
2. Schneidman-Duhovny D, Dror O, Inbar Y, Nussinov R, Wolfson HJ. PharmaGist: a webserver for ligand-based pharmacophore detection. Nucleic Acids Research 2008. [Abstract] [FREE Full Text]
Database search
3. Dror O, Schneidman-Duhovny D, Inbar Y, Nussinov R, Wolfson HJ. Novel approach for efficient pharmacophore-based virtual screening: method and applications. J Chem Inf Model. 2009 Oct;49(10):2333-43. [Abstract] [Full Text]

Fig. 4.94 Interface of PharmaGist

Input Molecules <u>view details: visualization of the detected features</u>

#	Molecule	Atoms	Features	Spatial Features	Aromatic	Hydrophobic	Donors	Acceptors	Negatives	Positives
1	NNRTI_1.mol2	30	8	8	1	4	1	2	0	0
2	NNRTI_2.mol2	34	9	9	2	3	1	3	0	0
3	NNRTI_3.mol2	60	14	13	3	3	3	4	0	1
4	NNRTI_4.mol2	43	14	14	3	4	2	5	0	0
5	NNRTI_5.mol2	39	9	9	2	3	2	2	0	0
6	NNRTI_6.mol2	46	15	15	3	6	2	4	0	0
7	NNRTI_7.mol2	41	13	12	2	4	1	5	0	1

Fig. 4.95 Detected pharmacophore features for the set of input molecules

Sort by score   Number of Aligned Molecules: 5

Score	Jmol	Features	Spatial Features	Aromatic	Hydrophobic	Donors	Acceptors	Negatives	Positives	Molecules
13.227	Jmol	5	5	1	3	0	1	0	0	NNRTI_6.mol2 NNRTI_2.mol2 NNRTI_4.mol2 NNRTI_5.mol2 NNRTI_7.mol2

Number of Aligned Molecules: 4

Click here

Score	Jmol	Features	Spatial Features	Aromatic	Hydrophobic	Donors	Acceptors	Negatives	Positives	Molecules
18.236	Jmol	5	5	2	3	0	0	0	0	NNRTI_4.mol2 NNRTI_2.mol2 NNRTI_6.mol2 NNRTI_7.mol2

Fig. 4.96 List of pharmacophore models obtained

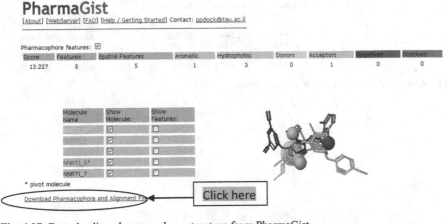

PharmaGist
[About] [WebServer] [FAQ] [Help / Getting Started] Contact: ppdock@tau.ac.il

Pharmacophore features: ☑

Score	Features	Spatial Features	Aromatic	Hydrophobic	Donors	Acceptors	Negatives	Positives
13.227	5	5	1	3	0	1	0	0

Molecule Name	Show Molecule:	Show Features:
	☑	☐
	☑	☐
	☑	☐
NNRTI_5*	☑	☐
NNRTI_7	☑	☐

* pivot molecule

Download Pharmacophore and Alignment File   **Click here**

Fig. 4.97 Downloading pharmacophore structure from PharmaGist

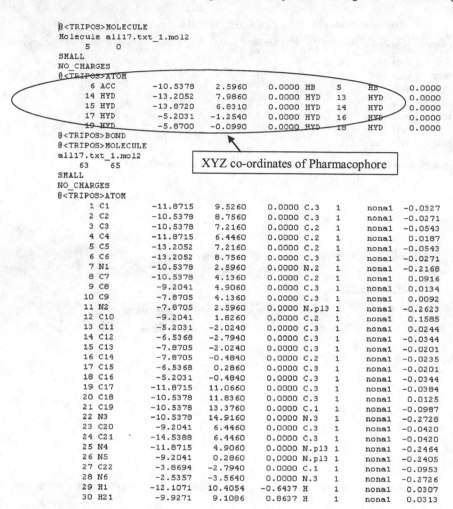

```
@<TRIPOS>MOLECULE
Molecule all17.txt_1.mol2
    5      0
SMALL
NO_CHARGES
@<TRIPOS>ATOM
      6 ACC      -10.5378     2.5960     0.0000 HB     5     HB      0.0000
     14 HYD      -13.2052     7.9860     0.0000 HYD   13     HYD     0.0000
     15 HYD      -13.8720     6.8310     0.0000 HYD   14     HYD     0.0000
     17 HYD       -5.2031    -1.2540     0.0000 HYD   16     HYD     0.0000
     19 HYD       -5.8700    -0.0990     0.0000 HYD   18     HYD     0.0000
@<TRIPOS>BOND
@<TRIPOS>MOLECULE
all17.txt_1.mol2
     63     65
SMALL
NO_CHARGES
@<TRIPOS>ATOM
      1 C1       -11.8715     9.5260     0.0000 C.3    1     nonal  -0.0327
      2 C2       -10.5378     8.7560     0.0000 C.3    1     nonal  -0.0271
      3 C3       -10.5378     7.2160     0.0000 C.2    1     nonal  -0.0543
      4 C4       -11.8715     6.4460     0.0000 C.2    1     nonal   0.0187
      5 C5       -13.2052     7.2160     0.0000 C.2    1     nonal  -0.0543
      6 C6       -13.2052     8.7560     0.0000 C.3    1     nonal  -0.0271
      7 N1       -10.5378     2.5960     0.0000 N.2    1     nonal  -0.2168
      8 C7       -10.5378     4.1360     0.0000 C.2    1     nonal   0.0916
      9 C8        -9.2041     4.9060     0.0000 C.3    1     nonal   0.0134
     10 C9        -7.8705     4.1360     0.0000 C.3    1     nonal   0.0092
     11 N2        -7.8705     2.5960     0.0000 N.pl3  1     nonal  -0.2623
     12 C10       -9.2041     1.8260     0.0000 C.2    1     nonal   0.1585
     13 C11       -5.2031    -2.0240     0.0000 C.3    1     nonal   0.0244
     14 C12       -6.5368    -2.7940     0.0000 C.3    1     nonal  -0.0344
     15 C13       -7.8705    -2.0240     0.0000 C.3    1     nonal  -0.0201
     16 C14       -7.8705    -0.4840     0.0000 C.2    1     nonal  -0.0235
     17 C15       -6.5368     0.2860     0.0000 C.3    1     nonal  -0.0201
     18 C16       -5.2031    -0.4840     0.0000 C.3    1     nonal  -0.0344
     19 C17      -11.8715    11.0660     0.0000 C.3    1     nonal  -0.0384
     20 C18      -10.5378    11.8360     0.0000 C.3    1     nonal   0.0125
     21 C19      -10.5378    13.3760     0.0000 C.1    1     nonal  -0.0987
     22 N3       -10.5378    14.9160     0.0000 N.3    1     nonal  -0.2728
     23 C20       -9.2041     6.4460     0.0000 C.3    1     nonal  -0.0420
     24 C21      -14.5388     6.4460     0.0000 C.3    1     nonal  -0.0420
     25 N4       -11.8715     4.9060     0.0000 N.pl3  1     nonal  -0.2464
     26 N5        -9.2041     0.2860     0.0000 N.pl3  1     nonal  -0.2405
     27 C22       -3.8694    -2.7940     0.0000 C.1    1     nonal  -0.0953
     28 N6        -2.5357    -3.5640     0.0000 N.3    1     nonal  -0.2726
     29 H1       -12.1071    10.4054    -0.6437 H      1     nonal   0.0307
     30 H21       -9.9271     9.1086     0.8637 H      1     nonal   0.0313
```

XYZ co-ordinates of Pharmacophore

Fig. 4.98 XYZ coordinates of a pharmacophore

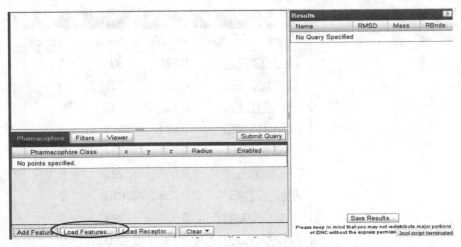

Fig. 4.99 Loading the pharmacophore features obtained from PharmaGist in ZINCPharmer

Pharmacophore	Filters	Viewer					Submit Query	
	Pharmacophore Class	x	y	z	Radius	Enabled		
>	HydrogenAcceptor	-5.28	-1.82	-2.68	0.50	☑	▾
>	Aromatic	-9.91	-3.76	0.95	1.10	☑	▾
>	Hydrophobic	-9.46	-1.68	0.58	1.00	☑	▾
>	Hydrophobic	-9.86	-5.75	-1.18	1.00	☑	▾
>	Hydrophobic	-9.33	-0.95	0.48	1.00	☑	▾

Fig. 4.100 Loaded pharmacophore feature

Fig. 4.101 Hits obtained after screening the ZINC database

Results			>
Name	RMSD	Mass	RBnds
ZINC00527894	0.675	301	3
ZINC00527894	0.623	301	3
ZINC35112180	0.727	248	9
ZINC04302453	0.674	313	4
ZINC55414757	0.701	304	7
ZINC02808677	0.856	298	5
ZINC14272433	0.786	310	2
ZINC02506209	0.695	299	9
ZINC00416398	0.737	320	5
ZINC68044181	0.743	308	11
ZINC77732800	0.732	308	11
ZINC80960302	0.795	304	8
ZINC02808551	0.737	312	6
ZINC21527941	0.695	310	7
ZINC72347048	0.832	318	5
ZINC11766014	0.742	304	10
ZINC16393265	0.647	301	4
ZINC02329987	0.619	315	4
ZINC69912371	0.730	306	13
ZINC63882269	0.681	289	9
ZINC02329987	0.670	315	4
ZINC04171154	0.836	300	6

<< < **1** 2 3 4 5 6 7 8 ≥ ≥≥

4,022 hits
70.513s

Save Results...

References

1. Taylor RD, Jewsbury PJ, Essex JW (2002) A review of protein-small molecule docking methods. J Comput-Aided Mol Des 16:151–166
2. Li H, Gao Z, Kang L et al (2006) TarFisDock: a web server for identifying drug targets with docking approach. Nucleic Acids Res 34:W219–W224. doi:10.1093/nar/gkl114
3. Halperin I, Ma B, Wolfson H, Nussinov R (2002) Principles of docking: an overview of search algorithms and a guide to scoring functions. Proteins 47:409–443
4. Bello M, Martínez-Archundia M, Correa-Basurto J (2013) Automated docking for novel drug discovery. Expert Opin Drug Discov 8:821–834
5. Glide, version 5.8, Schrödinger, LLC, New York, NY, 2012
6. http://www.nlm.nih.gov/medlineplus/druginfo/meds/a682550.html. Accessed 20 Oct 2013
7. http://www.nlm.nih.gov/medlineplus/druginfo/meds/a682401.html. Accessed 20 Oct 2013
8. http://www.nlm.nih.gov/medlineplus/druginfo/meds/a682402.html. Accessed 20 Oct 2013
9. Keith J, Ilari A, Savino C (2008) Protein structure determination by x-ray crystallography. In: Keith JM (ed) Bioinformatics, methods in molecular biology, vol 2. Humana Press, New York, pp 63–87
10. Schrödinger Suite (2012) Protein Preparation Wizard; Epik version 2.3, Schrödinger, LLC, New York, NY, 2012; Impact version 5.8, Schrödinger, LLC, New York, NY, 2012; Prime version 3.1, Schrödinger, LLC, New York, NY, 2012.
11. LigPrep, version 2.5, Schrödinger, LLC, New York, NY, 2012
12. http://www.ncbi.nlm.nih.gov/pccompound. Accessed 20 Oct 2013
13. Morris GM, Huey R, Lindstrom W et al (2009) AutoDock4 and AutoDockTools4: automated docking with selective receptor flexibility. J Comput Chem 30:2785–2791
14. Trott O, Olson AJ (2010) AutoDock Vina: improving the speed and accuracy of docking with a new scoring function, efficient optimization and multithreading. J Comput Chem 31:455–461
15. http://www.ebi.ac.uk/pdbsum/. Accessed 20 Oct 2013
16. The PyMOL Molecular Graphics System, Version 1.5.0.4 Schrödinger, LLC.
17. Bahm H-J (1992) The computer program LUDI: a new method for the de novo design of enzyme inhibitors. J Comput Aided Mol Des 6:61–78
18. Rarey M, Kramer B, Lengauer T, Klebe G (1996) A fast flexible docking method using an incremental construction algorithm. J Mol Biol 261:470–489
19. Kramer B, Rarey M, Lengauer T (1999) Evaluation of the FLEXX incremental construction algorithm for protein-ligand docking. Proteins 37:228–241
20. Barreca ML, Iraci N, De Luca L, Chimirri A (2009) Induced-fit docking approach provides insight into the binding mode and mechanism of action of HIV-1 Integrase Inhibitors. ChemMedChem 4:1446–1456
21. Schrödinger Suite (2012) Induced fit docking protocol; Glide version 5.8, Schrödinger, LLC, New York, NY, 2012; Prime version 3.1, Schrödinger, LLC, New York, NY, 2012.
22. Clauben H, Buning C, Rarey M, Lengauer T (2001) FlexE: efficient molecular docking considering protein structure variations. J Mol Biol 308:377–395
23. Hetenyi C, van der Spoel D (2002) Efficient docking of peptides to proteins without prior knowledge of the binding site. Protein Sci 11:1729–1737
24. Campbell SJ, Gold ND, Jackson RM, Westhead DR (2003) Ligand binding: functional site location, similarity and docking. Curr Opin Struct Biol 13:389–395
25. Sutherland JJ, Nandigam RK, Erickson JA, Vieth M (2007) Lessons in molecular recognition. 2. Assessing and improving cross-docking accuracy. J Chem Inf Model 47:2293–2302
26. Verdonk ML, Mortenson PN, Hall RJ et al (2008) Protein-ligand docking against non-native protein conformers. J Chem Inf Model 48:2214–2225
27. Wu G, Robertson DH, Brooks CL 3rd, Vieth M (2003) Detailed analysis of grid-based molecular docking: a case study of CDOCKER-A CHARMm-based MD docking algorithm. J Comput Chem 24(13):1549–1562

28. McGann M (2011) FRED pose prediction and virtual screening accuracy. J Chem Inf Model 51(3):578–596
29. http://www.eyesopen.com/. Accessed 20 Oct 2013
30. Jones S, Thornton JM (1996) Principles of protein-protein interactions. Proc Natl Acad Sci U S A 93:13–20
31. Davis C, Harris HJ, Hu K et al (2012) In silico directed mutagenesis identifies the CD81/claudin-1 hepatitis C virus receptor interface. Cellular Microbiol 14:1892–1903
32. Vincenzetti S, Pucciarelli S, Carpi FM et al (2013) Site directed mutagenesis as a tool to understand the catalytic mechanism of human cytidine deaminase. Protein Pept Lett 20:538–549
33. Keskin O, Ma B, Rogale K et al (2005) Protein-protein interactions: organization, cooperativity and mapping in a bottom-up systems biology approach. Phys Biol 2:S24–S35
34. Wendt MD (2012) Protein-Protein Interactions. doi:10.1007/978-3-642-28965-1
35. Villoutreix BO, Labbé CM, Lagorce D et al (2012) A leap into the chemical space of protein-protein interaction inhibitors. Curr Pharm Des 18:4648–4667
36. Xenarios I, Rice DW, Salwinski L et al (2000) DIP: the database of interacting proteins. Nucleic Acids Res 28:289–291. doi:10.1093/nar/28.1.289
37. Szklarczyk D, Franceschini A, Kuhn M et al (2011) The STRING database in 2011: functional interaction networks of proteins, globally integrated and scored. Nucleic Acids Res 39:D561–D568.
38. Chatr-aryamontri A, Breitkreutz BJ, Heinicke S et al (2013) The BioGRID interaction database: 2013 update. Nucleic Acids Res 41:D816–D823. doi:10.1093/nar/gks1158
39. http://string-db.org/. Accessed 20 Oct 2013
40. http://hexserver.loria.fr/. Accessed 20 Oct 2013
41. http://zdock.umassmed.edu/. Accessed 20 Oct 2013
42. Chen R, Li L, Weng Z (2003) ZDOCK: an initial-stage protein-docking algorithm. Proteins 52(1):80–87
43. Tovchigrechko A, Vakser IA (2006) GRAMM-X public web server for protein protein docking. Nucleic Acids Res 34:W310–W314
44. http://graylab.jhu.edu/docking/rosetta/. Accessed 20 Oct 2013
45. http://cluspro.bu.edu/login.php. Accessed 20 Oct 2013
46. Matsuzaki Y, Uchikoga N, Ohue M et al (2013) MEGADOCK 3.0: a high-performance protein-protein interaction prediction software using hybrid parallel computing for petascale supercomputing environments. Source Code Biol Med 8:18. doi:10.1186/1751-0473-8-18
47. http://www.ebi.ac.uk/msd-srv/capri/
48. Kozakov D, Brenke R, Comeau SR, Vajda S (2006) PIPER: an FFT-based proteindocking program with pairwise potentials. Proteins 65:392–406
49. Griffith R, Luu TTT, Garner J, Keller PA (2005) Combining structure-based drug design and pharmacophores. J Mol Graph Model 23:439–446.
50. Shin WJ, Seon BL (2013) Recent advances in pharmacophore modeling and its application to anti-influenza drug discovery. Expert Opin Drug Discov 8:411–426
51. Caporuscio F, Tafi A (2011) Pharmacophore modelling: a forty year old approach and its modern synergies. Curr Med Chem 18:2543–2553
52. Hecker EA, Duraiswami C, Andrea TA, Diller DJ (2002) Use of catalyst pharmacophore models for screening of large combinatorial libraries. J Chem Inf Comput Sci 42(5):1204–1211
53. Phase, version 3.4, Schrödinger, LLC, New York, NY, 2012
54. Coteron JM, Marco M, Esquivias J et al (2011) Structure-guided lead optimization of triazolopyrimidine-ring substituents identifies potent *Plasmodium falciparum* dihydroorotate dehydrogenase inhibitors with clinical candidate potential. J Med Chem 54:5540–5561. doi:10.1021/jm200592f
55. ACD/ChemSketch, version 12, Advanced Chemistry Development, Inc., Toronto, ON, Canada, http://www.acdlabs.com, 2013
56. Mills N (2006) ChemDraw Ultra 10.0. J Am Chem Soc 128:13649–13650

57. Chen IJ, Foloppe N (2008) Conformational sampling of druglike molecules with MOE and catalyst: implications for pharmacophore modeling and virtual screening. J Chem Inf Model 48:1773–1791. doi:10.1021/ci800130k
58. http://www.maybridge.com/default.aspx. Accessed 20 Oct 2013
59. http://bioinfo3d.cs.tau.ac.il/PharmaGist/. Accessed 20 Oct 2013
60. Schneidman-Duhovny D, Dror O, Inbar Y et al (2008) PharmaGist: a webserver for ligand-based pharmacophore detection. Nucleic Acids Res 36:W223–W228. doi:10.1093/nar/gkn187
61. Cavasotto CN, Orry AJW, Abagyan RA (2005) The challenge of considering receptor flexibility in ligand docking and virtual screening. Curr Comput Aided Drug Des 1:423–440
62. Vajda S, Hall DR, Kozakov D (2013) Sampling and scoring: a marriage made in heaven. Proteins 81:1874–1884
63. Seeliger D, Groot BL (2010) Conformational transitions upon ligand binding: holo structure prediction from Apo conformation. PLoS Comput Biol 6(1):e1000634

Chapter 5
Active Site-Directed Pose Prediction Programs for Efficient Filtering of Molecules

Abstract It is well known that the three-dimensional structure of a protein is a prerequisite in the field of structure-based drug discovery. Proteins are usually crystallized along with substrates (small molecules) and the site of binding is used for further computational study and virtual screening. Homology is a method that helps in modelling when a protein structure lacks co-crystallized ligands and requires knowledge of the binding site or the sequences which are yet to be crystallized, that require some structural understanding to correlate with biological functions. Homology modelling and active site prediction steps are discussed in detail using standard state-of-the-art software. Knowing the exact sites on a particular protein structure where other molecules can bind and interact is of paramount importance for any drug design effort. Having learnt the basic elements of docking, in this chapter we probe further into the binding sites and the specific properties that impart them the capability of getting bound by ligands. Active site-based features like topology, shape volume and amino acid composition all contribute to its preference for binding to a particular ligand molecule. Deducing this knowledge is the crux of an efficient active site-based screening of molecules. Active site information also helps in building a receptor-based pharmacophore query which can be applied as a constraint while screening molecular libraries. The later section therefore highlights some efforts towards active site-based virtual screening of molecules using an internally developed program which computes phi–psi-based fingerprints of proteins and binary fingerprints of ligands as a pre-filtering step for docking.

Keywords Active site · Homology modelling · Phi–psi fingerprints · Drug design

5.1 Introduction

There are several computational methods for the identification of binding sites. The different approaches discussed in the literature can be broadly classified as structure, sequence, knowledge or dynamics based. They can be further categorized as shown in Fig. 5.1:

M. Karthikeyan, R. Vyas, *Practical Chemoinformatics,* 271
DOI 10.1007/978-81-322-1780-0_5, © Springer India 2014

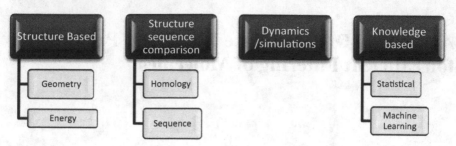

Fig. 5.1 Classification of known active site prediction methods

Geometric methods detect cavities and pockets on a protein's surface by employing a cubic grid strategy [1]. Energetic approaches are based on the interaction energy between protein and a van der Waals probe to identify energetically favourable binding sites.[2]. Homology-based methods primarily identify structures homologous to the target with a bound ligand [3]. Other methods in this category use sequence profiles, conserved features, motifs and descriptors to predict protein ligand binding site [4]. Using atomistic scale simulations, many active sites have been predicted especially for ion channel inhibitors [5]. Machine learning methods like the support vector machine (SVM) and artificial neural network (ANN) have been routinely used for predictive model building using amino acid features as inputs [6]. There are several free online software and tools that follow the above methods to find the active sites like LIGSITE [7], Pocket finder [8], Findsite [9], CASTp, etc. [10]. Commercial software such as Schrodinger [11], Molecular Operating Environment (MOE) [12], Discovery Studios [13], etc. also have binding site prediction modules in their toolkit. We will learn the use of some of these tools in the following sections.

5.2   A Practice Tutorial for Predicting Active Site Using SiteMap

Here, we predict the possible active sites for a three-dimensional (3D) protein using SiteMap module in Schrodinger. For our study, let us take a 3D protein from the Protein Data Bank (PDB) in its unbound state. The method used by the SiteMap is similar to the Goodfords' GRID algorithm [14]. It uses the interaction energies to locate the energetically favourable regions. It is necessary to remove the water molecules, cofactors or ligands (if any) present in the protein. The protein in our example does not have a bound ligand and it consists of say four chains (A, B, C and D) among which we considered chain A for this study. In the first step, the water molecules and extra chains are removed and the structure is saved in .mae format. In the next step, select the SiteMap option which initially traces the sites that include a set of site points on a grid. Then it creates the contour maps which generate

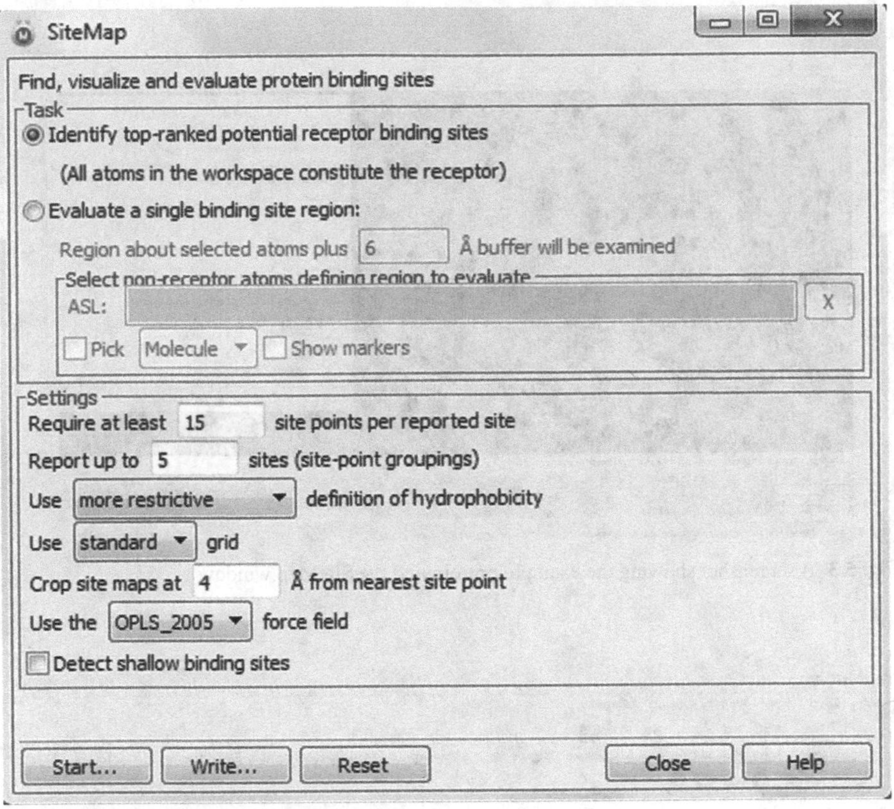

Fig. 5.2 The main SiteMap window

the hydrophilic and hydrophobic regions. Finally, each site is evaluated for various properties which are added to the project table (Fig. 5.2).

Evaluation of binding sites through SiteMap has two options. The first option is to identify top-ranked potential binding sites which cover the entire protein. The second is to find a single-binding site region where we can evaluate a single region for its hydrophilicity or hydrophobicity. To use this option, we have to select the active site residues or the co-crystallized ligand.

There is a facility of settings where the user can choose different options as per their requirement. The number of site points for a site (15 default) and the number of sites to be found (5 default) should be specified. Three types of grid are available—fine, standard and coarse—defined based on the distance between two points in the grid. Here, standard grid, the default option, is used. One can also choose between a more restrictive and less restrictive option for defining the hydrophobic regions. Two types of force fields are available of which Optimized Potentials for Liquid Simulations (OPLS) 2005 is the default one. In the given example, all default options are used (Fig. 5.3).

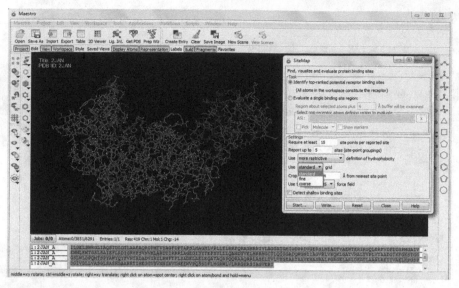

Fig. 5.3 A screenshot showing the example protein and the SiteMap window

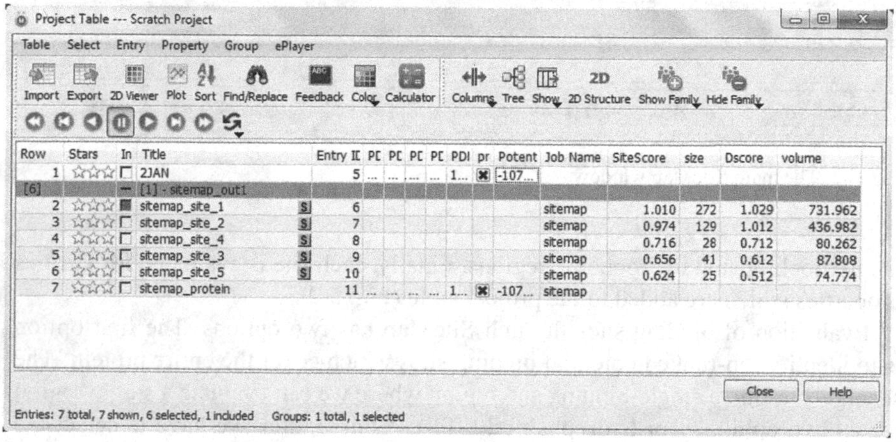

Fig. 5.4 The results as viewed in project table

Clicking on the start button launches the job, and it takes some time to calculate the score and the results are displayed in the project table (Fig. 5.4).

The site score value ranks the various binding sites. The sites are always placed in descending order of the site score values and hence the first site will be better than all other sites. It also gives the druggability score and volume of the active site.

The active site in the workspace appears in different colours and each colour represents a different property (Figs. 5.5 and 5.6).

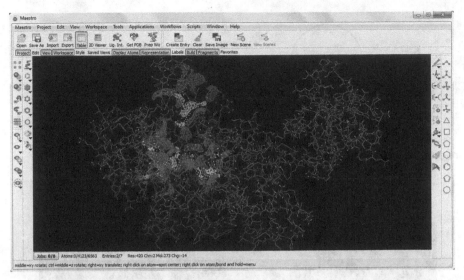

Fig. 5.5 The first active site as predicted by SiteMap

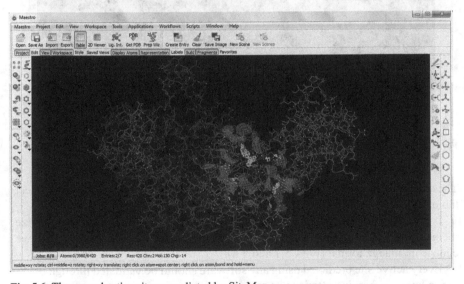

Fig. 5.6 The second active site as predicted by SiteMap

Hydrophobic map	Yellow mesh
Hydrophilic map	Green mesh
Hydrogen-bond donor map	Blue mesh
Hydrogen-bond acceptor map	Red mesh
Metal-binding map	Pink mesh
Surface map	Gray surface, 50 % transparency

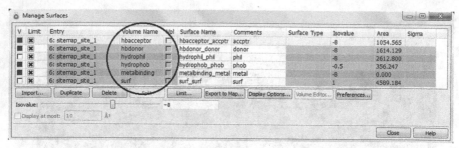

Fig. 5.7 The manage surfaces window

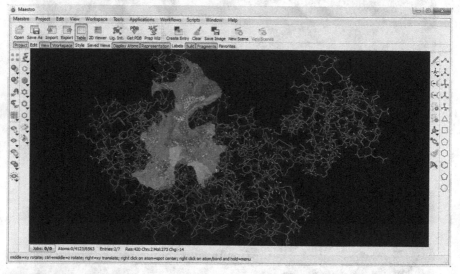

Fig. 5.8 The surface image of first active site

These can be visualized individually by clicking on the 'S' symbol in the project table.

It opens the manage surfaces window through which we can separately visualize the active site regions in workspace. The output file is generated in the folder we choose in .maegz format (Figs. 5.7 and 5.8).

Every active site predicted using any software should be validated through docking to know its binding efficiency.

5.3   A Practice Tutorial for Active Site Prediction Using MOE

This uses the geometric methods for searching the active sites different from the energy-based methods used by Schrodinger. In this, the relative positions and accessibility of receptor atoms are considered. We use the same protein example as above

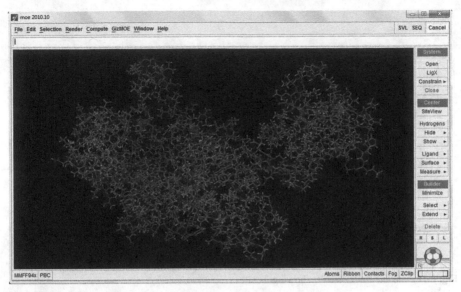

Fig. 5.9 The protein structure loaded in MOE workspace

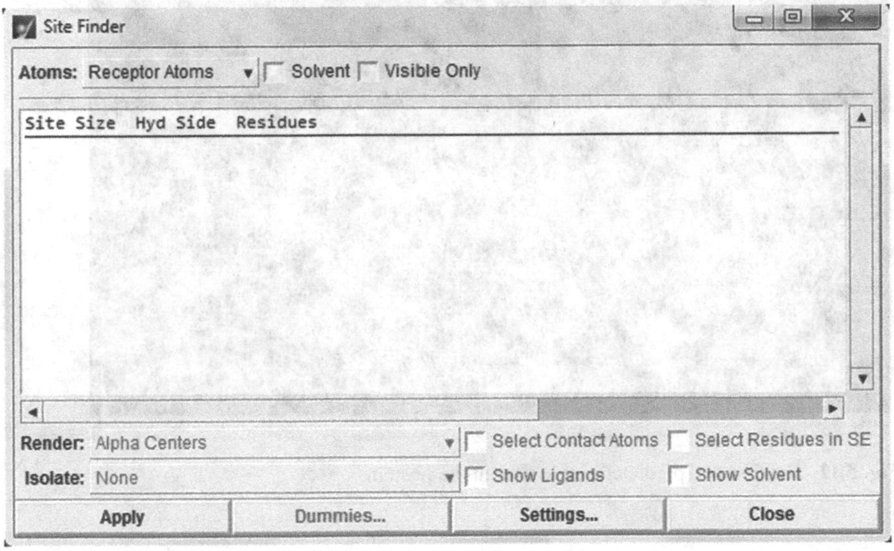

Fig. 5.10 The site finder MOE window

for this tutorial. Open the MOE and load the protein structure in to the workspace (Fig. 5.9).

Now go to the main *menu compute site finder*. It opens the window as shown in Fig. 5.10.

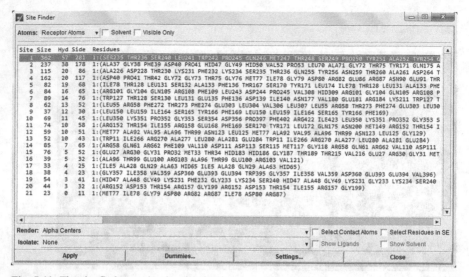

Fig. 5.11 The site finder results showing 21 sites

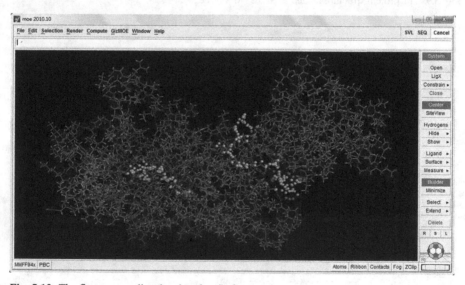

Fig. 5.12 The first two predicted active sites in the protein

Using the default settings, clicking on the start button will give the results showing the amino acid residue numbers in the window. For our example, it gave about 21 sites in descending order (Fig. 5.11).

The size column shows the number of atoms forming that site, the Hyd column shows the number of hydrophobic atoms involved and the side column shows the number of side-chain atoms involved (Fig. 5.12).

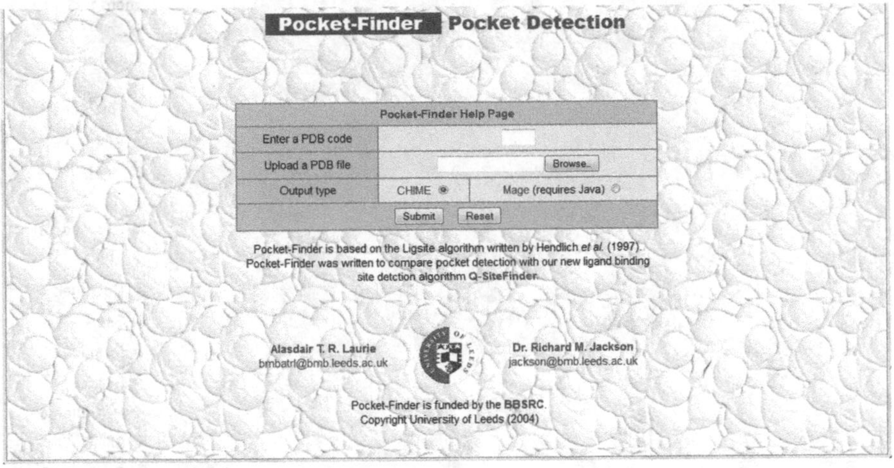

Fig. 5.13 The user interface of pocket finder server

5.4 Free Online Tools for Active Site Prediction

Pocket finder, an online tool, uses energy-based calculations to find the ligand-binding regions. The interface of the server is shown in Fig. 5.13. It follows the LIGSITE algorithm [16]. We can either give the PDB ID or browse the protein of interest. A few seconds after the submission of the job, we can see the results in the same page. It gives the best ten predicted sites.

It requires the Maze.java applet to visualize the protein. The result page has different boxes. It contains the viewer box, a box of display sites showing the ten different sites represented by different colours, a site info box which gives the volume of the first predicted site, a binding box around selected sites which gives the minimum and maximum coordinate values and a residues box which gives the list of residues occupying and surrounding all the sites (Figs. 5.14, 5.15, 5.16 and 5.17).

Q-SiteFinder [17], another tool, is similar to the pocket finder but the prediction accuracy is greater for Q-SiteFinder [18]. The tool has a simple interface. We can directly give PDB ID in the box or we can browse the protein of interest. Clicking on submit will give the result of top ten sites in a few seconds (Fig. 5.18).

We can see the results in the same page. Jmol [19] needs to be installed in the host computer to visualize the protein. The result page has different boxes (Fig. 5.19).

It contains the Jmol viewer box, a box to change the representation of the protein, a box of display sites showing the ten different sites represented by different colours, a site info box which gives the volume of the selected site, a binding box around selected sites which gives the minimum and maximum coordinate values and a residues box which gives the list of residues occupying and surrounding all the sites (Figs. 5.20, 5.21, 5.22 and 5.23).

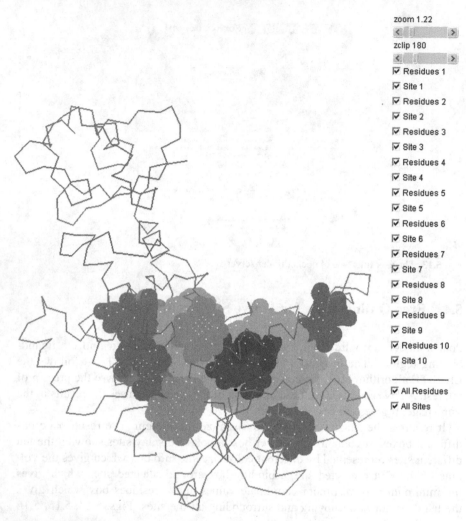

Fig. 5.14 A figure showing all the binding sites predicted

The coordinate values in both pocket finder and Q-SiteFinder can be further used to generate the grid for docking. Also, we can download the output in .pdb format and can visualize them in other tools.

Fig. 5.15 Top ten sites

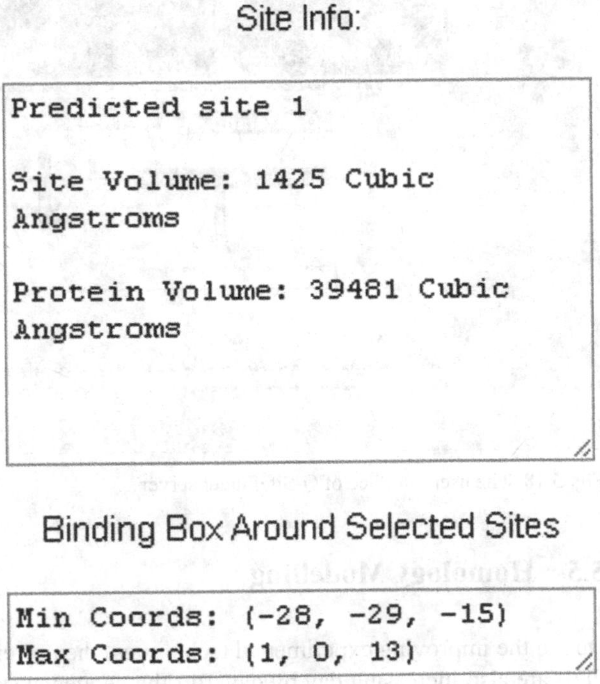

Site 1 ☑
Site 2 ☐
Site 3 ☐
Site 4 ☐
Site 5 ☐
Site 6 ☐
Site 7 ☐
Site 8 ☐
Site 9 ☐
Site 10 ☐

Help!
Download
Start Again

Fig. 5.16 Volume and coordinate details

Site Info:

```
Predicted site 1

Site Volume: 1425 Cubic
Angstroms

Protein Volume: 39481 Cubic
Angstroms
```

Binding Box Around Selected Sites

```
Min Coords: (-28, -29, -15)
Max Coords: (1, 0, 13)
```

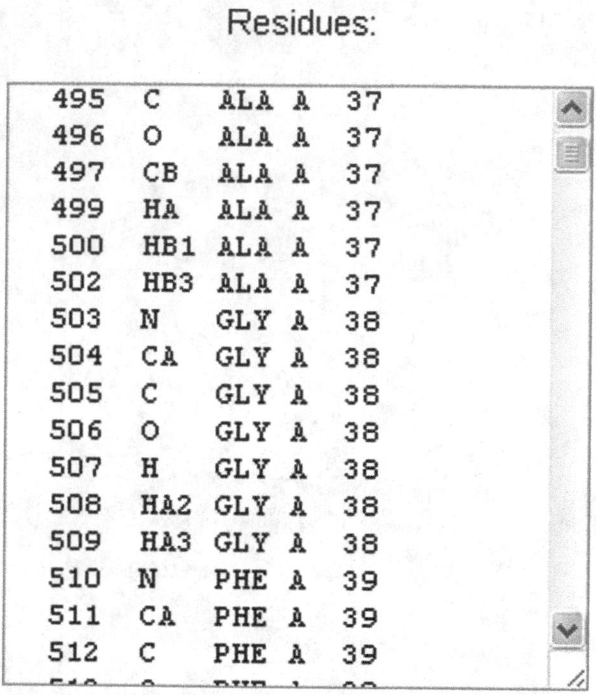

Fig. 5.17 Residue details

Residues:

```
495   C    ALA  A   37
496   O    ALA  A   37
497   CB   ALA  A   37
499   HA   ALA  A   37
500   HB1  ALA  A   37
502   HB3  ALA  A   37
503   N    GLY  A   38
504   CA   GLY  A   38
505   C    GLY  A   38
506   O    GLY  A   38
507   H    GLY  A   38
508   HA2  GLY  A   38
509   HA3  GLY  A   38
510   N    PHE  A   39
511   CA   PHE  A   39
512   C    PHE  A   39
```

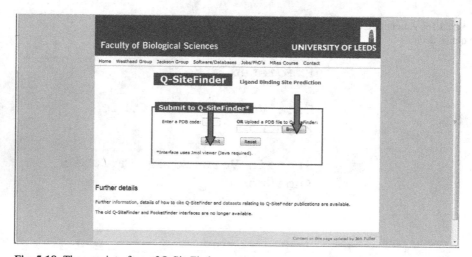

Fig. 5.18 The user interface of Q-SiteFinder server

5.5 Homology Modelling

Due to the improving experimental techniques, the protein structure deposits in the PDB are also increasing day by day. But the sequence structure gap is not reduced because of the fast sequencing techniques. Homology modelling helps to bridge

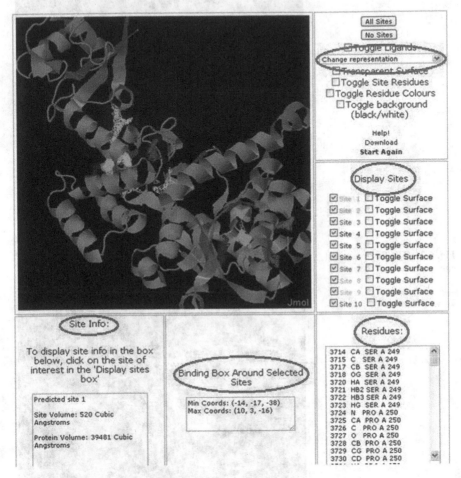

Fig. 5.19 The results page of Q-SiteFinder server

this gap [20]. Homology modelling is the procedure where we predict the 3D structure of the protein sequence computationally whose crystallized structure is not yet determined experimentally [21]. Though the accuracy rate of such models is still a matter of conflict, this is one of the wide research areas.

There are quite a good number of published papers about homology modelling of proteins [22–25]. For beginners, a brief view of homology modelling is provided. It basically includes the steps shown in Fig. 5.24 [26].

Even though every step is crucial in the generation of a better model, the first steps template recognition and initial alignment are considered as rate limiting. The

Fig. 5.20 All the predicted
sites seen in different colours

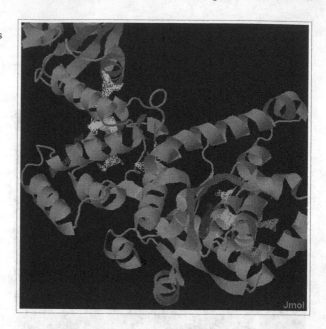

Fig. 5.21 The first predicted
site

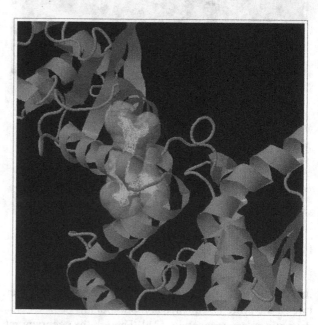

better the identity of the sequence with template the better will be the model [27].
Usually if the identity is less than 40 %, it is difficult to generate a good model. For
such cases, high computation techniques like threading [28] and ab initio techniques
[29] are used. Another more essential step is the model validation where the errors

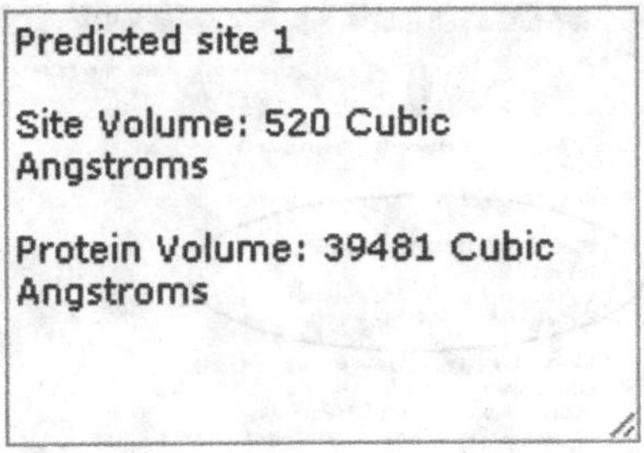

Fig. 5.22 Volume of first site

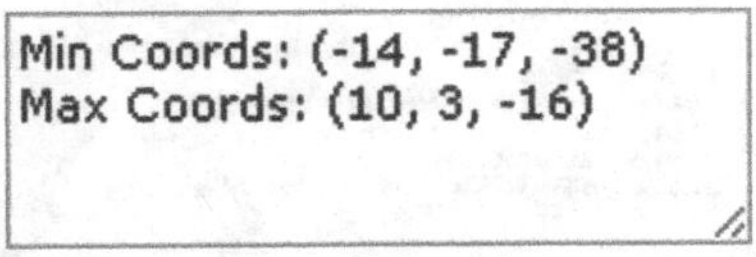

Predicted site 1

Site Volume: 520 Cubic
Angstroms

Protein Volume: 39481 Cubic
Angstroms

Fig. 5.23 Coordinates of the sites

Min Coords: (-14, -17, -38)
Max Coords: (10, 3, -16)

Fig. 5.24 Steps in homology modelling

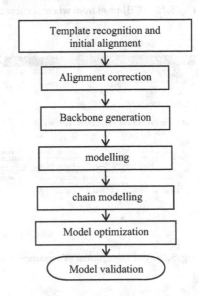

generated during the model generation are corrected. This is done by finding the model's energy using a force field and also the normality indices which examine the bond lengths and bond angles and the distribution of polar and nonpolar residues to detect the misfolded regions [30].

5.6 A Practice Tutorial for Homology Modelling

There are many commercial and free software through which modelling can be done. Modeller is a free tool for comparative modelling including loop modelling; some commercial GUI-based versions are also available [31]. We demonstrate an

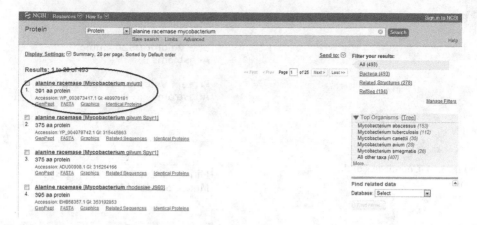

Fig. 5.25 NCBI page from where sequence was obtained

Fig. 5.26 The fasta format of sequence

example using the homology option in MOE [31]. The primary requirement for modelling is a protein sequence for which there is no crystallized structure. The sequence can be downloaded from databases like National Center for Biotechnology Information (NCBI; protein) [32], swiss-prot [33], uniprot [34], etc. In our example, the sequence was downloaded in fasta format from NCBI (Figs. 5.25 and 5.26).

Now, open the MOE window by double clicking the icon. Go to file open and browse the sequence in fasta format in to the sequence viewer. Click on the SEQ button present at the upside right corner of the MOE window as shown in figure. It opens the sequence viewer window (Fig. 5.27).

Then, go to the display option in the sequence editor window and check the compound name and single-letter residues boxes. This will display the name of the sequence and the single-letter representation of the sequence (Fig. 5.28).

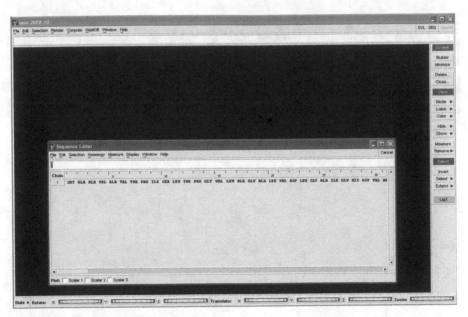

Fig. 5.27 The sequence in sequence editor

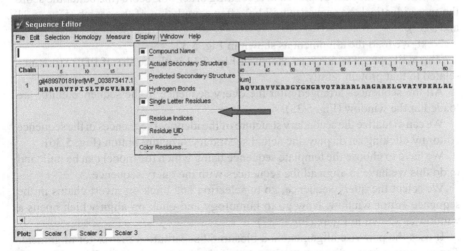

Fig. 5.28 The name and single-letter representation of the protein sequence

It is known that the next step is to search for a similar sequence for which a crystallized structure is available. For this, we have the PDB search option in MOE (similar to performing Basic Local Alignment Search Tool (BLAST)). Go to *homology PDB search* (Fig. 5.29).

This opens up the window that is shown in Fig. 5.30. Select the chain number one present at the upside right corner of the window. This will load the sequence in to the window as shown in Fig. 5.30.

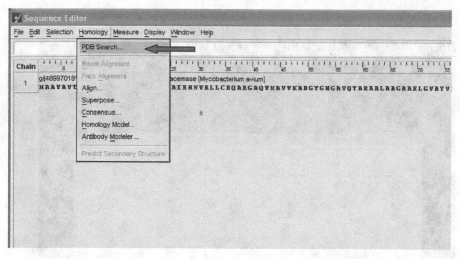

Fig. 5.29 The similarity search step

The search option will start the searching of the protein families for the sequence similarities and they are displayed in ascending order of Z score (the better the score the more identical the sequence will be) after completion of the search (Figs. 5.31 and 5.32).

Next, we load the alignment as shown in Fig. 5.33.

This will open a window as shown in Fig. 5.34 with the list of PDB structures related to that protein family.

All the sequences arranged after the query sequence in the sequence editor are loaded in the window (Fig. 5.35).

We can visualize the secondary structure of the identical sequences in the sequence editor by clicking on display, the actual secondary structure button (Fig. 5.36).

We have to choose the template sequence using which the model can be built and to do this we have to align all the sequences with the query sequence.

We select the query sequence, go to selection and click on invert chains in the sequence editor window. Now go to homology and click on align which opens a window as shown in figure (Fig. 5.37).

It uses the pairwise alignment and blosum62 substitution matrix for alignment. The alignment is the freeze button in chain selection and click ok. We can see the changes in the sequences in the sequence editor window (Fig. 5.38).

The pairwise sequence identity matrix is shown in the SVL command window (Fig. 5.39).

From the percentage, we observe that the protein with the 1XFC code has the highest identity. We can see the conserved residues between the query sequence and the selected sequence by clicking on selection, conserved residues and residue identity in the sequence editor window (Fig. 5.40).

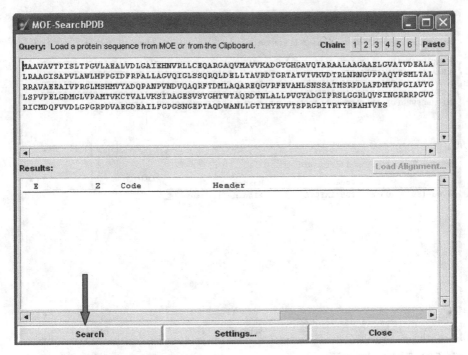

Fig. 5.30 The PDB search of the sequence

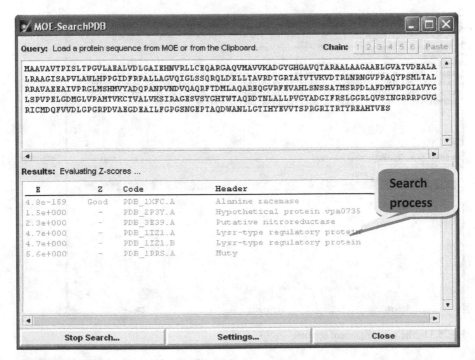

Fig. 5.31 The window during the search process

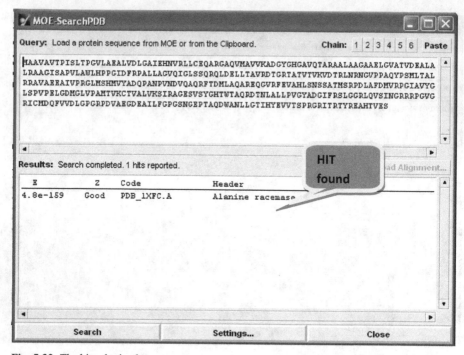

Fig. 5.32 The hits obtained

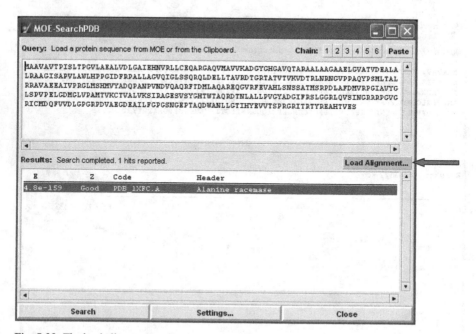

Fig. 5.33 The load alignment option

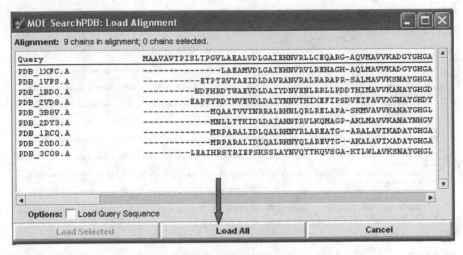

Fig. 5.34   Sequences related to that family (have crystallized structures)

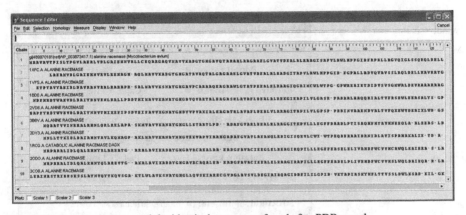

Fig. 5.35   Query sequence and the identical sequences found after PDB search.

This will highlight the conserved residues as shown in Fig. 5.41 and the colour of those residues can be changed by clicking the right button of mouse, selected residues and colour.

The next step is to build the homology model. In our example, the first sequence (next to the query sequence) is taken as the template to build the model. Click on homology and homology model (Fig. 5.42).

This opens up the model-building window (Fig. 5.43).

The selected sequence and the query sequence can be seen in models and templates division by default. The template sequence can be changed by clicking the drop-down menu and multiple sequences can also be selected. Specify the path for the output model to be saved by clicking on browse. The rotamer library and

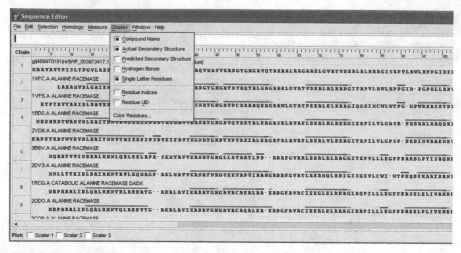

Fig. 5.36 The secondary structures of the identical sequences found after search

Fig. 5.37 The alignment window dialog box

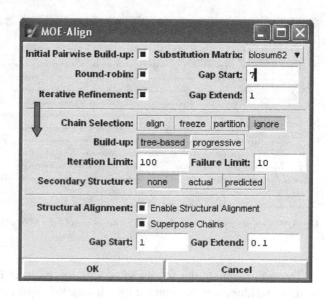

the loop library are selected by default in their respective divisions. In the model refinement division, choose the medium option for both intermediates and the final model. By default, the force field in the homology model of MOE is MMFF94× (distance dependent) which is not good for protein modelling. To select the specific force field, click on the potential setup button which opens a new window (Fig. 5.44).

Choose AMBER99 from the list, the solvation option and R-field. Save these parameters.

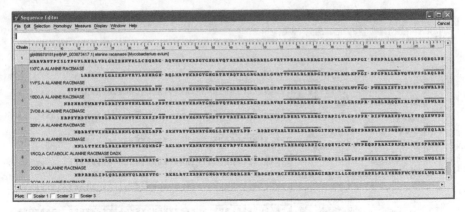

Fig. 5.38 The aligned sequences

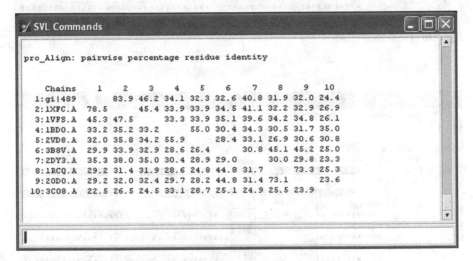

Fig. 5.39 The residue identity after alignment

The homology modelling process can be started now.

We can see the job running in the SVL command window as shown in Fig. 5.45.

MOE generates ten models in their ascending order of root-mean-square deviation (RMSD) value. Open the promodel.mdb file to visualize the models in the database viewer. The first model is displayed in Figs. 5.46 and 5.47.

To visualize the structure in a different format, go to render in MOE. Go to back bone and choose cartoon and then go to Render hide all (Figs. 5.48 and 5.49).

The next step is to evaluate the built homology model.

For this, we go to measure and then to protein geometry in the sequence editor window (Fig. 5.50).

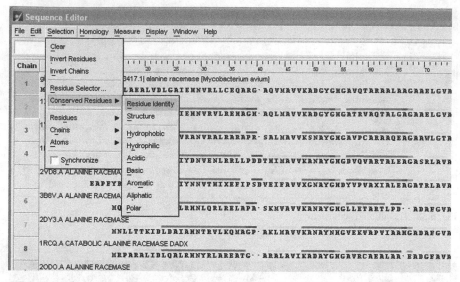

Fig. 5.40 The residue identity

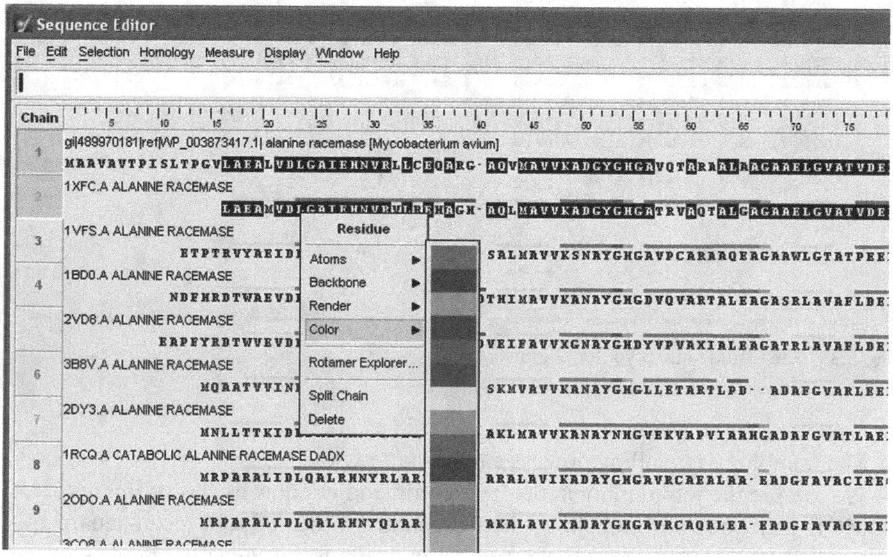

Fig. 5.41 Figure showing the conserved residues in different template proteins

It will open the protein geometry window where we can see the Ramchandran plot for the respective model. We can see the outliers by clicking on the data button at the upside right corner of the protein geometry window (Figs. 5.51 and 5.52).

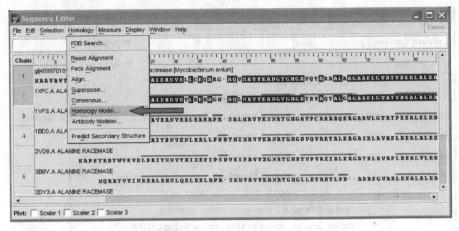

Fig. 5.42 The homology model option selection window

We can visualize the geometry of bond lengths, bond angles, dihedrals, etc. by choosing the required option from the drop-down menu of the check option (Figs. 5.53 and 5.54).

The report for the Ramchandran plot can be generated by clicking on the report button (Fig. 5.55).

The best model among the generated models can be saved separately by choosing the save option in the file menu of the MOE window. We can save the model in .pdb format.

This model can further be evaluated through the online available servers like procheck [35], what if [36], etc.

5.7   Model Validation Using Online Servers

Structural Analysis and Verification Server (SAVES) [37] is the widely used server for the validation of the protein models. It has options like ERRAT [38], VERIFY 3D [39], PROCHECK, etc. which have specific evaluating methods described in the website itself (Fig. 5.56).

We can choose file and browse the model in .pdb format that is to be validated and view the listed options, for this example we select PROCHECK.

The results will be displayed after few seconds as shown in Fig. 5.57. We can see each parameter by selecting the portable document format (PDF) or JPG present below each point. We can also see the Ramchandran plot by clicking the button present at the bottom of the page (Figs. 5.58 and 5.59).

Similarly, choose the pdb file and then ERRAT to view the results in the interface as shown in window (Figs. 5.60 and 5.61).

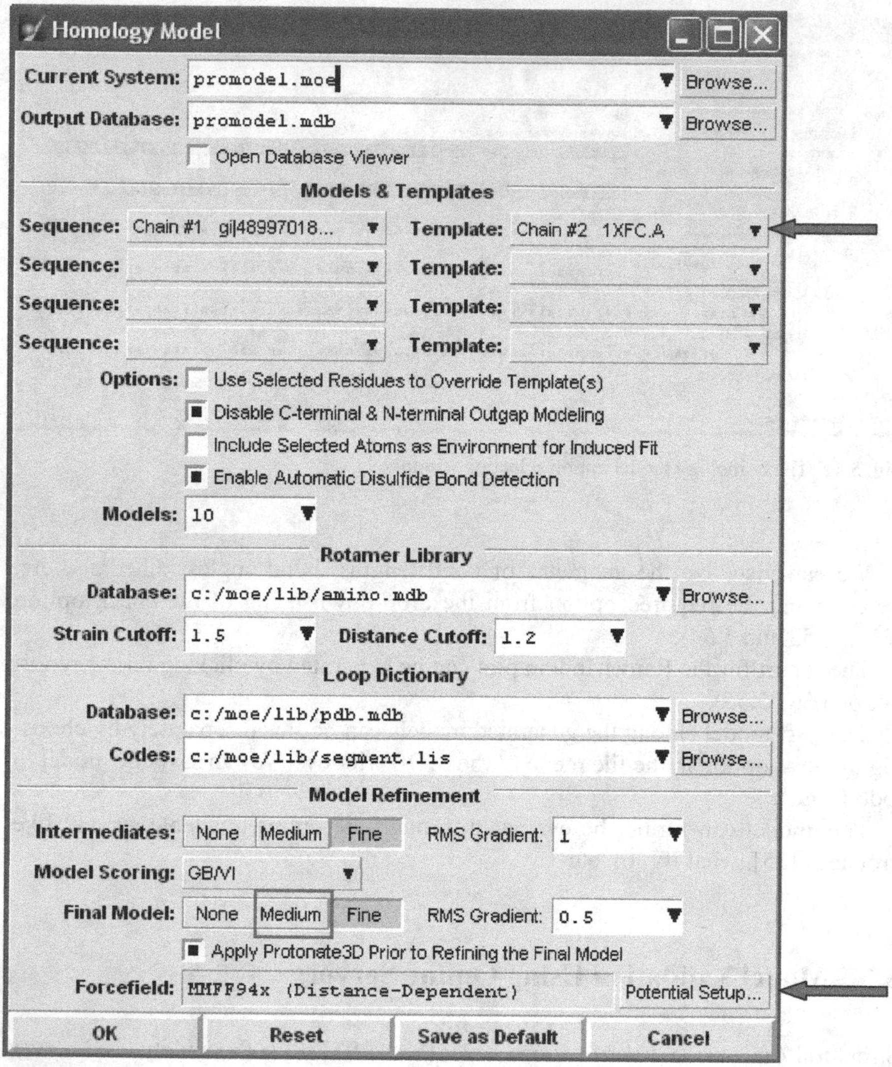

Fig. 5.43 The homology model window

5.8   Receptor-Based Pharmacophore

From the previous chapters, it is known that a pharmacophore is the group of features that is essential for the biological activity of a molecule and we have discussed it in the context of analogue or ligand-based virtual screening [40]. We can also build a pharmacophore query in the absence of ligands based only on the receptor information which is known as receptor-based pharmacophore modelling [41].

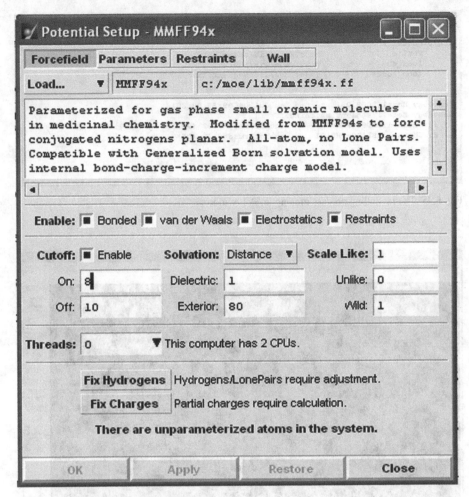

Fig. 5.44 The force field options

Usually, this is known as dynamic receptor-based pharmacophore as the pharmacophore is generated by considering the dynamic (different conformations) nature of the protein [42].

This usually includes four steps:

1. Structure quality assessment
2. Phase space sampling
3. Negative image construction
4. Hit analysis

In the third step (negative image sampling), the chemical features in the active site are known by molecular interaction field analysis and identification of excluded volumes [43]. It has its application in designing novel inhibitors in the drug discovery field [44].

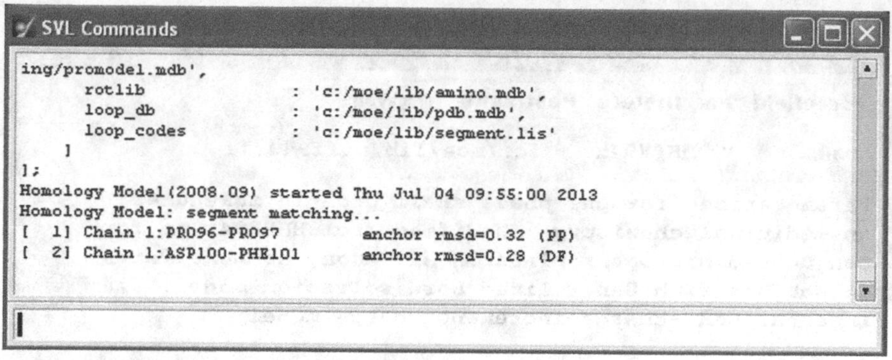

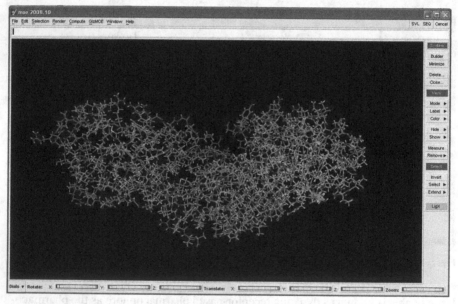

SVL Commands
ing/promodel.mdb',
rotlib : 'c:/moe/lib/amino.mdb',
loop_db : 'c:/moe/lib/pdb.mdb',
loop_codes : 'c:/moe/lib/segment.lis'
]
];
Homology Model(2008.09) started Thu Jul 04 09:55:00 2013
Homology Model: segment matching...
[1] Chain 1:PRO96-PRO97 anchor rmsd=0.32 (PP)
[2] Chain 1:ASP100-PHE101 anchor rmsd=0.28 (DF)

Fig. 5.45 The modelling process running in SVL window

moe 2008.10
File Edit Selection Render Compute GizMOE Window Help
SVL SEQ Cancel
System
Builder
Minimize
Delete...
Close...
View
Mode
Label
Color
Hide
Show
Measure
Remove
Select
Invert
Select
Extend
LigX
Dials Rotate: X Y Z Translate: X Y Z Zoom

Fig. 5.46 The model number 1 displayed in the MOE window

5.9 Studies on Active Site Structural Features

Identification, visualization and analysis of protein active site regions is the first and foremost mandatory step for any structure-based drug design program. Active site is the playing field where actual action takes place which could be either a catalytic activity of an enzyme or a drug action to modulate molecular processes. The residues at the catalytic active site are always conserved across families and even a small mutation at the active site can adversely impact the protein [45]. A

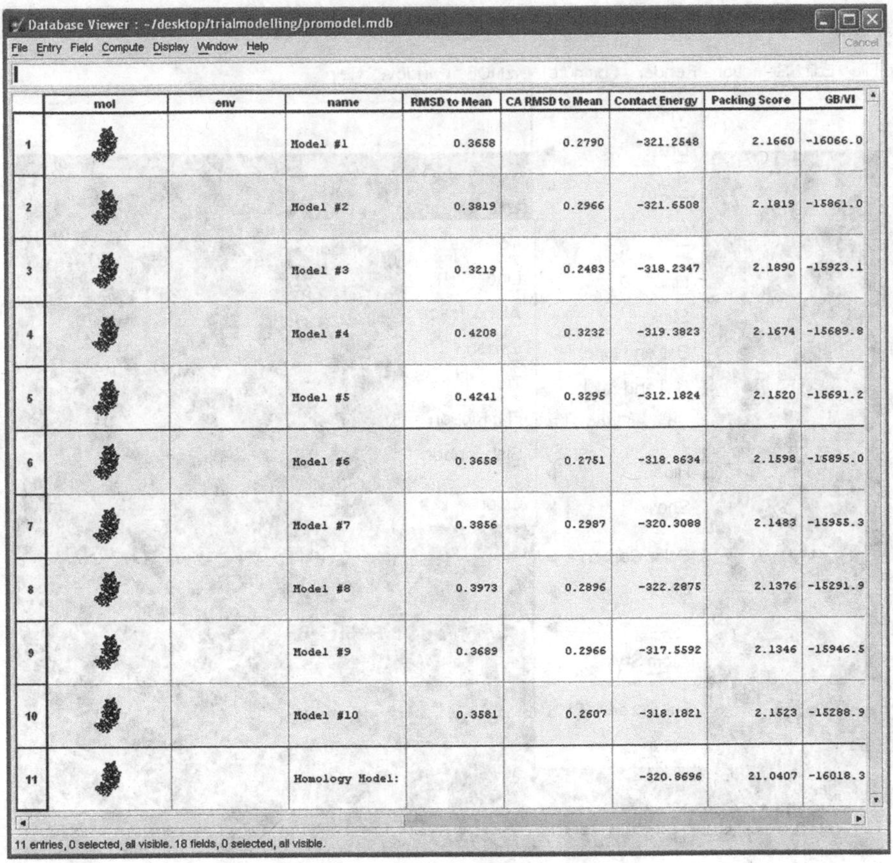

	mol	env	name	RMSD to Mean	CA RMSD to Mean	Contact Energy	Packing Score	GB/VI
1			Model #1	0.3658	0.2790	-321.2548	2.1660	-16066.0
2			Model #2	0.3819	0.2966	-321.6508	2.1819	-15861.0
3			Model #3	0.3219	0.2483	-318.2347	2.1890	-15923.1
4			Model #4	0.4208	0.3232	-319.3823	2.1674	-15689.8
5			Model #5	0.4241	0.3295	-312.1824	2.1520	-15691.2
6			Model #6	0.3658	0.2751	-318.8634	2.1598	-15895.0
7			Model #7	0.3856	0.2987	-320.3088	2.1483	-15955.3
8			Model #8	0.3973	0.2896	-322.2875	2.1376	-15291.9
9			Model #9	0.3689	0.2966	-317.5592	2.1346	-15946.5
10			Model #10	0.3581	0.2607	-318.1821	2.1523	-15288.9
11			Homology Model:			-320.8696	21.0407	-16018.3

Fig. 5.47 The saved models in the database viewer window of MOE

prerequisite for a bioactive molecule is that it should be able to locate and fit into the buried active site region of the target protein. The change in protein conformation upon ligand binding in the active site to effect a particular biological response gives important clues about the protein function [46]. Detecting and characterizing the active site therefore assumes tremendous importance in the arena of drug design. Active site features for example topology, electronic environment, energy, shape, size, volume, chirality, hydrophobicity, salt bridges, solvation, electrostatic potential, surface accessibility, secondary structural elements and chemical fragment interactions, etc. enable a ligand to bind to a protein in a biological system [47]. The analysis of physiochemical properties of binding sites helps us in the design of high-affinity ligands. Computationally intensive density functional theory (DFT) methods have been employed to study active site structural features using first principles [48]. Machine learning algorithms have also been used to generate atom-based fingerprints of the ligand binding sites in a protein [49]. The algorithms

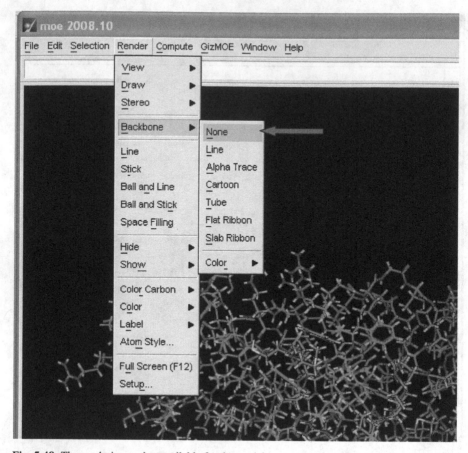

Fig. 5.48 The rendering option available for the model

used in these prediction programs are based on residue information and surface/ volume properties of the active site.

5.9.1   Application of Active Site Features in Chemoinformatics

Target active site identification is easy but predicting protein druggability is difficult [50]. Machine learning approaches such as random forest [51], SVM [52] have been attempted for discriminating druggable and non-druggable sites based on pocket attributes. Though several docking methods are available to score a large molecular database for complementarity to a protein active site, they usually yield hundreds of hits. So, there is a need to reduce the initial hit list without losing information about potential ligands by applying some efficient pre-docking filters. Moreover, docking protocols usually provide interaction energies between protein and ligand but do not

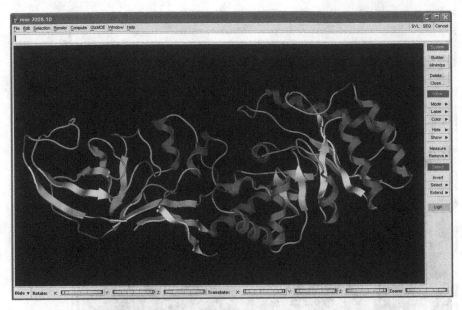

Fig. 5.49 The cartoon form of the built model

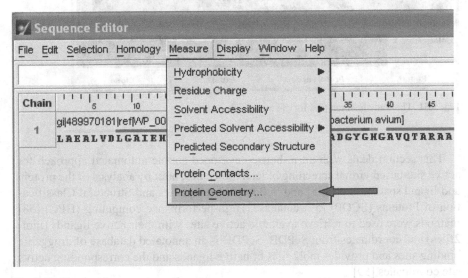

Fig. 5.50 The protein geometry option

directly facilitate the design of new ligands [53]. Molecular dynamics methods are increasingly being used in drug discovery approaches in identifying cryptic binding sites, active site dynamics and free energies but suffer from two limitations—force fields that need to be refined and high computation demand [54].

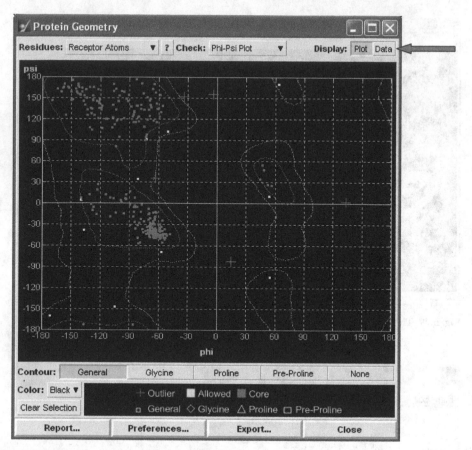

Fig. 5.51 The Ramchandran plot for the model

This section deals with an in-house-developed simple automated approach for active site-based virtual screening of lead molecules, built by analysis of the protein and ligand space of Pdb [55] and ScPdb [56] complexes and Structural Classification of Proteins (SCOP) [57] database. High-performance computing (HPC) [58] methods were used to retrieve available active sites with their native ligands (mol-2files) and coordinates from ScPDB. ScPDB is an annotated database of druggable binding sites and provides mol2 files of native ligands and the corresponding active site coordinates [59].

5.9.1.1   Code for Getting the Protein Coordinates from PDB File Format (X, Y, Z values)

```java
public double[][] getProtCoord(String fname) {
    String coord = "";
    int aid = 1;
    int haid = 1;
    int lacnt = 0;
    int pacnt = 0;
    int max = 200000; // at-least 200k atoms - Change here as per
your needs
    double[][] pcoor = new double[max][3];
    String pdbid = fname;
    try {
        FileInputStream fStream = new FileInputStream(fname);
        BufferedReader in = new BufferedReader(new
InputStreamReader(fStream));
        String b = "";
        int chk = 0;
        while ((b = in.readLine()) != null && pacnt < max && lacnt
< 999 && chk == 0) {
            if (b.startsWith("ATOM    ")) //
            {
                String[] e = stringToArray(b);
                if (e.length == 12) {
                    pcoor[pacnt][0] = Double.valueOf(e[6]);//for 7
column X
                    pcoor[pacnt][1] = Double.valueOf(e[7]);//for 8
column Y
                    pcoor[pacnt][2] = Double.valueOf(e[8]);//for 9
column Z
                    String at = e[11];//for 11 column
                } else if (e.length != 12) {
                    chk++;
                }
                pacnt++;
            }  //atom
        }//while
        in.close();//in object is close
    } catch (IOException e) {
        System.out.println("File input error");
    }
    double[][] pcoor1 = new double[pacnt][3];
    for (int v = 0; v < pacnt; v++) {
        pcoor1[v] = pcoor[v];
    }
    return pcoor1;
}
```

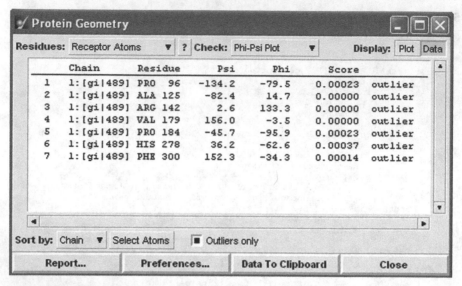

Fig. 5.52 The outliers data of the model

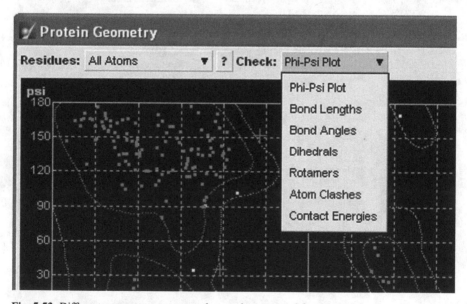

Fig. 5.53 Different geometry parameters that can be measured for the model

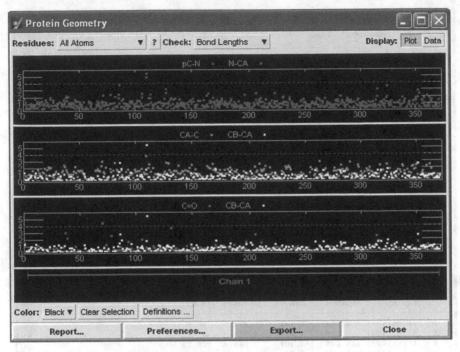

Fig. 5.54 The geometry of the bond lengths

5.9.1.2   Code for Obtaining the Ligand Coordinates (X, Y, Z values) from PDB (Protein–Ligand complex)

```
public double[][] getLigCoord(String fname) {
    int aid = 1;
    int haid = 1;
    int lacnt = 0;
    double[][] lcoor = new double[1000][3];
    String pdbid = fname;
    try {
        FileInputStream fStream = new FileInputStream(fname);
        BufferedReader in = new BufferedReader(new InputStreamReader(fStream));
        String b = "";
        int chk = 0;
        while ((b = in.readLine()) != null && lacnt < 999 && chk == 0) {
            if (b.startsWith("HETATM  ")) {
                String[] e = stringToArray(b);
                if (e.length == 12) {
                    lcoor[lacnt][0] = Double.valueOf(e[6]);//for 7 column
                    lcoor[lacnt][1] = Double.valueOf(e[7]);//for 8 column
                    lcoor[lacnt][2] = Double.valueOf(e[8]);//for 9 column
                    String at = e[11];//for 11 column
                } else if (e.length != 12) {
                    chk++;
                }
                lacnt++;
            }
        }
    } catch (IOException e) {
        System.out.println("File input error");
    }
    double[][] lcoor1 = new double[lacnt][3];
    for (int v = 0; v < lacnt; v++) {
        lcoor1[v] = lcoor[v];
    }
    return lcoor1;
}
```

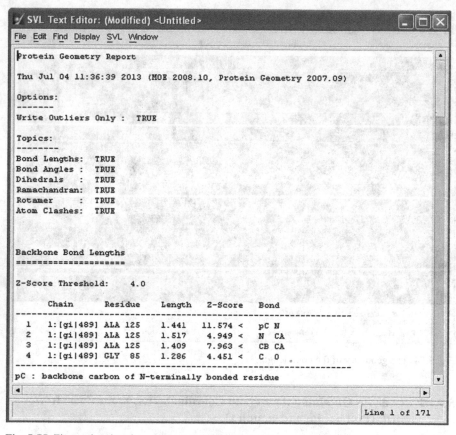

Fig. 5.55 Figure showing the protein geometry report of the model

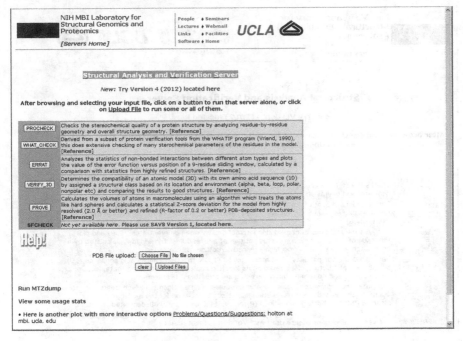

Fig. 5.56 The SAVES server interface

Get Coordinates (XYZ values) from MOL2 format

```
double[][] getCoordMol2(String fname) {
        double[][] out = new double[10000][3];
        System.out.println(fname);
        String param = "";
        int acnt = 0;
        try {
                BufferedReader br = new BufferedReader(new FileReader(new
File(fname)));
                String s = "";
                int start = 0;
                while ((s = br.readLine()) != null) {
                    if (s.contains("@<TRIPOS>ATOM")) {
                        start = 1;
                    }
                    if (start == 1 && s.length() > 50) {
                        String[] a = stringToArray(s.substring(18,
48).trim().replaceAll("   ", " "));
                        out[acnt][0] = Double.valueOf(a[0]);
                        out[acnt][1] = Double.valueOf(a[1]);
                        out[acnt][2] = Double.valueOf(a[2]);
                        param += "   " + a[3] + "\t" + out[acnt][0] + "\t" +
out[acnt][1] + "\t" + out[acnt][2] + "\n";
                        acnt++;
                    }
                    if (s.contains("@<TRIPOS>BOND")) {
                        start = 0;
                    }
                }
                br.close();
        } catch (Exception e) {
                System.out.println(e);
        }
        double[][] d1 = new double[acnt][3];
        for (int i = 0; i < acnt; i++) {
            d1[i] = out[i];
        }
        return d1;
    }
```

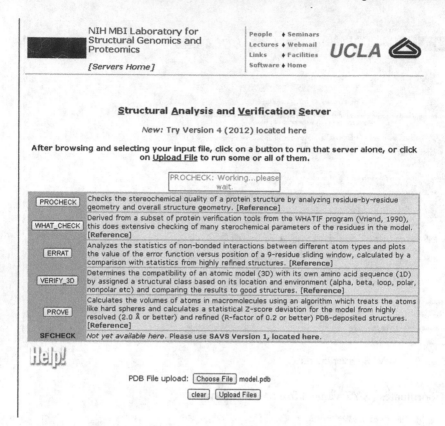

Fig. 5.57 Processing by PROCHECK

Each atom of every individual amino acid present in the complexes was processed to extract the associated sets of phi–psi angles to generate a statistically significant cumulative Ramchandran plot for chain A of all proteins [60] (Fig. 5.62).

For proteins without co-crystallized ligands, fingerprints corresponding to distinct protein classes were created by identifying distinguishing features. Ligand characterization in binding site is very important to understand the intermolecular interactions leading to the desired biological effect. At the time of molecular docking, the force field of proteins opposes the force field of ligands. The magnitude of this force field depends upon the active site environment and the ligand which has to displace water molecules in the active site. Calculating the force field is time consuming and requires more precision. However, if we understand the active site very well, then we know what molecular fragments to put therein and thus avoid intensive force field computations. Once we know in advance the fragments to put, we can place the interacting residues and remaining linkers in the active site. This is the core concept of de novo drug design or fragment-based drug design (FBDD) [61].

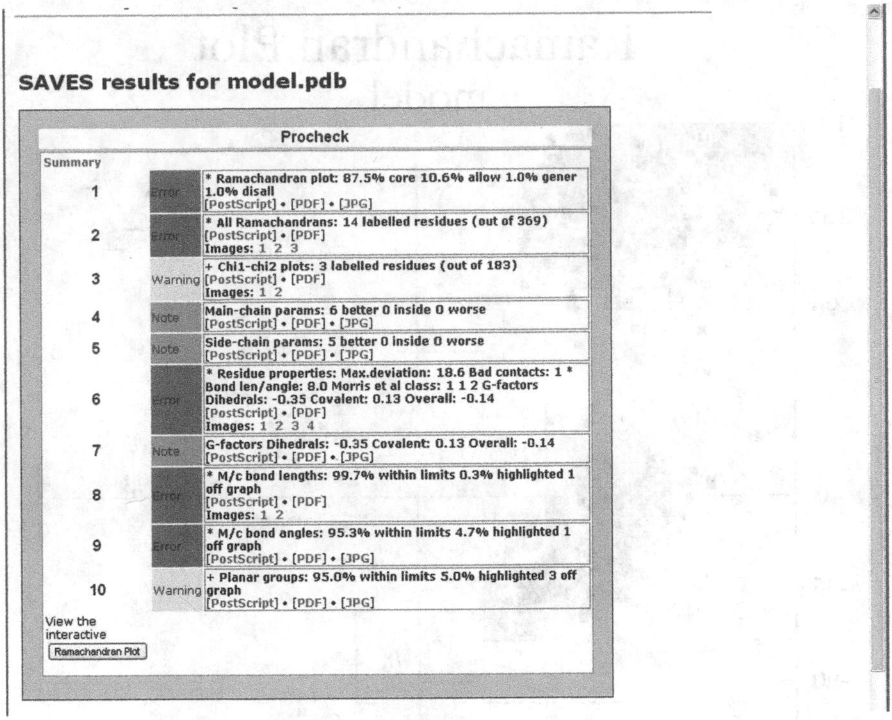

SAVES results for model.pdb

		Procheck
Summary		
1	Error	* Ramachandran plot: 87.5% core 10.6% allow 1.0% gener 1.0% disall [PostScript] • [PDF] • [JPG]
2	Error	* All Ramachandrans: 14 labelled residues (out of 369) [PostScript] • [PDF] Images: 1 2 3
3	Warning	+ Chi1-chi2 plots: 3 labelled residues (out of 183) [PostScript] • [PDF] Images: 1 2
4	Note	Main-chain params: 6 better 0 inside 0 worse [PostScript] • [PDF] • [JPG]
5	Note	Side-chain params: 5 better 0 inside 0 worse [PostScript] • [PDF] • [JPG]
6	Error	* Residue properties: Max.deviation: 18.6 Bad contacts: 1 * Bond len/angle: 8.0 Morris et al class: 1 1 2 G-factors Dihedrals: -0.35 Covalent: 0.13 Overall: -0.14 [PostScript] • [PDF] Images: 1 2 3 4
7	Note	G-factors Dihedrals: -0.35 Covalent: 0.13 Overall: -0.14 [PostScript] • [PDF] • [JPG]
8	Error	* M/c bond lengths: 99.7% within limits 0.3% highlighted 1 off graph [PostScript] • [PDF] Images: 1 2
9	Error	* M/c bond angles: 95.3% within limits 4.7% highlighted 1 off graph [PostScript] • [PDF] • [JPG]
10	Warning	+ Planar groups: 95.0% within limits 5.0% highlighted 3 off graph [PostScript] • [PDF] • [JPG]

View the interactive
[Ramachandran Plot]

Fig. 5.58 The PROCHECK results for the model

Ligands are generally computationally processed as fingerprints which are binary bit string representations of molecular structure and properties of a molecule and often employed in chemical similarity searching methods [62]. The in-house program generates ligand-based fingerprints by considering two important properties—the topology and charge present on each ligand. Structure data file (SDF) of ligands were input into the program named LIGBIT. The length (l), breadth (b) and height (h) dimensions of the ligands were used to obtain the centre of the active site grid box using a Java-based script. The size of the unit cell is complementary to active site dimension of the protein. The ligands were sorted based on fingerprints. The program can also calculate the volume of active site with volume of ligand for comparison. Rigid body transformations such as rotation and translation can be carried out to simulate actual molecular recognition in a biological system. The ligand molecule was rotated in the box by 5 in x, y and z directions and translated in the centre. This method was able to generate 216 poses for a small molecule. The pose, orientation and interaction of the 7,211-pdb ligands in active sites were similarly studied and predicted. The program was applied for generating and screening poses of acquired immunodeficiency syndrome (AIDS) inhibitors available in the National Cancer Institute (NCI) [60] (Fig. 5.63).

PROCHECK

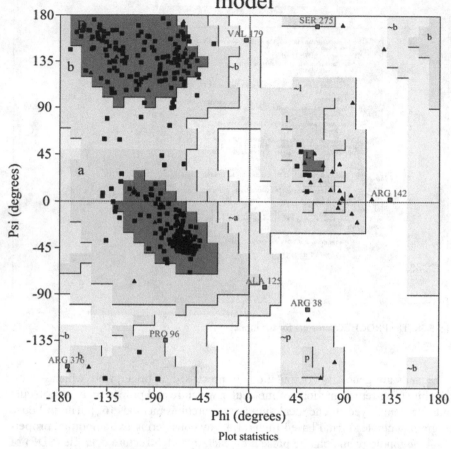

Ramachandran Plot
model

Plot statistics

Residues in most favoured regions [A,B,L]	272	87.5%
Residues in additional allowed regions [a,b,l,p]	33	10.6%
Residues in generously allowed regions [~a,~b,~l,~p]	3	1.0%
Residues in disallowed regions	3	1.0%
	----	------
Number of non-glycine and non-proline residues	311	100.0%
Number of end-residues (excl. Gly and Pro)	2	
Number of glycine residues (shown as triangles)	36	
Number of proline residues	22	

Total number of residues	371	

Fig. 5.59 The pdf of the Ramchandran plot by PROCHECK

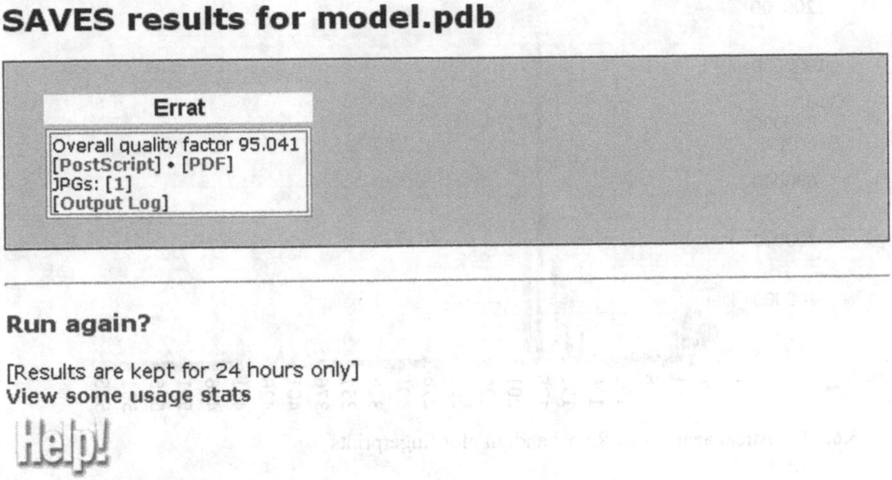

SAVES results for model.pdb
Errat
Overall quality factor 95.041
[PostScript] • [PDF]
JPGs: [1]
[Output Log]
Run again?
[Results are kept for 24 hours only]
View some usage stats
Help!

Fig. 5.60 Results for ERRAT

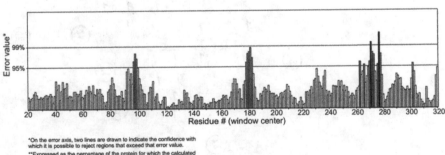

Program: ERRAT2
Chain#:1
Overall quality factor**: 95.041
Error value*
99%
95%
20
40
60
80
100
120
140
160
180
200
220
240
260
280
300
320
Residue # (window center)
*On the error axis, two lines are drawn to indicate the confidence with which it is possible to reject regions that exceed that error value.
**Expressed as the percentage of the protein for which the calculated error value falls below the 95% rejection limit. Good high resolution structures generally produce values around 95% or higher. For lower resolutions (2.5 to 3Å) the average overall quality factor is around 91%.

Fig. 5.61 The results of ERRAT in PDF format

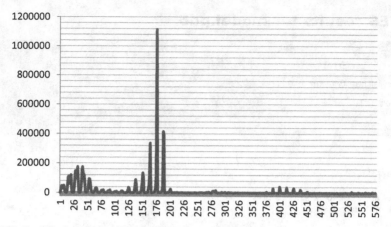

Fig. 5.62 Statistical analysis of Ramchandran plot fingerprints

Fig. 5.63 Active site ligand
fingerprints developed
by LIGBIT program for a
molecule

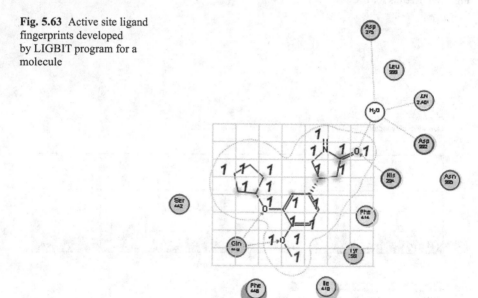

5.10   Thumb Rules for Active Site Identification and Homology Modelling

- The initial target structure obtained from a crystallographic database always needs to be energy minimized to make it energetically reasonable.
- If the sequence similarity of the target and the template is below 30%, then opt for other methods like threading and ab initio

- The quality of the homology model has to be checked thoroughly before subjecting it to further modelling studies like docking

5.11 Do it Yourself Exercises

Perform the following tasks using the protein sequence of-galactopyranose mutase of Leishmania major organism.

a. Download the fasta sequence and perform BLAST.
b. Generate two good homology models using two different templates (use MOE software).
c. Validate the model and examine the structural differences between the two models.
d. Find the active sites in the two models (use SiteMap (Schrodinger) and Site-Finder (MOE)).
e. Examine the differences between the active sites generated within the model and also between the two models (e.g. residues involved, volume of cavity, nature of residues, number of residues, etc.).

5.12 Questions

Q1. What is an active site? Discuss its importance in drug designing efforts.
Q2. What are different methods used to find the active site in a protein structure? Give one example of each.
Q3. What is sequence structure gap? Explain how homology modelling helps to bridge that gap?
Q4. Why is it always necessary to validate a homology model?
Q5. What information can be inferred from a Ramchandran plot of a protein?

References

1. Dai T, Liu Q, Gao J, Cao Z, Zhu R (2011) A new protein-ligand binding sites prediction method based on the integration of protein sequence conservation information. BMC Bioinformatics, 12(Suppl 14):9
2. Jain T, Jayaram B (2005) An all atom energy based computational protocol for predicting binding affinities of protein–ligand complexes. FEBS Lett 579:6659–6666
3. Wass MN, Sternberg MJE (2009) Prediction of ligand binding sites using homologous structures and conservation at CASP8. Proteins 77(Suppl 9):147–151
4. Henschel A, Winter C, Kim WK, Schroeder M (2007) Using structural motif descriptors for sequence-based binding site prediction. BMC Bioinforma 8(Suppl 4):5

5. Schmidt MR, Stansfeld PJ, Tucker SJ, Sansom MS (2013) Simulation-based prediction of phosphatidylinositol 4,5-bisphosphate binding to an ion channel. Biochemistry 52(2):279–281
6. Wang X, Mi G, Wang C, Zhang Y, Li J, Guo Y, Pu X, Li M (2012) Prediction of flavin mononucleotide binding sites using modified PSSM profile and ensemble support vector machine. Comput Biol Med 42(11):1053–1059
7. Hendlich M, Rippmann F, Barnickel G (1997) LIGSITE: automatic and efficient detection of potential small molecule-binding sites in proteins. J Mol Graph Model 15(6):359–363
8. http://www.modelling.leeds.ac.uk/pocketfinder/help.html
9. Brylinski M, Skolnick J (2008) A threading-based method (FINDSITE) for ligand-binding site prediction and functional annotation. Proc Natl Acad Sci 105:129–134
10. http://sts.bioengr.uic.edu/castp/
11. http://www.schrodinger.com/
12. http://www.chemcomp.com/
13. http://accelrys.com/products/discovery-studio/structure-based-design.html
14. Bitetti-Putzer R, Joseph-McCarthy D, Hogle JM, Karplus M (2001) Functional group placement in protein binding sites: a comparison of GRID and MCSS. Comput Aided Mol Des 15(10):935–960
15. Laurie ATR, Jackson RM (2006) Methods for the prediction of protein-ligand binding sites for structure-based drug design and virtual ligand screening. Curr Protein Pept Sci 7(5):395–406
16. Zhang Z, Li Y, Lin B, Schroeder M, Huang B (2011) Identification of cavities on protein surface using multiple computational approaches for drug binding site prediction. Bioinformatics 27(15):2083–2088
17. http://www.modelling.leeds.ac.uk/qsitefinder/
18. Laurie ATR, Jackso RM (2005) Q-SiteFinder: an energy-based method for the prediction of protein–ligand binding sites. Structural bioinformatics 21(9):1908–1916
19. http://jmol.sourceforge.net/
20. Krieger E, Nabuurs SB, Vriend G (2003) Homology modeling In: Bourne PE, Weissig H (eds) Structural Bioinformatics. Wiley, Liss, pp 507–521
21. Marti-Renom MA, Stuart AC, Fiser A, Sanchez R, Melo F, Sali A (2000) Comparative protein structure modeling of genes and genomes. Annu Rev Biophys Biomol Struct 29:291–325
22. Parulekar RS, Barage SH, Jalkute CB, Dhanavade MJ, Fandilolu PM, Sonawane KD (2013) Homology modeling, molecular docking and DNA binding studies of nucleotide excision repair uvrc protein from M. tuberculosis. Protein 32(6):467–476
23. Cashman DJ, Ortega DR, Zhulin IB, Baudry J (2013) Homology modeling of the CheW coupling protein of the chemotaxis signaling complex. PLoS One 8(8):e70705
24. Pang C, Cao T, Li J, Jia M, Zhang S, Ren S, An H, Zhan Y (2013) Combining fragment homology modeling with molecular dynamics aims at prediction of Ca^{2+} binding sites in CaBPs. J Comput Aided Mol Des (in press)
25. Wang P, Zhu BT (2013) Usefulness of molecular modeling approach in characterizing the ligand-binding sites of proteins: experience with human PDI, PDIp and COX. Curr Med Chem (in press)
26. Holtje H-D, Sippl W, Rognan D, G Folkers (2008) Molecular modeling. Wiley, Weinheim
27. Tramontano A (1998) Homology modeling with low sequence identity. Methods 14(3):293–300
28. Brylinski M (2013) eVolver: an optimization engine for evolving protein sequences to stabilize the respective structures. BMC Res Notes 6:303. doi:10.1186/1756-0500-6-303
29. Zhang Y (2013) Interplay of I-TASSER and QUARK for template-based and ab initio protein structure prediction in CASP10. Proteins (in press)
30. Kihara D, Chen H, Yifeng D Yang YD (2009) Quality assessment of protein structure models. Curr Protein Pept Sci 10:216–228
31. http://www.chemcomp.com/journal/provalid.htm
32. http://www.ncbi.nlm.nih.gov/

33. Eswar N, Marti-Renom, MA, Webb B, Madhusudhan, MS, Eramian, D, Shen, Pieper MU, Sali A (2006) Comparative protein structure modeling with MODELLER. In: Coligan JE, Dunn BM, Speicher DW, Wingfield PT (eds) Current protocols in bioinformatics. Wiley, New York

34. http://web.expasy.org/groups/swissprot/

35. http://www.uniprot.org/

36. http://www.ebi.ac.uk/thornton-srv/software/PROCHECK/

37. http://swift.cmbi.ru.nl/servers/html/index.html

38. http://nihserver.mbi.ucla.edu/SAVES/

39. http://nihserver.mbi.ucla.edu/ERRATv2/

40. http://nihserver.mbi.ucla.edu/Verify_3D/

41. Wermuth CG, Ganellin CR, Lindberg P, Mitscher LA (1998) Glossary of terms used in medicinal chemistry (IUPAC Recommendations 1998) Pure Appl Chem 70 (5):1129–1143

42. Chen J, Lai L (2006) Pocket v.2: further developments on receptor-based pharmacophore modeling. J Chem Inf Model 46(6):2684–2691

43. Deng J, Lee KW, Sanchez T, Cui M, Neamati N, Briggs JM (2005) Dynamic receptor-based pharmacophore model development and its application in designing novel HIV-1 integrase inhibitors. J Med Chem 48(5):1496–1505

44. Ebalunode JO, Dong X, Ouyang Z, Liang J, Eckenhoff RG, Zheng W (2009) Structure-based shape pharmacophore modeling for the discovery of novel anaesthetic compounds. Bioorganic Med Chem 49(10):2333–2343

45. Dror O, Schneidman-Duhovny D, Inbar DY, Nussinov R, Wolfson HJ (2002) A novel approach for efficient pharmacophore-based virtual screening: method and applications. J Mol Biol 324:105–121

46. Bartlett GJ, Porter CT, Borkakoti N, Thornton JM (2002) Analysis of catalytic residues in enzyme active sites. J Mol Biol 324(1):105–121

47. Cuneo MJ, Beese LS, Hellinga HW (2008) Ligand-induced conformational changes in a thermophilic ribose-binding protein. BMC Struct Biol 8:50

48. Gliubich F, Gazerro M, Zanotti G, Delbono S, Bombieri G, Berni R (1996) Active site structural features for chemically modified forms of rhodanese. J Biol Chem 271(35):21054–21061

49. Li S, Hall MB (2001) Modeling the active sites of metalloenzymes. 4. predictions of the unready states of [NiFe] desulfovibrio gigas hydrogenase from density functional theory. Inorg Chem 40(1):18–24

50. Kumar A, Chaturvedi V, Bhatnagar S, Sinha S, Siddiqi MI (2009) Knowledge based identification of potent anti-tubercular compounds using structure based virtual screening and structure interaction fingerprints. J Chem Inf Model 49(1):35–42

51. Desaphy J, Azdimousa K, Kellenberger E, Rognan D (2012) Comparison and druggability prediction of protein-ligand binding sites from pharmacophore-annotated cavity shapes. J Chem Inf Model 52(8):2287–2299

52. Zhang N, Li B-Q, Gao S, Ruan J-S, Cai Y-D (2012) Computational prediction and analysis of protein [gamma]-carboxylation sites based on a random forest method. Mol Bio Syst 8:2946–2955

53. Somarowthu S et al (2011) High-performance prediction of functional residues in proteins with machine learning and computed input features. Biopolymers 95:390–400

54. Ewing TJA, Kuntz ID (1997) Critical evaluation of search algorithms for automated molecular docking and database screening. J Comp Chem 18:1175–1189

55. Marx D, Hutter J (2000) Ab initio molecular dynamics: theory and Implementation In: Grotendorst J (ed.) Modern methods and algorithms of quantum chemistry. John von Neumann Institute for Computing, Jülich, pp 301–449

56. http://www.rcsb.org/pdb/home/home.do

57. http://cheminfo.u-strasbg.fr:8080/scPDB/2012/db_search/acceuil.jsp?uid=5563865723544052736

58. http://scop.mrc-lmb.cam.ac.uk/scop/

59. http://www.nvidia.com/object/tesla-supercomputing-solutions.html

60. Meslamani J, Rognan D, Kellenberger E (2011) sc-PDB: a database for identifying variations and multiplicity of 'druggable' binding sites in proteins. Bioinformatics 27(9):1324–1326
61. Unpublished results
62. Wenying Y, Hui X, Jiayuh L, Chenglong L (2013) Discovery of Novel STAT3 small molecule inhibitors via in silico site-directed fragment-based drug design. J Med Chem 56(11):4402–4412
63. Putta S, Lemmen C, Beroza P, Greene J (2002) A novel shape-feature based approach to virtual library screening. J Chem Information Comp Sci 42(5):1230–1240

Chapter 6
Representation, Fingerprinting, and Modelling of Chemical Reactions

Abstract Designing a better molecule is just one aspect of computational research, but getting it synthesized for biological evaluation is the most significant component in a drug discovery program. A molecule can be formed by a number of synthetic routes. Manually keeping track of all the available options for a product formation in various reaction conditions is a herculean task. Chemoinformatics comes to the rescue by providing a number of computational tools for reaction modelling, albeit less in number than structure property prediction software. The current computational tools help us in modelling various aspects of a given organic reaction—synthetic feasibility, synthesis planning, transition state prediction, the kinetic and thermodynamic parameters, and finally mechanistic features. Several methods like empirical, semiempirical, quantum mechanical, quantum chemical, machine learning, etc. have been developed to model a reaction. The computational approaches are based on the concept of rational synthesis planning, retro-synthetic approaches, and logic in organic synthesis. In this chapter, we begin with reaction representation in computers, reaction databases, free and commercial reaction prediction programs, followed by reaction searching methods based on ontologies and reaction fingerprints. The commonly employed quantum mechanics (QM) and quantum chemistry (QC)-based methods for intrinsic reaction coordinate (IRC) and transition state (TS) determination using the B3LYP/6–31G* scheme are described using simple name reactions. Most of the computational reaction prediction programs such as CHAOS/CAOS are based on the identification of the strategic bonds which are likely to be cleaved or formed during a certain chemical transformation. Accordingly, an algorithm has been developed to identify more than 300 types of unique bonds occurring in chemical reactions. The effect of implicit hydrogens on chemical reactivity modelling is discussed in the context of bioactivity spectrum for structure–activity relationship studies. Other parameters affecting reactivity such as solvent polarity, thermodynamics etc. are also briefly highlighted for frequently used name reactions, hazardous high-energy reactions, as well as industrially important reactions involving bulk chemicals.

Keywords Chemical reaction modelling · Chemoinformatics · Retro-synthesis · Artificial intelligence · Ontologies

M. Karthikeyan, R. Vyas, *Practical Chemoinformatics,* 317
DOI 10.1007/978-81-322-1780-0_6, © Springer India 2014

6.1   Introduction

Synthesis of new molecules involves general chemical reactions, for example, oxidation, reduction, esterification, hydrolysis, etc., which constitute important biochemical processes for sustaining life. Typically, there are nine trillion amazing reactions per cell per day taking place in the human body [1]. A simple process as breathing, in human aerobic respiration, requires a host of chemical reactions, the key reaction being succinic acid dehydrogenase (SDH) enzyme-catalyzed removal of hydrogen atoms from succinic acid, the substrate in the Krebs cycle [2]. Another important chemical reaction in nature is photosynthesis, the most critical reaction on the planet for production of chemical energy [3]. Today, a chemical biologist can design selective chemical coupling reactions that proceed in cells without affecting cellular chemistry to understand the chemical mechanism in biological systems [4]. Knowledge about reactions in signaling pathways helps us in understanding cellular communication better [5]. A holistic study of all these reactions provides a broad perspective on many areas of active research at the interface of chemistry and biology, such as understanding the effects of drug administration on biological systems [6]. Chemical reactions are carried out essentially via functional groups attached to the carbon backbone of organic molecules and their interconversions [7]. Functional groups are the reaction centers in a molecule and determine its characteristics including chemical reactivity. The common biologically important functional groups are hydroxyl, carboxyl, carbonyl, amine, ester, amide, disulfide, and phenyl.

6.2   Reaction Representation in Computers

A reaction is a collection of reagents, products, and agents. It is represented by a reaction data file (RDF) or a reaction file (Rxn) [8]. The reagent, product and agent elements are molecule objects embedded into a reaction data file. The type of an element is defined by its relative position with regard to the reaction arrow. Here, let us take the example of Diels–Alder reaction which is a simple 4+2 cycloaddition reaction of an alkene and a conjugated diene to form a cyclohexene ring system ([9]; Figs. 6.1, and 6.2).

6.3   Computational Methods in Reaction Modelling

Computational chemistry provides a host of methods for molecular modelling of reactions [10]. A brief overview of these methods is provided here for a clear understanding of their underlying basic differences. The references cited in this section should be referred to for obtaining an in-depth analysis.

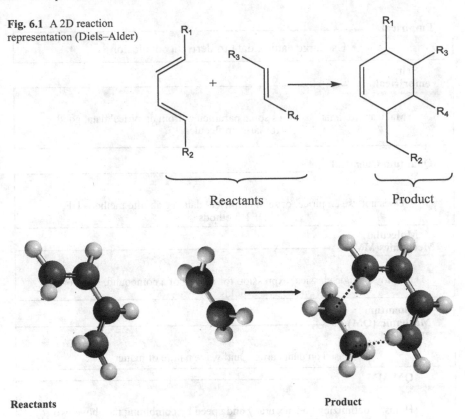

Fig. 6.1 A 2D reaction representation (Diels–Alder)

Fig. 6.2 A 3D Diels–Alder reaction representation (view generated using Spartan)

6.3.1   *Empirical and Semiempirical Methods*

Semiempirical calculations are set up with the same general structure as a Hartree–Fock calculation. Within this framework, certain pieces of information, such as two-electron integrals, are approximated or completely omitted. In order to correct the errors introduced by omitting these parts of the calculation, the method is parameterized, by curve fitting in a few parameters or numbers, in order to give the best possible agreement with experimental data. The semiempirical calculations are much faster than the ab initio calculations; however, the results can sometimes be erratic. If the molecule under study is similar to molecules in the database used to parameterize the method, then the results may be very good. If this molecule is significantly different from anything in the parameterization set, the answers may be poor. Semiempirical calculations have been very successful in computational organic chemistry, where only a few elements are used extensively and the molecules are of medium size (Fig. 6.3).

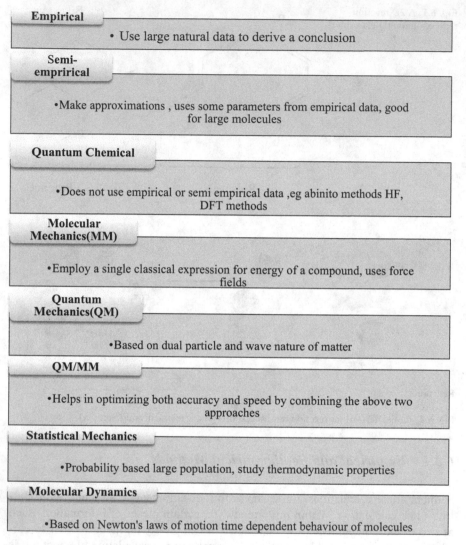

Empirical
• Use large natural data to derive a conclusion

Semi-emprirical
•Make approximations , uses some parameters from empirical data, good for large molecules

Quantum Chemical
•Does not use empirical or semi empirical data ,eg abinito methods HF, DFT methods

Molecular Mechanics(MM)
•Employ a single classical expression for energy of a compound, uses force fields

Quantum Mechanics(QM)
•Based on dual particle and wave nature of matter

QM/MM
•Helps in optimizing both accuracy and speed by combining the above two approaches

Statistical Mechanics
•Probability based large population, study thermodynamic properties

Molecular Dynamics
•Based on Newton's laws of motion time dependent behaviour of molecules

Fig. 6.3 Computation approaches for modelling of chemical reactions

6.3.2   Molecular Mechanics Methods

Molecular mechanics methods are good for modelling big molecule systems where it is computationally expensive to employ quantum mechanics. These methods employ a molecular force field which is   potential energy as a function of all atomic positions. It is used to study the molecular properties without any need for computing a wave function or total electron density [11]. The force field expression consists of simple classical equations, such as the harmonic oscillator equation to describe the energy associated with bond stretching, bending, rotation, and intermolecular forces, such as van der Waals interactions and hydrogen bonding of a molecule. All the constants in these equations are obtained from experimental data or an ab initio calculation. In a molecular mechanics method, the database of compounds used to parameterize the method is crucial to its success. A semiempirical method may be parameterized against a specific set of molecules but a molecular mechanics method is parameterized against a specific class of molecules, such as proteins [12]. As molecular mechanics can model enormous molecules, such as proteins and segments of DNA, it is the primary tool of computational biochemists. However, there are many chemical properties that are not defined within the method, such as treatment of electronic excited states. Generally the molecular mechanics software packages have the powerful and easy to use graphical interfaces. Because of this, mechanics is often used because it is easy, even though it may not be a good way to completely describe a system.

6.3.3 Molecular Dynamics Methods

Molecular dynamics consists of examining the time-dependent characteristics of a molecule, such as vibrational motion or Brownian motion within a classical mechanical description [13]. Molecular dynamics when applied to solvent/solute systems allow the computation of properties such as diffusion coefficients or radial distribution functions for use in statistical mechanical treatments. In this calculation a number of molecules are given some initial position and velocity. New positions are calculated a short time later based on this movement, and the process is iterated for thousands of steps in order to bring the system to an equilibrium. Next the data are Fourier transformed into the frequency domain. A given peak can be chosen and transformed back to the time domain, to see the motion at that frequency.

6.3.4 Statistical Mechanics and Thermodynamics

Statistical mechanics is the mathematical means to extrapolate thermodynamic properties of bulk materials from a molecular description of the material [14]. Statistical mechanics computations are often performed at the end of ab initio calculations for gas-phase properties. For condensed-phase properties, often molecular dynamics calculations are necessary in order to do a computational experiment. Thermodynamics is one of the best-developed physical theories and it gives a good theoretical starting point for the analysis of molecular systems.

6.3.5   The Quantum Mechanical/molecular Mechanical Approach

Proper description of a chemical reaction requires a quantum mechanical (QM) treatment, as electronic rearrangements are involved. The basic idea of combining QM and molecular mechanical (MM) potentials into a hybrid QM/MM description of enzymes was developed in the pioneering study of lysozyme by Warshel and Levitt in 1976 [15]. Warshel and Levitt recognized that QM calculations, especially at that time, were feasible only for small chemical systems, enzymatic reactions generally, represent a small fraction and thus an oversimplified model of the real enzyme–substrate system. The region further away from the reacting groups provides mainly conformational and nonbonded contributions. These contributions can be adequately described by (classical) molecular mechanics (MM) and the electrostatic interaction of classical particles with the reacting QM system. Therefore, the system can be divided into a small region around the active site to be described quantum mechanically, while the surrounding protein can be adequately represented by simpler molecular mechanics. Thus, the total energy of the whole system (i.e. the enzyme as well as surrounding solvent) could be decomposed into

$$V = \text{Classical} + \text{Quantum} + \text{Quantum/classical},$$

where the first two terms describe the MM and QM regions and the latter term represents the interactions between the two. With the development of better computers, quantum chemical methods, and MM force fields for proteins, the usefulness of this principle became widely recognized and QM/MM methods were further developed. Singh and Kollman presented a QM/MM method in 1986, based on ab initio QC (Gaussian 03) [16].

6.3.6   Modelling the Transition State of Reactions

The transition state (TS) of a biological reaction or a chemical reaction is a particular configuration along the reaction coordinate (bond length or bond angle). It is defined as *the state corresponding to the highest potential energy along the reaction coordinate*. It is also referred to as saddle point. At this point, energy is higher and the reaction is perfectly irreversible (Fig. 6.4).

The activated complex of a reaction can refer to either the TS or other states along the reaction coordinate between reactants and products, especially those which are close to the TS. A collision between reactant molecules may or may not result in a successful reaction. The outcome depends on factors such as the relative kinetic energy, relative orientation, and internal energy of the molecules. Even if the collision partners form an activated complex, they are not bound to go on

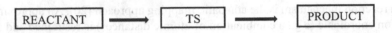

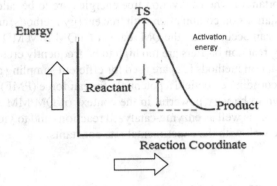

Fig. 6.4 First-order saddle point is transition state between two local minima (for example, reactant and product of a chemical reaction)

Fig. 6.5 Transition state of a reaction

and form products, and instead the complex may fall apart back to the reactants (Fig. 6.5).

TS structures can be determined by searching for first-order saddle points on the potential energy surface (PES). Such a saddle point is a point where there is a minimum in all dimensions but one. Almost all quantum chemical methods can be used to find TS. However, locating them is often difficult and there is no method guaranteed to find the right TS. There are many different methods of searching for TS and different QC program packages include different ones. Many methods of locating TS also aim to find the minimum energy pathway (MEP) along the PES. Each method has its advantages and disadvantages depending on the particular reaction under investigation.

To characterize a reaction pathway on a potential energy surface, in principle, the reaction intermediates (minima) and the TS (saddle points) connecting those intermediates need to be identified. A common approach in gas-phase reaction modelling is to optimize the relevant TS and perform a subsequent calculation of the intrinsic reaction coordinate (IRC) [17] towards the intermediates on both sides of the barrier. Vibrational analysis of the intermediates and TS can be used to derive thermodynamic contributions to the energetics of the reaction. However, for the larger QM/MM models, these methods are generally too expensive or impractical [18]. The TS optimization is based on the Hessian for the core degrees of freedom only while the "environment" is kept at its minimum at every TS optimization step. This method has successfully been applied to analyze enzyme reactions [19]. A very efficient, but more approximate, method to scan the potential energy surface for a given reaction mechanism is the adiabatic mapping approach [20]. An approximate reaction coordinate (e.g., a combination of atomic distances) is restrained and used to drive the system stepwise from reactants to products. Simple geometry optimization is performed at every step. This method has proven to be very useful in calculations on enzymes. Other approximate methods have been developed which optimize an entire pathway as a whole, involving multiple intermediates and TS and without expensive calculations of second derivatives [21]. The methods listed above are very useful when a single protein conformation is expected to adequately represent the reacting enzyme. When more extensive conformational sampling is important or when activation free energies are to be calculated molecular dynamic simulations, in combination with free energy methods, are required. In theory, reactions can occur within the QM region of a QM/MM MD simulation, but in practice, many reaction barriers are too high to be frequently crossed. Therefore, free energy simulation methods [22] are used for efficient sampling along an approximate reaction coordinate, to yield a potential of mean force (PMF). These methods have been shown to be very powerful in the context of QM/MM simulations of reactions in solution as well as enzyme-catalyzed reactions and to yield free energy barriers that agree well with the experimental rate constants.

6.4   TS Modelling of Organic Transformations

Some organic transformations are frequently used in chemical synthesis in both laboratory and industry. These are termed as name reactions [23]. Here, we will discuss a few important name reactions and provide detailed reaction modelling steps for the Diels–Alder reaction which is a typical carbon–carbon bond-forming cycloaddition transformation that proceeds with high stereocontrol.

6.4.1   Name Reactions

Aldol Reaction (Condensation) [24] Traditionally, it is the acid- or base-catalyzed condensation of one carbonyl compound with the enolate/enol of another, which may or may not be the same, to generate a β-hydroxy carbonyl compound—an aldol. The method is composed of self-condensation, polycondensation, generation of regioisomeric enols/enolates, and dehydration of the aldol followed by Michael addition, q.v. The development of methods for the preparation and use of preformed enolates or enol derivatives that dictate specific carbon–carbon bond formation have revolutionized the coupling of carbonyl compounds (Fig. 6.6):

Cope rearrangement [25] The highly stereoselective [3, 3] sigmatropic rearrangement of 1,5 dienes is called as Cope rearrangement. When the R group is an alcohol, it is called as oxy-Cope rearrangement (Fig. 6.7).

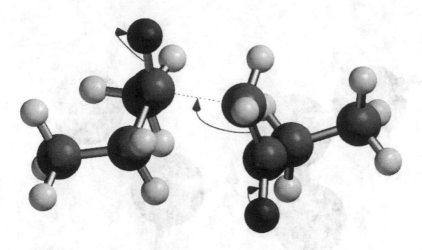

Fig. 6.6 Aldol condensation: transition state of two butanone determined using C1_AM1 method

Claisen Condensation (Acetoacetic Ester Condensation) [26] Base-catalyzed condensation of an ester containing an α-hydrogen atom with a molecule of the same ester or a different one to give β-keto esters (Fig. 6.8):

$$CH_3COOC_2H_5 + CH_3COOC_2H_5 \xrightarrow{C_2H_5O^-} CH_3COCH_2COOC_2H_5 + C_2H_5OH$$

Chugaev elimination [27] This reaction involves the formation of alkenes through pyrolysis of the corresponding xanthates via *cis* elimination:

There is no rearrangement of the carbon skeleton of the substrate molecules. Mechanistic studies have revealed a concerted cyclic mechanism. It is considered a very useful reaction for transformation of alcohols to olefins (Fig. 6.9).

Markovnikov addition reaction [28] This reaction involves addition of a protic acid HX to an olefin wherein the acidic hydrogen adds to the carbon with fewer alkyl substituents and the halide becomes attached to the carbon with more alkyl substituents, the mechanistic reason being the formation of a stable carbocation during the addition process (Fig. 6.10):

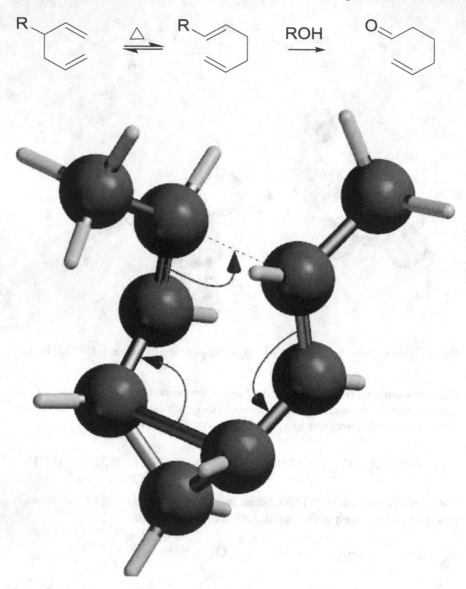

Fig. 6.7 Cope rearrangement: transition state of Cope rearrangement of *cis*-dipropenyl cyclopropane generated with AM1 method

6.4.2 *A Practice Tutorial for Transition State and Intrinsic Reaction Coordinate Modelling*

Here, we will demonstrate the steps for modelling the Diels–Alder reaction using the Gaussian software.

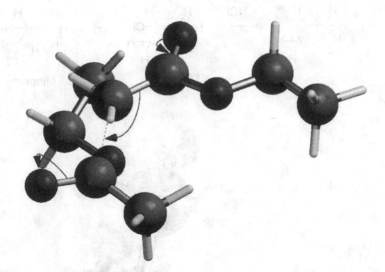

Fig. 6.8 Claisen condensation: transition state of ethyl acetate generated using AM1 method in Spartan

Fig. 6.9 Chugaev elimination: transition state of xanthate ester generated using the AM1 method

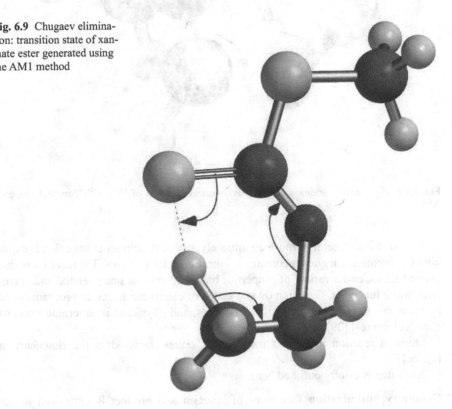

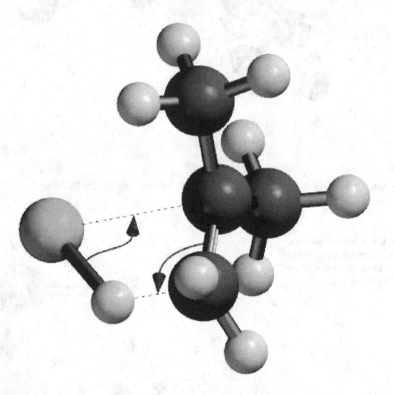

Fig. 6.10 Markovnikov addition: TS of Markovnikov addition of HCl with 2-methyl propene AM1

Gaussian is a general purpose ab initio electronic structure package that is capable of computing energies, geometries, vibrational frequencies, TS, reaction paths, excited states, and a variety of properties based on various uncorrelated and correlated wave functions. Gaussian 09 is a series of electronic structure programs used by chemists, chemical engineers, biochemists, and physicists in emerging areas of chemical interest [29].

Steps in reaction modelling involve the stages depicted in the flowchart in Fig. 6.11.

Each step is briefly outlined here.

Geometry optimization Geometry of reactant and product is optimized to get equilibrium geometry. The energy is obtained at minima, that is, minimum energy

Fig. 6.11 General steps for reaction modelling

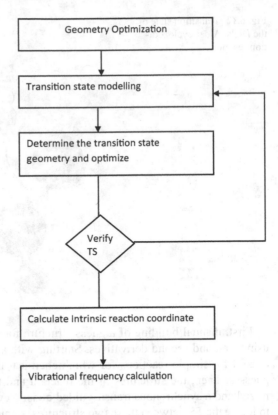

conformation. This is done to adjust the bond length and angles according to the standards mentioned. This is then used for TS calculation.

TS reaction modelling TS corresponds to the saddle point on the potential energy surface. Like minima, saddle points are stationary points with all forces zero. Unlike minima, one of the second derivatives in the saddle point is negative. The negative eigen value corresponds to the reaction coordinate. TS search thus locates points having one negative eigen value. The first thing in TS search is to identify the reaction mode and maximize energy along this mode, while minimizing energy in all other directions. (Fig. 6.12).

TS is a state in the course of reaction when one bond breaks and a new bond forms. This state is imaginary which cannot be isolated. TS cannot be found in an experiment as it is short lived, so it is not possible to view how the TS looks like experimentally. In quantum chemical reaction modelling detailed information about the geometry of TS and other physical properties associated with the TS can be found and also activation energy and activation entropy can be calculated which tells us that the energy of TS is more than the reactant.

Calculation of TS geometry and optimization Here are some approaches to locate TS structures for chemical reactions:

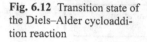

Fig. 6.12 Transition state of the Diels–Alder cycloaddition reaction

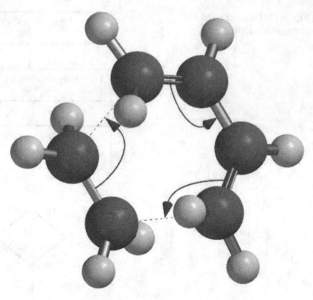

First, manual building of a guess structure for the TS and optimization is done using first and second derivatives. Starting with a guess TS structure is often successful for simple reactions for which chemical intuition provides reasonable TS guesses. Then, the structures of the reactant and the product were built and optimized and a synchronous transit-guided quasi-Newton approach (QST2) was used to locate the TS between these two structures. Again, structures of the reactant complex, the product complex, and a guess for the TS were built, and a synchronous transit-guided quasi-Newton approach (QST3) was used to optimize the TS. Then, reaction path was scanned to identify saddle points.

The scanning approach is effective when there is only one reaction coordinate, as in the case of transitions between conformational isomers.

Verifying calculated TS geometry There are two ways to verify that a particular geometry corresponds to a saddle point (TS) and further this saddle point connects potential energy minima corresponding to reactant and product. We should verify that the Hessian matrix (matrix of second-energy derivatives with respect to coordinates) yields one and only one imaginary frequency. For this, it is required that vibrational frequency be obtained for the proposed TS. Frequency calculation should be carried out using the same method that was used to find TS. The imaginary frequency will be in the range of 400–2,000 cm^{-1}. We should verify that a normal coordinate corresponding to imaginary frequency connects reactant and product; this can be done by animating the normal coordinate corresponding to imaginary frequency. Optimization subject to fixed position on the reaction coordinate can be done by IRC; this is the pathway linking reactant, TS, and product together.

IRC calculation TS geometry may be connected to the ground state geometry by IRC calculation. In this path followed moving from TS towards product in the forward direction and from TS towards reactant in the reverse direction.

```
chk=opt_r.chk                           checkfile name
%mem=6MW
%nproc=1
# opt pm3 geom=connectivity                        command section

Title Card Required
                                      Charge and spin
0 1
 C
 O                    1            B1
 H                    2            B2    1         A1
 C                    1            B3    2         A2    3        D1
 H                    4            B4    1         A3    2        D2
 H                    4            B5    1         A4    2        D3
 O                    4            B6    1         A5    2        D4
 H                    7            B7    4         A6    1        D5
 O                    1            B8    2         A7    4        D6

    B1              1.43000000
    B2              0.96000000
    B3              1.54000000
    B4              1.07000000
    B5              1.07000000
    B6              1.43000000         geometry specification
    B7              0.96000000
    B8              1.25840000
    A1            109.50000006
    A2            119.88652694
    A3            109.47120255
    A4            109.47120255
    A5            109.47123134
    A6            109.50000006
    A7            120.22694612
    D1             -0.11110000
    D2             -0.00460740
    D3            119.99540740


 1 2 1.0 4 1.0 9 2.0
 2 3 1.0
 3
 4 5 1.0 6 1.0 7 1.0
 5
```

Fig. 6.13 Format of a gif file showing details of the molecular structure and the connectivity

TS and reactant structures from AM1, 3–21G, and 6–31G* calculations have been used for activation energy calculations.

Vibrational frequency calculation This is done to verify if the TS structure is correctly modeled or not as TS is found at negative imaginary frequency and negative eigen value.

6.4.2.1   IRC Calculation Using Gaussian Program

Gaussian basically takes ".gjf" files as input which mainly has the structures of the compounds. The structure of a typical gif file is highlighted in Fig. 6.13 (Figs. 6.14, and 6.15).

If we open the ".gjf" file in Notepad/WordPad, then the details of the structure along with the connectivity table and coordinates appear.

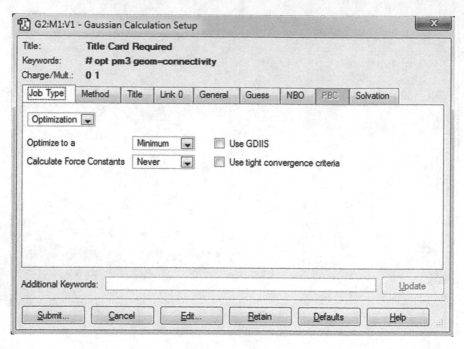

Fig. 6.14 A snapshot of reactant optimization calculation

Fig. 6.15 Representation of a
compound in the ".gif" file in
Gauss View

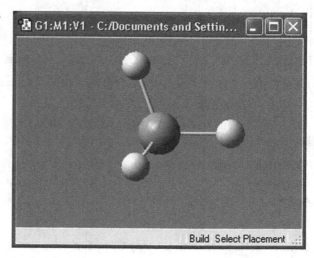

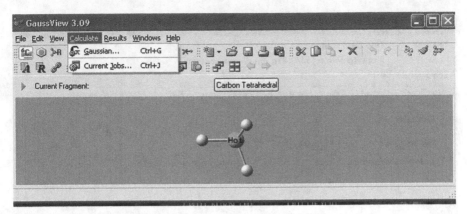

Fig. 6.16 Gaussian calculation is carried out from the Calculate tab

Following are some Gaussian keywords which are used during the calculation of TS:

1. **Opt = TS:** This is used for optimization to a TS rather than a local minimum using the Berny's optimization method.
2. **QST2:** This requires the reactant and product structures as input, specified in two consecutive groups. This mainly generates a guess for the transition structure that is something midway between the reactants and products.
3. **QST3:** This searches for a transition structure using the synchronous transit-guided quasi Newton method, which is used for locating the transition structure. This mainly requires three molecules: reactants, products, and an initial structure for the TS.

Steps for calculating the TS for the Diels–Alder reaction with QST2:

1. Geometry optimization:

Gauss View program has to be used in order to perform geometry optimization of the reactants and products (1,3-butadiene, ethene, cyclohexane).

We perform three separate calculations for each molecule to carry out the geometry optimization:

a. First, create the structures of all the three compounds involved in the reaction, in the Gauss View program with the help of the toolbar in the Gauss View program.
b. To carry out geometry optimization, open any of the molecules drawn in the Gauss View program, go to the Calculate tab in the program, and select Gaussian (Fig. 6.16).

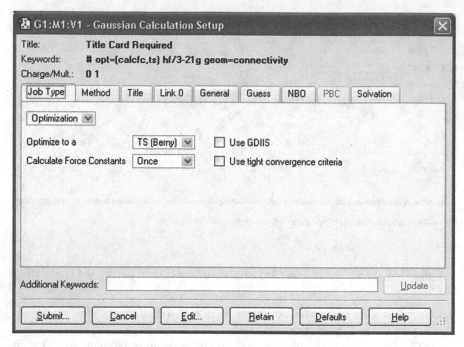

Fig. 6.17 A screen capture showing the Gaussian calculation setup

c. Select the Gaussian tab, a window appears in which we are supposed to set the parameters. In Job type, select optimization to a: TS (Berny) and force constants = Once; in method, select the appropriate method with the appropriate basis set, and click on the submit button (Figs. 6.17 and 6.18).

When the job is submitted, a window appears which monitors the job.

2. After the geometry optimization is done, further calculations can be carried out in Gauss View itself. To carry out the calculations, one should create a file which has two sheets. In the first sheet, one should have the reactants with appropriate distance and the other should have the product/s.

a. Open the optimized structure of 1,3-butadiene in Gauss View, press Edit, and select Copy
b. Press the File button, press the Edit button, and select Paste and Append Molecule and name it as QST2.
c. Open the optimized structure of ethane, press Edit, and select Copy.
d. Open the QST2 file, press Edit, and select Paste and Append Molecule.
e. Both the reactants are now in one sheet. Maintain a distance of 3A between the reactants.
f. Open the optimized cyclohexane structure along with the QST2 file. Highlight the cyclohexane structure, press Edit, and select Copy.

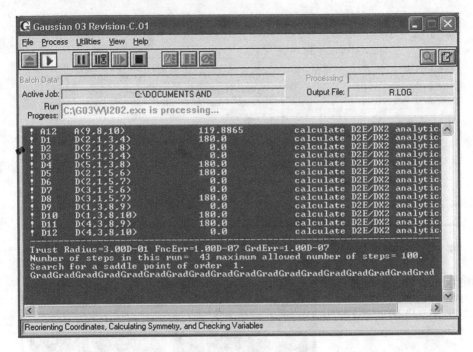

Fig. 6.18 The Gaussian process window

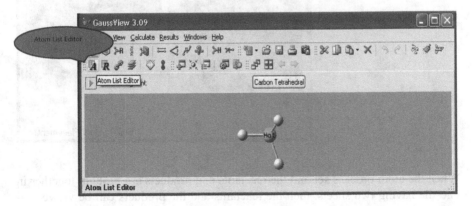

Fig. 6.19 The callout showing the Atom List Editor

g. Highlight the QST2 structure, press Edit, select Paste, and Add to Molecular group
h. Make sure that the order of the atoms in reactants and the products are the same. (Press the Atom List Editor button. A table displaying the Atom order appears). (Figs. 6.19 and 6.20)

G1:M1:V1 - Atom List Editor

File   Edit   Rows   Columns   Help

Highlight	#	Symbol	NA	NB	NC	Bond	Angle	Dihedral	X	Y	Z
⊙	1	C							0.000000	0.000000	0.000000
⊙	2	H	1			1.070000			1.070000	0.000000	0.000000
⊙	3	C	1	2		1.540000	119.886527		-0.767357	1.335201	0.000000
⊙	4	H	3	1	2	1.070000	119.886527	180.000000	-1.837357	1.335201	-0.000000
⊙	5	C	1	3	4	1.355200	119.886527	0.000000	-0.682243	-1.170944	-0.000000
⊙	6	H	5	1	3	1.070000	120.226946	0.000000	-1.752243	-1.170944	-0.000000
⊙	7	H	5	1	3	1.070000	119.886527	180.000000	-0.149080	-2.098649	-0.000000
⊙	8	C	3	1	5	1.355200	120.226946	180.000000	-0.092083	2.510179	0.000000
⊙	9	H	8	3	1	1.070000	120.226946	0.000000	0.977898	2.516536	0.000000
⊙	10	H	8	3	1	1.070000	119.886527	-180.000000	-0.630749	3.434700	0.000000
Add	.	?	8	3	1	2.511867	31.987411	0.000000	0.000000	0.000000	0.000000

Active Sublist Filters: None

Fig. 6.20 Atom List Editor window

Fig. 6.21 Reactants and products brought together in one file in two separate sheets

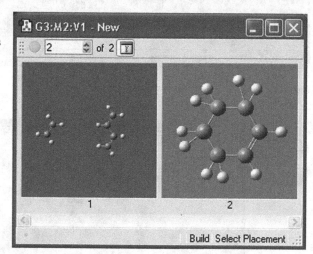

G3:M2:V1 - New

2 ⇕ of 2

1   2

Build Select Placement

When the atom order is set and the reactants and products are brought together in one file having two sheets, then the reactants and the products can be viewed together in one file itself in two different sheets (Fig. 6.21).

3. QST2 calculations:

   a. Highlight the QST2 file having both reactants and products, press Calculate, and select Gaussian.

   b. In the Job type, select Optimization to TS (QST2) (Fig. 6.22).

   c. In the method box, select Ground State, Restricted Hartree–Fock, with Basis set being set to 6-31G, Charge to 0, and spin multiplicity to Singlet (Fig. 6.23).

   d. Submit the calculation.

   e. Open the output file in Notepad and verify that the calculation terminated successfully and that convergence was accomplished.

   f. Open the output file in Gauss View in order to observe the TS of the reaction.

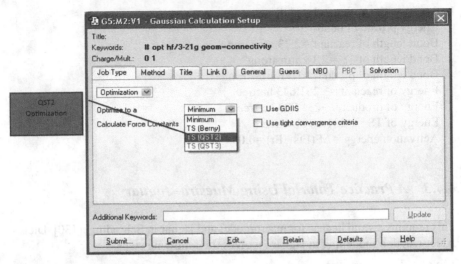

Fig. 6.22 Optimization using TS (QST2)

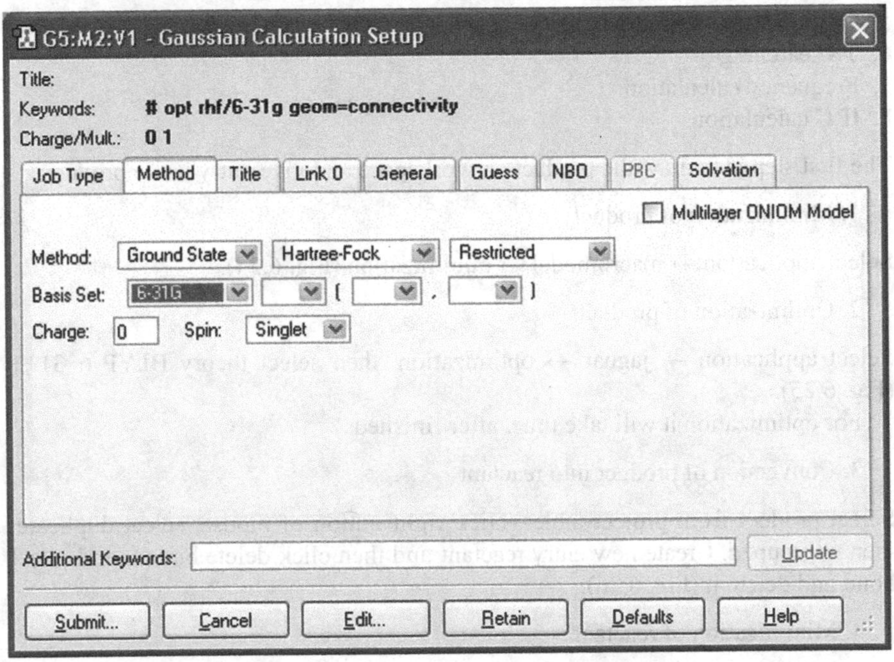

Fig. 6.23 Method dialog box

In this case, the optimum bond distance are-
Energy profile of bond-
Bond length of reactant $= 2.23$ armstrong
Bond length of TS $= 2.18$ armstrong
Bond length of product $= 1.51$ armstrong
Energy of reactant $= -231.643$ hartree
Energy of product $= -231.723$ hartree
Energy of TS $= -231.604$ hartree
Activation energy $= \Delta E(Ets - Er) = 0.039$ hartree

6.4.3 A Practice Tutorial Using Maestro–Jaguar

This requires a valid license for macromodel and jaguar in Schrodinger [30]. Diels–Alder reaction modelling in jaguar involves the following steps:

1. Minimization of product
2. Optimization of product
3. Conversion of product to reactant
4. Minimization of reactant
5. Optimization of product
6. TS searching
7. Frequency calculation
8. IRC calculation

The first step is to draw the product on workspace and give entry name product:

1. Minimization of product

Select application → macromodel → minimization (Fig. 6.24).

2. Optimization of product

Select application → jaguar → optimization, then select theory BLYP 6–31++ (Fig. 6.25).
For optimization it will take time, after finished.

3. Conversion of product into reactant

Select product from project table—click right button of mouse, select duplicate, then ungrouped. Create new entry reactant and then click delete button, and select bond and delete it (Fig. 6.26).

4. Minimization of reactant

Select application → macromodel → minimization.

5. Optimization of reactant

Select application → jaguar → optimization.

Select application - macromodel—minimization

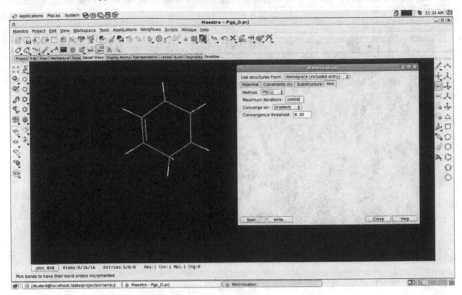

Fig. 6.24 A screen shot of the Jaguar module in Schrodinger

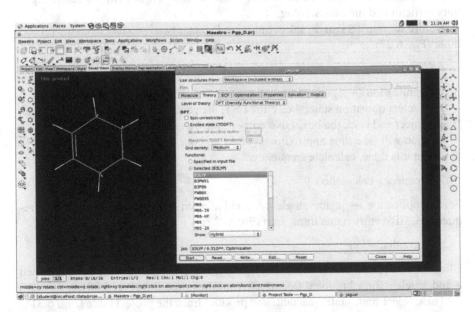

Fig. 6.25 Basis set selection in Jaguar

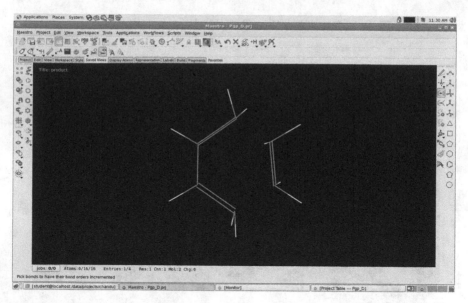

Fig. 6.26 Product selection in Jaguar

It will also take time, whenever it is finished, start TS search; TS searching is very difficult and time consuming.

Before starting TS search, select optimized reactant and product from the project table (Fig. 6.27).

6. TS searching

Select application → jaguar → transition state searching.

First select transition state, then click LST, then choose both reactant and product from project and click the box. Give entry name trans_state (Fig. 6.28).

The job will take time approximately half an hour depending on your system.

When it is done, calculate frequency.

7. Frequency calculation

Select application → jaguar-single point calculation→properties→vibrational frequencies. Give entry name trans_freq (Fig. 6.29).

8. IRC calculation

Click read button on the bottom of the jaguar panel and select trans-freq 01.in, open it and unselect the vibrational frequency from the properties (Fig. 6.30).

Then, select trans-state, reactant, and product from the project table (Fig. 6.31). Now

Select application → reaction coordinate → IRC and choose all three TS, reactant, and product. Here, there is no need to click on the box (Fig. 6.32).

When it is done, open the project table and see the reaction coordinate and energy and make a reaction plot.

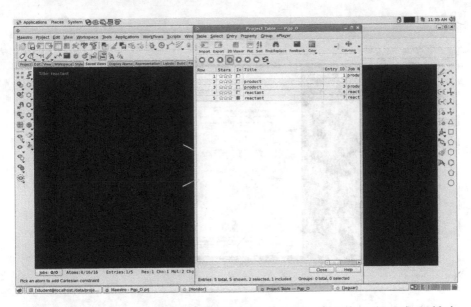

Fig. 6.27 Step showing selection of the optimized product and reactant from the project table in Jaguar

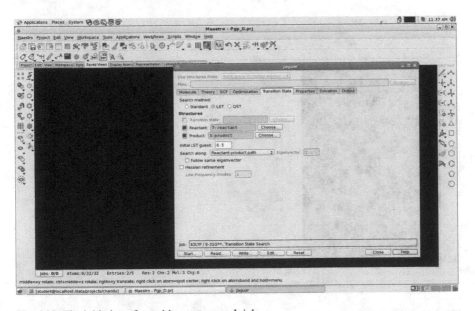

Fig. 6.28 The initiation of transition state search job

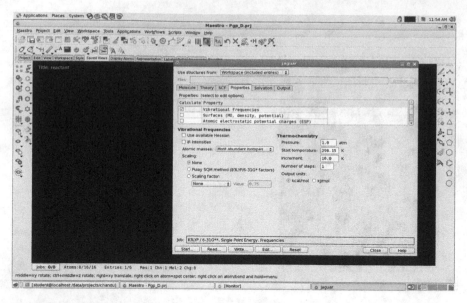

Fig. 6.29 Property calculation option in Jaguar, here vibrational frequencies will be computed

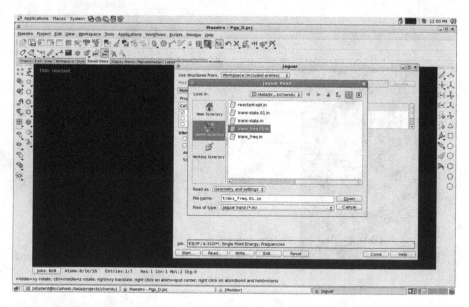

Fig. 6.30 Selection of the transition state-frequency input file

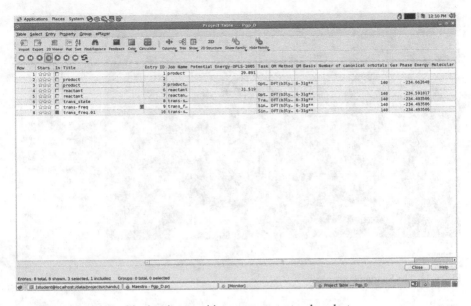

Fig. 6.31 The project table showing transition state, reactant and product

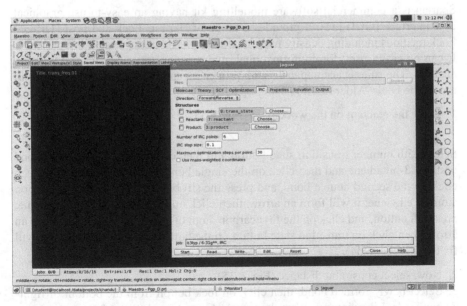

Fig. 6.32 Screen shot showing selection of the IRC (Intrinsic reaction coordinate)

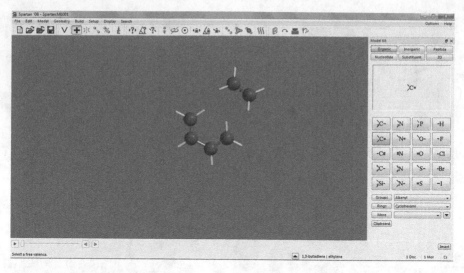

Fig. 6.33 The Spartan homepage

6.4.4 A Practice Tutorial Using Spartan

Spartan is a commercial software modelling kit having an easy-to-use graphical user interface (GUI) [31]. It provides a range of Hartree–Fock and post-Hartree–Fock methods including density functional theory. The latest version is Spartan14. It can be easily used for conformational analysis, spectral analysis, and reaction analysis. The suite is accompanied with properties and spectral databases. We will model the Diels–Alder reaction using Spartan 08:

1. Draw the reactant on the workspace (Fig. 6.33).

Select 25th number transition state button and then click on the first double bond on the 1,3-butadiene and then click on the single bond; it will form an arrow, then click on the second double bond, and press the sift button and click on the carbon atom of ethylene; it will form an arrow, then click on the double bond of ethylene, press sift button, and click on the first carbon atom of 1,3-butadiene; it will form an arrow, then click on transition state button on the bottom of right-hand side. It will fix the TS (Fig. 6.34).

2. Now, select constraint distance (Fig. 6.35).
3. Select newly formed bond, then click on lock button from the bottom of right-hand side. Likewise, select another newly formed bond (Fig. 6.36).
4. Now, click on selected bond; it will form a brown color arrow. Likewise, select second arrow (Fig. 6.37).

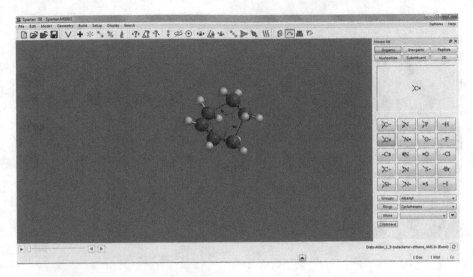

Fig. 6.34 Transition state Diels–Alder reaction

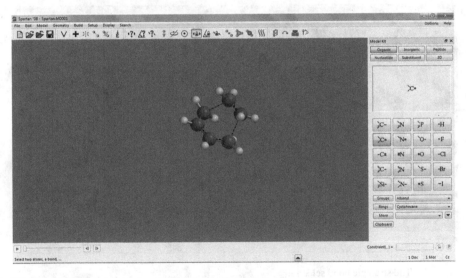

Fig. 6.35 Constraint distance specification in Spartan

5. Click on display button and select properties, then click on dynamic box, then change the value from 1.3 to 3.5 and steps 30. Do not forget to press enter button after putting each value (Fig. 6.38).
6. Next select calculation from setup, with energy profile on the ground state and semiempirical method PM3, and then click submit (Fig. 6.39).

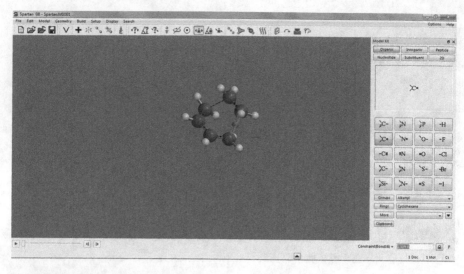

Fig. 6.36 The newly formed bonds are selected

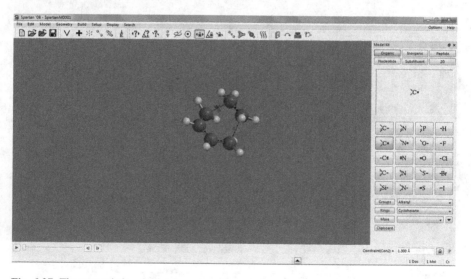

Fig. 6.37 The strategic bond selection

7. When it will complete, form prof file, click display, select spreadsheet, click add button, select energy, then click ok (Fig. 6.40).
8. click constraint distance button, then click on bond, and then click p on the bottom of right-hand side. It will add another column (Fig. 6.41).
9. Using display button, select plot, then select constraint on x-axis and energy on y-axis, then click ok. It will form a plot for reaction (Fig. 6.42).

   Reaction path (Fig. 6.43)

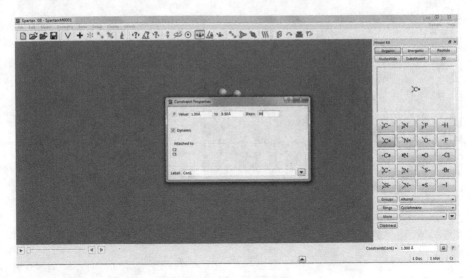

Fig. 6.38 Constraint property value selection in Spartan

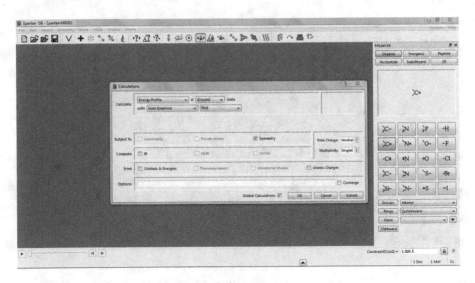

Fig. 6.39 The options provided in the calculation setup menu

6.5 Reaction-Searching Approaches and Tools

Simplified Molecular-Input Line-Entry System (SMILES) that are used to represent chemical reactions digitally are called reaction SMILES. SMIRKS [32] is a hybrid language of SMILES and SMARTS [33] and is also used for reaction expressions.

Reaction SMILES representation contains three parts: Reactant, Agent, and Product which are separated by ">" that represents the arrow "→" in a reaction.

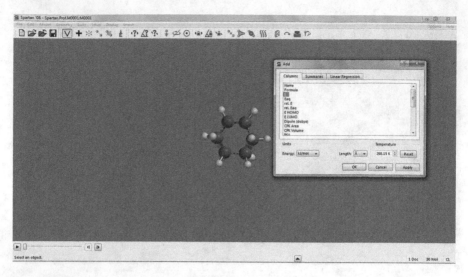

Fig. 6.40 Select the energy option in columns tab

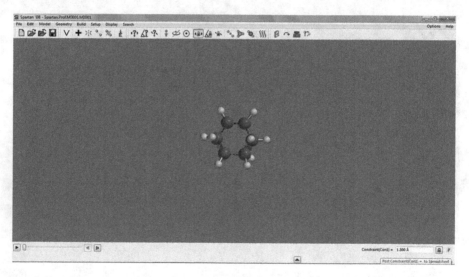

Fig. 6.41 Adding a new column using the post button provided in properties dialog box

Reactant: A substance participating in a chemical reaction, especially a directly reacting substance present at the initiation of the reaction and participates in it by contributing one or more atoms to the products. This can be a compound or multiple compounds or molecules.

SMILES are used for representing reactants in reaction SMILES.

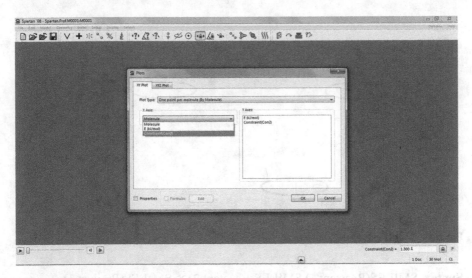

Fig. 6.42 Step for generating the reaction plot in Spartan

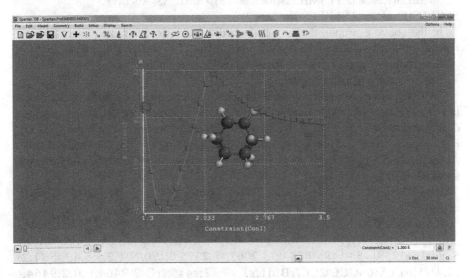

Fig. 6.43 The reaction path plot showing constraint on x axis and energy on y -axis

Agent: They are substances that act as catalysts or solvents and do not directly participate in contributing or accepting atoms during the reaction. They are represented as SMILES between two ">."

Product: Molecules which are the final results of the reaction are called products. They are also represented as SMILES.

Reaction SMILES: $C=CC=C.C=C \gg Cl=CCCCCl$

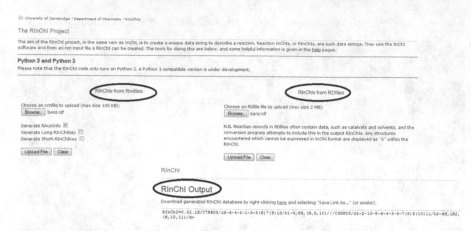

Fig. 6.44 The RInchi Project homepage

Syntax SMILES(Reactant 1).SMILES(Reactant 2) ≫ SMILES(Product 1). SMILES(Product 2)

SMILES(Reactant 1).SMILES(Reactant 2) > SMILES(Agent1) > SMILES(Product 1).SMILES(Product 2)

In the above example, the entities before ">" are the reactants. There are two reactants in the reaction given and are added by a "." in the Reaction SMILES.

Entities between ">" and ">" are the agents and are added by "."

Entities after ">" is the product.

SMIRKS have been recently used for searching chemical reactions in electronic laboratory notebooks [34]. Another identifier used is RInchi which creates a unique data string to describe a reaction based on Inchi software and a rxn input file [35]. For instance, the RInchi output generated by submitting the rxn file for the Diels–Alder reaction at the RInchi project server [36] shows the RAuxInfo and long and short RInChiKeys (Fig. 6.44).

RInChI =0.02.1S/C2H4/c1-2/h1-2H2//C4H6/c1-3-4-2/h3-4H,1-2H2///C6H10/ c1-2-4-6-5-3-1/h1-2H,3-6H2/d+

RAuxInfo=0.02.1/0/N:1,2/E:(1,2)/rA:2nCC/rB:d1;/rC:-2.2393,-.6777,0;-1.5248,-.2652,0;//0/N:2,4,1,3/E:(1,2)(3,4)/rA:4nCCCC/rB:d1;s1;d3;/rC:-5.9148,0,0;-5.2003,.4125,0;-5.9148,-.825,0;-5.2003,-1.2375,0;///0/N:1,2,3,5,4,6/E:(1,2) (3,4)(5,6)/rA:6nCCCCCC/rB:d1;s1;s3;s2;s4s5;/rC:2.9464,0,0;2.9464,-.825,0;3.6609,.4125,0;4.3754,0,0;3.6609,-1.2375,0;4.3754,-.825,0;

Long-RInChIKey = bSA-FEANN-VGGSQFUCUMXWEO-UHFFFAOY-N-KAKZBPTYRLMSJV-UHFFFAOY-N–HGCIXCUEYOPUTN-UHFFFAOY-N

Short-RInChIKey = bSA-FEANN-EAILMCWWNJ-CCFCLFGEWB-EANNAT-PGMB-NEANN-NEANN-NEANN

A set of components (CMLReact) for managing chemical and biochemical reactions have been added to Chemical Markup Language (CML) which can be combined to support most of the strategies for the formal representation of reactions

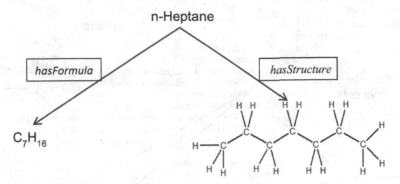

Fig. 6.45 Some simple chemical concepts and their relationships

[37]. Reaction signatures consisting of a simple linear string of letters suitable to index every reaction in a reaction database for computer access have also been developed [38].

6.5.1 Chemical Ontologies Approach for Reaction Searching

Ontology is defined as "basic terms and relations comprising the vocabulary of a topic area as well as the rules for combining terms and relations to define extensions to the vocabulary" [39]. Though ontologies are proposed mainly for the purpose of knowledge sharing historically, in the modern information age, the term ontology is viewed from the perspective of artificial intelligence (AI) with an objective to achieve better information organization and effective retrieval of useful information knowledge sharing across community. In the domain of computer science, ontology refers to an engineering artifact, constituted by a specific vocabulary used to describe a certain reality, plus a set of explicit assumptions captures domain knowledge in a generic way. It also provides common vocabulary and agreed understanding of a domain and makes the domain knowledge to be reused and shared across applications and groups. Accordingly, a chemical ontology tries to conceptualize the chemical knowledge in a narrow or broader perspective, depending on the granularity level of formalization [40, 41]. It is used to describe chemical objects and relationships for enabling the search across multiple data sources bridging some of the graphical and linguistic representations (Fig. 6.45).

Generally, the domain knowledge can be formalized in an ontology with three fundamental components, namelyclass/concept, relation/property, and instance/individual. The concepts identified in the domain are classes of ontology and they are usually organized in taxonomies. A concept/class can be anything about which something is said; it can be a material, nonmaterial, strategy, process, reasoning process, etc. The interaction between the concepts in taxonomy is relation and the instances are specific examples of concepts. Apart from these three fundamental

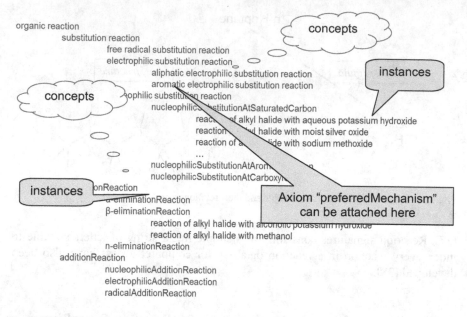

Fig. 6.46 Components of ontology in concept taxonomy

components, ontologies are often organized with two more components, namely function and axiom. Function is a special case of relation between more than two concepts. The axioms are used to model sentences that are always true. For example, a substrate with a halogen atom attached on primary carbon is expected to react with a strong nucleophile in nonpolar condition preferring SN_2 mechanism. This condition can be modeled as an axiom and conveniently attached to an appropriate concept in the taxonomy in the ontology (Fig. 6.46).

Identifying the type of relationship between the concepts within the same ontology is an important task. A *subclassOf* [42] relation is used to relate the concepts having parent–child relationship and it is traditionally named as *isA* relationship. In Fig. 6.47, the concepts substitutionReaction and additionReaction are subclasses of organic reaction and can be related with *isA* relationship. A *subclassPartitionOf* relates a parent concept with a set of child concepts which are mutually disjoint. The concepts nucleophilicSubstitutionAtSaturatedCarbon, nucleophilicSubstitutionAtAromaticCarbon and nucleophilicSubstitutionAtCarboxylCarbon are mutually disjoint and subclasses of nucleophilicSubstitutionReaction. An *exhaustiveSubclassOf* relation relates a parent concept with a set of mutually disjointed subclasses covering the entire parent concept. The concepts nucleophilicSubstitutionReaction, electrophilicSubstitutionReaction, and radicalSubstitutionReaction may be considered as examples of having *exhaustiveSubclassOf*.

Transformation of the conceptual model into an implemented one involves the transformation of ontological representations into formal machine-readable specifications using ontology representation language. For this purpose, selection of a

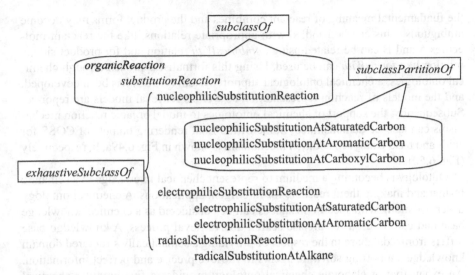

Fig. 6.47 Classification showing types of relationships

```
<?xml version="1.0" encoding="UTF-8"?>
<!DOCTYPE organicReaction SYSTEM "rxnontoDTD.dtd" >
<organicReaction xmlns = "" id="orc" title="Reaction">
        <substitutionReaction id="sub" title="substitution">
            <nucleophilicSubstitutionReaction id="sub/nuc" title="">
                    <nucleophilicSubstitutionAtSaturatedCarbon id="sub/nuc/saC-ind"
                    title="Nucleophilic Substitution Reaction at Saturated Carbon
        Atom">
                        <reactionInstance id="sub/nuc/saC-ind001"
                                title="reaction of alkyl halide with aqueous
                                potassium hydroxide"/>
                        <reactionInstance id="sub/nuc/saC-ind002"
                                title="reaction of alkyl halide with moist silver
oxide"/>
                    </nucleophilicSubstitutionAtSaturatedCarbon>
                </nucleophilicSubstitutionReaction>
        </substitutionReaction>
...
</organicReaction>
```

Fig. 6.48 Part of reaction ontology specified in XML

knowledge representation language is important. At present, Extensible Markup Language (XML) [43] is the state-of-the-art ontology specification technology. OWL [44] is W3C standard which is an XML-based ontology specification language. A part of reaction ontology specified in XML is shown in Fig. 6.48.

Ontology specification in XML

The taxonomies of different ontologies can also be related with specific relations. Such relations play a crucial role when the ontologies are integrated with some applications. For example, a concept of "aliphatic nucleophilic substitution reaction" can be related with a concept of "nucleophilic reagent" using *hasReagentClass* in one direction and also with a reverse relation as *reagentClassIn* relation. In a general reaction like A+B → C+D, if formalized along with the plus symbols, the fundamental meaning of reactant combines and the product forms may become ambiguous. This can be handled with appropriate relations, like the reactant molecules A and B can be related using *combinesWith* relation and for product side, a relation like *formsWith* can be used. Using this formalist approach through chemical ontologies, a chemical ontological support system (COSS) has been developed, and the models of reaction representation as well as retrieval models are reported. Subsequently, the support of chemical ontologies to model organic reaction mechanisms can also be demonstrated. Some of the final rendering outputs of COSS for acid- and base-catalyzed addition mechanism is shown in Fig. 6.49a, b, respectively (Fig. 6.50).

Ontology is becoming a medium to represent chemical knowledge in a semantic format and making them reusable for intelligent applications. A chemical ontology described along with specific instances can be considered as a chemical knowledge base and can be used for precise search and retrieval process. A knowledge base differs from a database in the respect of providing a semantically structured domain knowledge-supporting software agents to retrieve precise and perfect information. An exhaustive or elaborate chemical ontology provides a fine granular chemical knowledge enabling deeper semantics, whereas the reverse results in a description with shallow semantics. In recent years, reaction-specific chemical ontologies have started evolving. However, the reaction representation and its description, in a more meaningful way, can be achieved through the development of appropriate ontologies developed on the components of reaction, starting from atoms, groups, functional groups, etc. [45, 46] and associating them intermediately with chemical transformation and then ultimately with reaction.

6.5.2 Reaction Searching Using Fingerprints-Based Approach

In the previous chapter, we learnt the use of fingerprints for searching chemical structures. Of equal importance is the need for searching chemical reactions to estimate their similarity using computational tools. In this section, we will therefore learn how to use reaction fingerprints for searching in databases and online servers. Similar techniques are employed in both cases. Structural properties that are present in the reaction are used for estimation of reaction similarity. Two reactions can be considered similar if their product side and/or reactant side are similar. With this consideration, reaction similarity is reduced to molecular or *structural similarity*.

An alternative approach is to characterize the reaction transformation carried out by identifying the *changing atoms* and the *changing bonds* in the reaction with respect to the reactants and the product structures. An atom is changing if either of the conditions is met:

1. One or more of its bond is changed (i.e., the bond is different on the left side compared to the right side) or
2. It is present only on one side of the reaction and it has a non-changing atom neighbor.

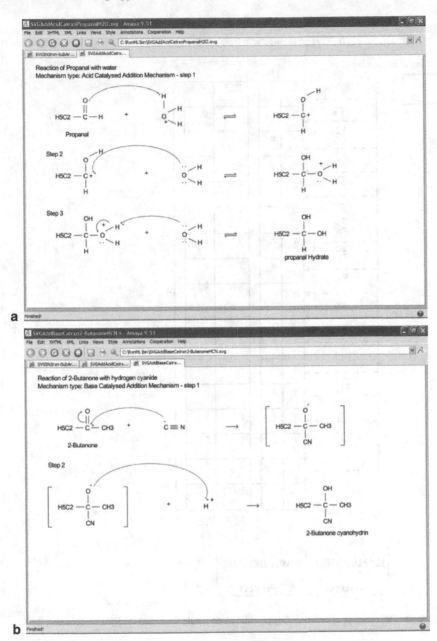

Fig. 6.49 Rendering of acid-catalyzed addition reaction mechanism (**a**) and base–catalyzed (**b**) with the support of chemical ontological support system (COSS)

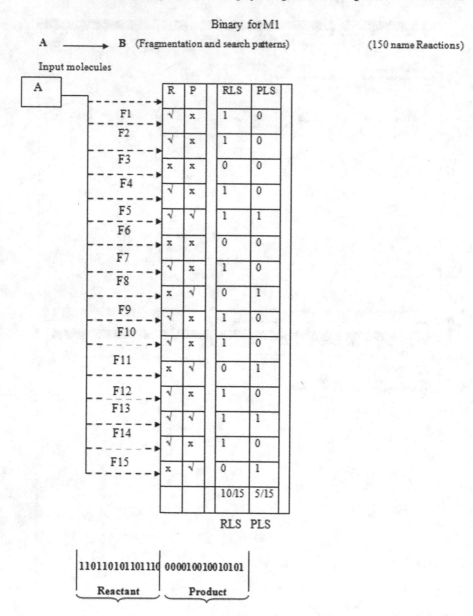

Fig. 6.50 The concept of reaction fingerprints to compute RLS and PLS for molecules

Changing atoms and changing bonds define the "reacting center" of the reaction. The reacting center is specific to a particular type of reaction. Nevertheless, another type of reaction similarity can be introduced by focusing on the reacting center of the reaction. This transformational similarity is less influenced by the particular reactant and product present in a reaction, but it is dominated by the reaction mechanism. Both of these types of reaction similarity are found to be useful in comparing and matching reactions.

6.5.2.1   Tools Available with an Academic License

In the ChemAxon reaction module, similarity-searching program [47] reaction fingerprints are used. The structure of the reaction fingerprint is composed of eight segments including chemical fingerprint (CFp) of the reactant or reactants and reagent, the product, the reactant side of the reaction center, the product side of the reaction center, the reactant side of the reaction center including its 1 bond neighborhood, the product side of the reaction center including its first bond neighborhood, the reactant side of the reaction center including its 2 bond neighborhood, and the product side of the reaction center including its second bond neighborhood.

The total length of the reaction fingerprint is 2,048 bits. The above-defined eight segments of the reaction fingerprint are laid out in the schema below (segment sizes given in number of bits):

512	512	128	128	128	128	256	256

This reaction fingerprint enables both types of reaction similarity calculations, and with the expense of some extra storage space, it makes the transformational similarity calculation efficient in all three predefined levels of coarseness.

Two types of reaction similarity calculations have been introduced: structural and transformational. Structural distinguishes the reactant and the product sides, while transformational relates to three levels of coarseness. With these considerations, five metrics need to be introduced to efficiently estimate the five different categories of reaction similarity. These metrics are as follows:

- ReactantTanimoto
- ProductTanimoto
- StrictReactionTanimoto
- MediumReactionTanimoto
- CoarseReactionTanimoto

All of these metrics are based on the Tanimoto metric; consequently, the degree of similarity is expressed from 0 to -1. ReactantTanimoto considers only the first quarter of the reaction fingerprint that represents the reactants in the reaction and ignores the rest of the reaction fingerprint. Therefore, it estimates the structural similarity of the reactants only. ProductTanimoto takes the second quarter of the fingerprint that is associated with the products. StrictReactionTanimoto takes the last two segments of the reaction fingerprint that represents the reacting center of both the reactant and the product side of the reaction with the broadest neighborhood and ignores the first 3/4 of the reaction fingerprint. Similarly, MediumReactionTanimoto applies the Tanimoto metric to the fifth and sixth segments, while CoarseReactionTanimoto takes the third and the fourth segments that encode the reacting center of the reactant and the product side, respectively.

6.5.2.2   In-House Developed Fingerprint-Generating Program

In the traditional fragment-based fingerprint approaches based on a pattern of say five atoms or four atoms, the algorithm will search around the molecule in all possible ways traversed through the molecules to detect presence or absence of patterns in that molecule, and the reaction similarity is based on a Tanimoto metric as discussed above. In our method, a chemical structure is stored using 16 fingerprints. Sixteen numbers of 4 bytes each are optimum for screening 16 integers with a capacity of 4 bytes each. Thus, in total, we allocate 64 bytes per structure. To store one reaction fingerprint, 512 bytes are required. Although there are a number of named organic reactions reported in chemistry, yet for the present work, we restricted ourselves to 150 name reactions having 150 reactants and 150 products. Their binary reaction fingerprints were computed and the complete data with reactant-like scores (RLS) and product-like scores (PLS) are available at the moltable server [48]. From this, the most frequently used reaction fingerprints and distinct species involved in these reactions, i.e., during conversion from reactant to product were identified. Out of 1,000 species, 450 distinct species were found to occur frequently. Each species was mapped to name reactions to confirm whether it occurred as a reactant or product molecule. On this basis, we computed 305 binary reaction fingerprints for each molecule search (hit=1, no hit=0).

We took a functional class of compounds and obtained the cumulative fingerprints based on reactant/product likeliness. For example, given ethyl alcohol CH_3CH_2OH as a query molecule the algorithm will check whether it is a reactant or a product. The OH functional group can be reactant in an esterification step or act as a product in a hydrolysis step. Likewise, an ester group is a product but can be a reactant in hydrolysis process. Alcohol functional group can be converted to aldehyde, acid, or ester; the aldehyde group can further undergo oxidation, reduction, or condensation process. So, it is impossible for a chemist to manually look at all possible options. Using our methodology, we can provide a fragment-based alert as to what reactions it is likely to undergo and what are the anticipated products, whether a new molecule can be formed. This prior knowledge will aid decision making in synthesis design. As the total number of products and reactants are fixed, an RLS and a PLS can be given to the query molecules submitted by user to determine the percentage of reactive functional groups of reactants or products present in it. Since we are considering the reaction as a whole, the solvent, catalyst, and reaction conditions present therein are integrated inherently in the reaction information itself (Fig. 6.50).

6.5.2.3   Code to Obtain Reactions

```
public int[] getReactReactionId(String[][] smartsData,String fp,char a){
        int[] temp=new int[250];
        int cnt=0;
        for (int i = 0; i < fp.length(); i++) {
                if(fp.charAt(i)==a){
                        temp[cnt]=Integer.parseInt(smartsData[i][2]);
                        cnt++;
                }
        }
        int[] out=new int[cnt];
        for(int i=0;i<cnt;i++){
                out[i]=temp[i];
                //System.out.println("react "+out[i]);
        }
        return out;
}

    public int[] getProReactionId(int totelReact,String[][] smartsData,String
fp,char a){
        int[] temp=new int[250];
        int cnt=0;
        for (int i = 0; i < fp.length(); i++) {
                if(fp.charAt(i)==a){
                        temp[cnt]=Integer.parseInt(smartsData[totelReact+i][2]);
                        cnt++;
                }
        }
        int[] out=new int[cnt];
        for(int i=0;i<cnt;i++){
                out[i]=temp[i];
                //System.out.println("pro "+out[i]);
        }
        return out;
}
```

We built a matrix of 29 × 305 reaction fingerprints for molecules belonging to 29 therapeutic categories; the objective was to deduce the difference in selectivity pattern within the class of drugs or leads and identify distinct fingerprints representative of the classes. A cumulative reaction fingerprints spectrum of 4,000 molecules present in 29 drug classes can be used for annotating a query molecule with any of the distinct 29 classes of drugs/leads. Given any molecule, one can predict whether that molecule falls in any of the 29 therapeutic classes based on the availability of diverse patterns/fingerprints and PLS and RLS (Table 6.1).

6.5.3   Tools for Reaction Searching

Scifinder is a commercial tool provided by Chemical Abstracts Service (CAS) [49]. It is enabled with advanced reaction-searching feature that includes assigning roles for reaction participants, substructure drawing features, and filtering results using any of the desired attributes. After logging into Scifinder in the *Explore Reactions* screen, the user can specify sites where bonds are changed, include functional

Table 6.1 Binary reaction fingerprints of common drugs

S.No.	Drug name	RLS	PLS	Reaction binary fingerprints
1	Lipitol	41	28	00000000001010000000000011001110010000010000000
				00100000000000000000000100000000001000010 1000
				10100010000101000000001011001010100000100010 0000
				11110010110000100000010010000000101000100000 0110
				01101000000000010110100001100000000010000000 0000
				00010000011000000001100101000100100000001 0001
				000000100001101001000000000011
2	Levobunalol	45	28	010000000000010100000000001000111001000011010 00
				100100000000000000001000001010000001011000 1010
				00001000100000010010000010110010101000100 0100
				00010100010110000100000010010000001010000 0000
				01100010000000110001001010000110001100000 0000
				00000000101000001000010001101100100100000 0000
				01010000000000001111000100000000000001
3	Sildenafil	38	27	000000000000010000010000010001110010000010 0100
				00000000000000000000000001000000000001000010 00
				00101000100001000010000010110010000000100001 011
				01011010010110000100000010110000000000010000 00
				01100110000000100001001010000110000000001000 00
				000000001000000010000000011000010001001010100
				01000000000010000111100100000000000011
4	Ibuprofen	25	16	00000000000010000010000000011100100000100000 0
				00000000000000000000000001000000000000000000 00
				00101000100001000000000101000100000001000000 0
				00011010000110000000000100000000000000010000 00

Table 6.1 (continued)

S.No.	Drug name	RLS	PLS	Reaction binary fingerprints
5	Aspirin	26	20	01100110000000000000000010000110000000100000
				00000000000000000000000000100000100010010000000
				010000000000000000110000100000000000011
				00000000000010000000000000000011000100000100000
				00000000000000000000000000100000000000000001010
				00001000100001000000000001010001000000001000000
				00011110010110000000000001000000000000000100000
				01100110000000000000000101000011000000001000000
				00000000100000000000000000010001010001001000000
				010000000000000000111000100000000000011

groups, map atoms, assign roles, etc. The reaction structure drawing tool can be used to create a reaction query or upload a formerly saved query in the .cxf format.

Both exact structure- and substructure-searching options are available. Functional groups or atoms can be locked to prevent substitution or ring fusion. The answer sets are determined by the Tanimoto similarity metric. A reaction search for a typical Diels–Alder reaction between a diene and a dienophile having a cyano functional group yielded 253 reactions (Fig. 6.51).

The results can be sorted by relevance, accession number, product yield, etc. To further narrow the search results, advanced search options can be specified such as number of steps, source, publication years, solvents, and nonparticipating functional groups (Fig. 6.52).

As seen in Fig. 6.53 by restricting the publication years to the past 5 years, the search results retrieved 36 reactions. The results can be further refined by additional criteria like reaction structure, product yield, reaction classification, etc. Moreover, one can also analyze by catalyst, available detailed experimental procedure, journal name, etc. All entries in the reaction results are connected to the CAS substance database records, which is highlighted by placing the cursor over the structure, and substance information regarding commercial source, synthetic procedure and regulatory information is revealed. Stepwise tutorials for reaction searching in Scifinder

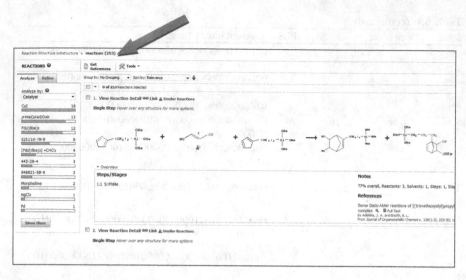

Fig. 6.51 Scifinder substructure search for the Diels–Alder reaction reaction-based query

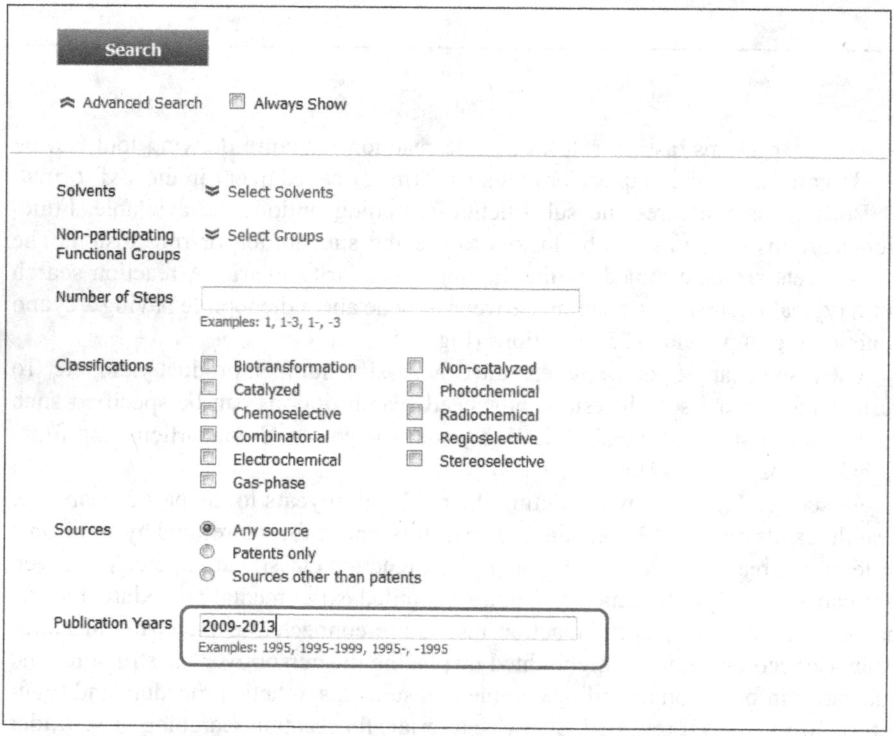

Fig. 6.52 Advanced search option specifying the publication years to the past 5 years

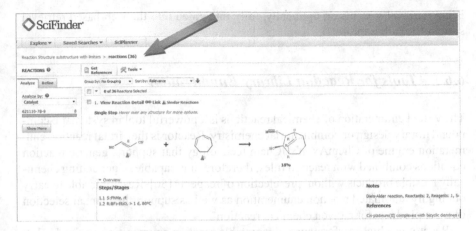

Fig. 6.53 The hits narrow down to 36 using advanced search options

are available at https://scifinder.cas.org/help/scifinder/R18/index.htm#reactions/search_by_reaction_structure.htm.

6.6 Reaction Databases

For knowledge-based approach, there exist a host of chemical reaction databases free as well as proprietary. The Beilstein Information System is the world's largest collection of chemical properties of organic compounds [50]. CASREACT, an online database, provides access to chemical reactions reported in the journal literature, maintained by CAS with substructure-based reaction-retrieval capabilities [51]. ChemInform Reaction Library (CIRX) is another source of information that enables chemists to predict the suitability of synthetic methods for the design of novel molecules focusing on the latest novel reactions and methods for organic synthesis [52]. Database information on chemo- and regio-selective reactions makes it especially useful in identifying viable routes to novel compounds. ChemReact68 contains essential information on 68,000 reactions referenced in literature published from 1974 to 2001 [53]. While the above-discussed databases are proprietary, there are some good online sources also available, such as Chemogenesis, Organic Syntheses (ORGSYN), SyntheticPages, Synthesis Protocols, chemical methodology and library development (CMLD), The Chemical Thesaurus, and web reactions [54].

BioPath is a database of biochemical pathways that provides access to metabolic transformations and cellular regulations derived from the Roche Applied Science "Biochemical Pathways" wall chart [55]. In the current version, BioPath also provides access to biological transformations reported in the primary literature. The BioPath database is available in Symyx MOL/RDF format for integration into existing retrieval systems or, optionally, fully integrated into the web-based retrieval system BioPath.Explore.

6.6.1 Tools for Reaction Library Enumeration

The virtual enumeration of chemical reactions is a powerful tool in systematic compound library design or combinatorial chemistry. Reactor is the virtual reaction enumeration engine of ChemAxon's JChem technology that supports generic reaction equations combined with reaction rules; therefore, it is capable of generating chemically feasible products without preselection of reagents [56]. Reactor is able to carry out highly automated reaction enumeration as well as support the manual selection of main products for a given chemical reaction.

Reactor is a high-performance, integratable reaction enumeration engine [57]. It works with generic reaction equations that can be defined and imported in various formats, including among others SMIRKS/SMARTS strings, RDF, RXN, and MRV files, or be drawn in MarvinSketch.

Reagent(s) are processed according to the given reaction schema; if the reaction is in RDF or MRV format, reaction rules, reactant standardization, and some additional properties are also possible to set in RDF/MRV tags.

Reaction schemes can include stereochemical information. Reactor is capable of handling both tetrahedral and double bond stereochemistry flexibly; inversion and retention centers as well as *cis–trans* configuration changes can be determined within Reactor's smart reaction schemes. Prochiral reaction schemes are also supported since version 5.5, allowing the user to manage syn/anti additions.

Reactor can be set up to carry out simple sequential enumeration, combinatorial enumeration, generating combinatorial virtual synthetic libraries. Users also have the option to exclude unwanted products from the enumeration results manually, restricting the outcome of the reaction enumeration process to the desired main products only. Reactor supports the generation of product or reaction libraries in a large variety of different output formats.

It has the option to copy arbitrary property fields from the input reactant files to the results. These can include, e.g., solubility or availability information of the reactants. Also, Reactor can generate synthesis codes for each reaction in the enumeration process containing selected information from the reaction scheme and the reactants. The stand-alone version of Reactor has a GUI for configuring the reaction enumeration process. The Reactor GUI leads users step by step through the whole configuration process of the virtual chemical reaction. Reactor has also been integrated into Instant JChem and JChem for Excel. It is also available in the workflow management tools KNIME and Pipeline Pilot. In its stand-alone version, it can be used through the GUI, as a command line application and also through a full featured Java API. Reactor offers full platform independence; it is equally available for Microsoft Windows, Mac OS, and Linux platforms.

Reactor has an integrated reaction sketcher and editor tool. Users can create their own reaction schemes and add corresponding reaction rules to them using

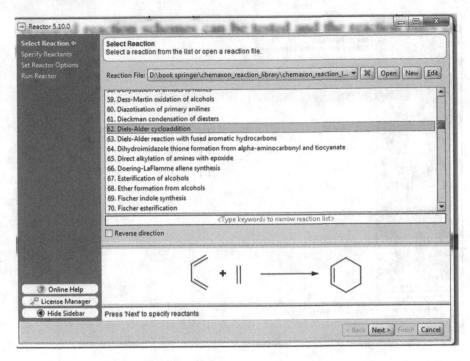

Fig. 6.54 Next step is to define the reactants

the Chemical Terms language. The prepared reaction schemes can be tested and the reaction rules can be validated using the integrated reaction-testing tool of Reactor.

6.6.2 A Practice Tutorial

The Reactor package includes a large and constantly increasing library of organic chemical reactions that can be used directly without any further configuration. The list of available reactions of ChemAxon's reaction library is provided on their website [58]. Here, we are selecting the Diels–Alder cycloaddition reaction mentioned in the previous section (Fig. 6.54).

Using the generic reaction equations, virtual synthetic compound libraries can be generated under full manual control. When doing so, users have the opportunity to draw and edit reactants directly and to select chemically meaningful products from the output of the enumeration process by using their chemical intuition on the fly. This approach is particularly advantageous for enumerating small, focused libraries. We define reactants 1 and 2 of the Diels–Alder reaction (Fig. 6.55).

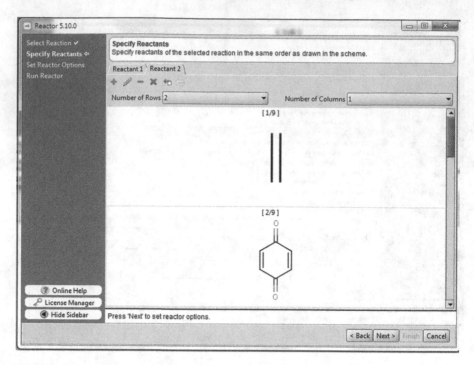

Fig. 6.55 Reactant 2 is specified

The next step is to get the synthesizable molecules by proper reaction rules defined in Chemical Terms, ChemAxon's scripting language that is designed to add chemical intelligence to chemoinformatics applications. Through Chemical Terms, a large number of calculated properties can be included in the reaction rules to produce valid compound libraries. Besides calculating physicochemical properties on the fly, Chemical Terms language also supports importing of arbitrary fields from the input reactant files to be used for the evaluation of the reaction rules (Figs. 6.56, 6.57 and 6.58).

6.7   Artificial Intelligence in Chemical Synthesis

To assist rational synthetic planning by a chemist, a number of computer programs to suggest viable chemical routes have been developed. The known general computational approaches are empirical, semiempirical, and knowledge based, all of them drawing their inspiration from the well-established reactions and certain principles of organic synthesis [59]. Empirical approaches can theoretically provide millions of reactions, but they may or may not be synthetically possible in the laboratory [60]. Quantum chemical approaches involve studying the ground state of individual atoms and molecules, the excited states, and the TS that occur during chemical reactions [61]. Quantum to molecular mechanics (Q2MM) methods allow application

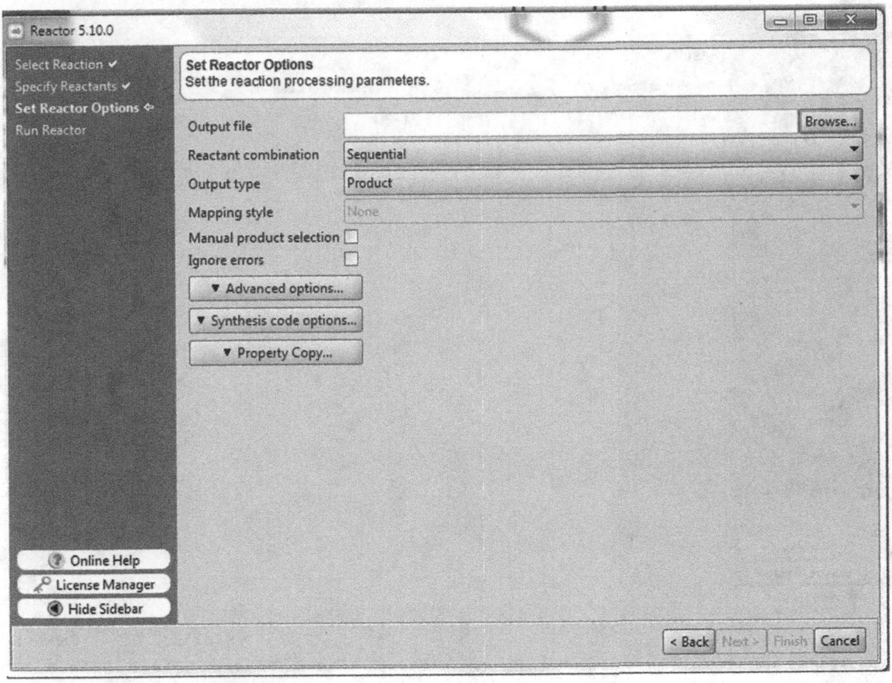

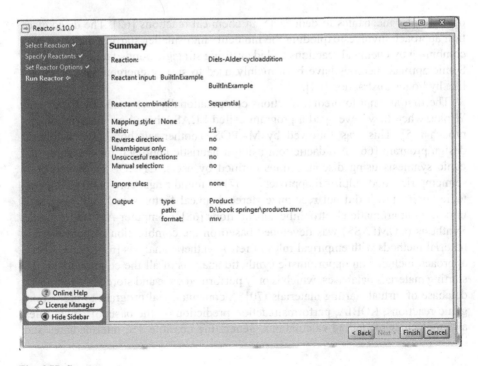

Fig. 6.56 Reaction processing parameters setup screen

Fig. 6.57 Summary page

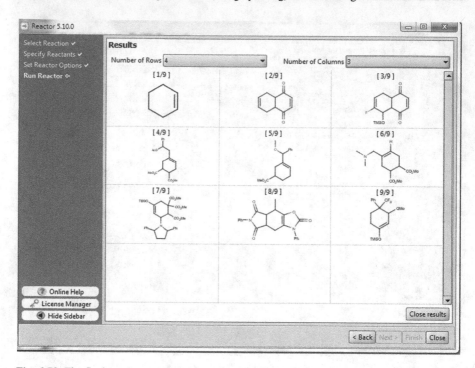

Fig. 6.58 The final results step retrieves various products of Diels–Alder reaction

of molecular mechanics to deduce TS in chemical reactions [62]. The disconnection approach involves exploding the molecule into smaller starting materials and combining by chemical reactions and identifying strategic bonds/facile bonds [63]. Semiempirical methods have been mainly used to survey energetics of reactions like hydrogen abstraction [64].

The first attempt to predict reactions computationally was made by Corey and Wipke, when they developed a program called LHASA based on the synthon approach [65]. This was followed by MAPOS, another synthon-based synthesis design program [66]. A didactic tool using a heuristic approach for designing organic synthesis using disconnections defined by users, CHAOS employed both semiempirical and empirical approaches [67]. It found rings, core bonds, and strategic bonds, but it did not recognize stereochemical features and could not take into account aromatic electrophilic substitutions [68]. Computer-Assisted Organic Synthesis (COMPASS) was developed based on the combination of pure combinatorial methods with empirical rules of retro-synthetic analysis [69]. The CAESA approach included an opportunistic synthetic analysis of all the compounds in the starting materials databases, which is only performed once and stored in a relational database of virtual starting materials [70]. A computational program to predict organic reactions, ROBIA, performs reaction prediction on the basis of coded rules and molecular modelling calculations, generating possible TS, intermediates and products given the starting material and reaction conditions [71]. Recent reaction prediction programs include Reaxys which has over 400 indexed fields of experimentally validated data extracted from journals and patents, important chemistry-related literature, and patent sources [72]. SYLVIA, a commercial program, rapidly evaluates the synthetic accessibility score of organic compounds on a scale of 1 for straightforward synthesis to 10 for complex and challenging synthesis by employing various structure- and reaction-based parameters ([73]; Table 6.2).

To build intelligence into the reaction prediction programs, the strategic bonds which are cleaved or formed during a reaction are identified. A disconnection approach is used to reveal the *synthon* and *retron* for a reaction which helps the user in designing synthetic routes for a molecule of interest.

6.8   Modelling Enzymatic Reactions

An important range of drug–host interactions involves covalent binding or chemical reactions, which are often catalyzed by enzymes. Prediction of these metabolic processes requires detailed insight into the mechanisms of the reactions involved, as well as computational methods that account for the reactivates of the compounds and proteins involved. Theoretical models of enzyme reactions are becoming increasingly important in applied areas for making predictions of biochemical conversions or designing bioactive compounds with desired chemical properties [75].

The most important class of enzymes is the family of cytochrome P450s, which is capable of catalyzing a variety of reactions, mainly oxidations, of a broad range of compounds [76]. Their catalytic flexibility is based on the heme cofactor that is present in the active site and has exceptional catalytic properties. Other enzymes include flavin-dependent monooxygenases, dehydrogenases, esterases, and peptidases. Many reactions are catalyzed by several enzyme families, e.g., epoxide hydrolases, glutathione S-transferases, glucuronyl transferases, etc. [77]. Resistance against antibiotics involves the occurrence of enzymes in the target microorganism that specifically convert the antibiotic to a non-antibiotic metabolite [78]. In some drug design strategies, prodrugs are metabolized to the active drug specifically in the target tissue, e.g., tumor tissue only [79]. In the future, a number of techniques for enzyme reaction modelling need to be developed for applications in studies of drug metabolism.

6.9   Thumb Rules for Performing Reaction Representation, Fingerprints, and Modelling

- The TS observed in a modelling process should be checked for correct geometry, calculated bond orders, and vibrational frequencies.

S. No.	Name reaction	Reactants	Products
1	Aldol Condensation		
2	Claisen condensation		
3	Oxy Cope rearrangement		
4	Diels Alder reaction		
5	Mannich reaction		
6	Birch Reduction		

Table 6.2 Strategic key bond formation in few examples of name reactions [74]

6.10 Do it Yourself

1. Search esterification reaction in the various online reaction sources
2. Model any name reaction of your choice using Gaussian program and interpret the results

6.11 Questions

i. What are the ways of representing reactions in computers?
ii. Write a short note on reaction file formats.
iii. Highlight the various methods used in reaction modelling.
iv. What do you understand by the term artificial intelligence in organic synthesis? Elaborate giving examples.
v. What are the challenges in modelling enzymatic reactions?

References

1. Knowles JR (1980) Enzyme-catalyzed phosphoryl transfer reactions. Annu Rev Biochem 49:877–919
2. Rich PR (2003) The molecular machinery of Keilin's respiratory chain. Biochem Soc Trans 31(6):1095–1105
3. McMurry J, Begley TP (2005) The organic chemistry of biological pathways. In: Lehninger (ed) Principles of biochemistry. Roberts and Company Englewood, Colo
4. Sorensen SD, Nicole O, Peavy RD, Montoya LM, Lee CJ, Murphy TJ, Traynelis SF, Hepler JR (2003) Common signaling pathways link activation of murine PAR-1, LPA, and S1P receptors to proliferation of astrocytes. Mol Pharmacol 64(5):1199–1209
5. Li Y, Agarwal PA (2009) Pathway-based view of human diseases and disease relationships. PLoS One 4(2):e4346
6. http://www.biocarta.com/genes/catMetabolism.asp. Accessed 17 Oct 2012
7. Timberlake KC (2009) General, organic, and biological chemistry: structures of life, 3rd edn. Prentice Hall
8. Bersohn M, Esack A (1977) A computer representation of synthetic organic reactions. Comput Chem 1(2):103–107
9. Nicolaou KC, Snyder SA, Montagnon T, Vassilikogiannakis G ChemInformA: the Diels–Alder reaction in total synthesis. Cheminform 33(36).
10. Yilmazer ND, Korth M (2013) Comparison of molecular mechanics, semi-empirical quantum mechanical, and density functional theory methods for scoring protein ligand interactions. J Phys Chem B 117(27):8075–8084
11. Ferguson DM, Gould IR, Glauser WA, Schroeder S, Kollman PA (1992) Comparison of ab initio, semiempirical, and molecular mechanics calculations for the conformational analysis of ring systems. J Comput Chem 13(4):525–532

12. Nagaoka M, Okuno Y, Yamabe T (1991) The chemical reaction molecular dynamics method and the dynamic transition state: proton transfer reaction in formamidine and water solvent system. J Am Chem Soc 113(3):769–778

13. Myrvold WC Statistical mechanics and thermodynamics: a Maxwellian view. Studies in History and Philosophy of Science Part B: Studies in History and Philosophy of Modern Physics 42(4):237–243

14. Amara P, Field MJ (2002) Combined quantum mechanical and molecular mechanical potentials. Encyclopedia of computational chemistry. Wiley

15. Singh UC, Kollman PA (1986) A combined ab initio quantum mechanical and molecular mechanical method for carrying out simulations on complex molecular systems: applications to the CH3Cl+ Cl– exchange reaction and gas phase protonation of polyethers. J Comput Chem 7(6):718–730

16. Crehuet R (2005) The reaction path intrinsic reaction coordinate method and the Hamilton Jacobi theory. J Chem Phys 122:234105

17. van der Kamp MW, Mulholland AJ Combined quantum mechanics/molecular mechanics (QM/MM) methods in computational enzymology. Biochemistry 52(16):2708–2728

18. Hao HL, Zhenyu MP, Jerry KB, Weitao Y (2008) Quantum mechanics/molecular mechanics minimum free-energy path for accurate reaction energetics in solution and enzymes: sequential sampling and optimization on the potential of mean force surface. J Chem Phys 128(3):034105

19. Kirchner B, Vrabec J, Jaramillo-Botero A, Nielsen R, Abrol R, Su J, Pascal T, Mueller J, Goddard W III First-principles-based multiscale, multiparadigm molecular mechanics and dynamics methods for describing complex chemical processes. Multiscale molecular methods in applied chemistry. Springer Berlin Heidelberg, pp 1–42.

20. Moreland DW, Dauben WG (1985) Transition-state modeling in acyclic stereoselection. A molecular mechanics model for the kinetic formation of lithium enolates. J Am Chem Soc 107(8):2264–2273

21. Schwartz SG, Henkelman GS, Jahannesson JH (2002) Methods for finding saddle points and minimum energy paths. Theoretical methods in condensed phase chemistry. Springer, Netherlands, pp 269–302

22. Wang Z. Barton-Kellogg Olefination. Comprehensive organic name reactions and reagents. Wiley

23. Bernardi AA, Capelli M, Gennari C, Goodman JM, Paterson I (1990) Transition-state modeling of the aldol reaction of boron enolates: a force field approach. J Org Chem 55(11):3576–3581

24. Jones M, Fleming S (2010) "Organic Chemistry", Norton, 4th edn. Chapter 19, p 932–946, 965–985.

25. Berson J, Jones M (1964) 86, 5019

26. http://www.organic-chemistry.org/namedreactions/claisen-condensation.shtm. Accessed 17 Oct 2012

27. Puy CHDe (1960) Chem Rev 60:444

28. Hughes P (2006) Was Markovnikov's rule an inspired guess?. J Chem Educ 83(8):1152

29. Gaussian 03, Revision C.02, Frisch MJ, Trucks GW, Schlegel HB, Scuseria GE, Robb MA, Cheeseman JR, Montgomery JA Jr, Vreven T, Kudin KN, Burant JC, Millam JM, Iyengar SS, Tomasi J, Barone V, Mennucci B, Cossi M, Scalmani G, Rega N, Petersson GA, Nakatsuji H, Hada M, Ehara M, Toyota K, Fukuda R, Hasegawa J, Ishida M, Nakajima T, Honda Y, Kitao O, Nakai H, Klene M, Li X, Knox JE, Hratchian HP, Cross JB, Bakken V, Adamo C, Jaramillo J, Gomperts R, Stratmann RE, Yazyev O, Austin AJ, Cammi R, Pomelli C, Ochterski JW, Ayala PY, Morokuma K, Voth GA, Salvador P, Dannenberg JJ, Zakrzewski VG, Dapprich S, Daniels AD, Strain MC, Farkas O, Malick DK, Rabuck AD, Raghavachari K, Foresman JB, Ortiz JV, Cui Q, Baboul AG, Clifford S, Cioslowski J, Stefanov BB, Liu G, Liashenko A, Piskorz P, Komaromi I, Martin RL, Fox DJ, Keith T, Al-Laham MA, Peng CY, Nanayakkara A, Challacombe M, Gill PMW, Johnson B, Chen W, Wong MW, Gonzalez C, Pople JA, Gaussian Inc., Wallingford CT (2004)

30. Jaguar, version 7.9 (2012) Schrödinger, LLC, New York, NY
31. Spartan'10 Wavefunction, Inc.Irvine, CA
32. http://www.daylight.com/dayhtml/doc/theory/theory.smirks.html. Accessed 17 Oct 2012
33. http://www.daylight.com/dayhtml/doc/theory/theory.smarts.html. Accessed 17 Oct 2012
34. Christ CD, Zentgraf M, Kriegl JM (2012) Mining electronic laboratory notebooks: analysis, retrosynthesis, and reaction based enumeration. J Chem Inf Model 52(7):1745–1756
35. Goodman JM RInChIs and reactions abstracts of papers, 242nd ACS National Meeting & Exposition, Denver, CO, United States, August 28–September 1, 2011 (2011), CINF–40
36. http://www-rinchi.ch.cam.ac.uk/
37. Holliday GL, Murray-Rust P, Rzepa HS (2006) Chemical markup, XML, and the world wide web. 6. CMLReact, an XML vocabulary for chemical reactions. J Chem Inf Model 46(1):145–157
38. Hendrickson JB (2010) Systematic signatures for organic reactions. J Chem Inf Model 50(8):1319–1329
39. Neches R, Fikes RE, Finin T, Gruber TR, Patil R, Senator T, Swartout WR (1991) Enabling technology for knowledge sharing. AI Magazine 12(3):16–36
40. Sankar P, Aghila G (2006) Design and development of Chemical ontologies for reaction representation. J Chem Inf Model 46:2355–2368
41. Sankar P, Aghila G (2007) Ontology aided modeling of organic reaction mechanisms with flexible and fragment based XML markup procedures. J Chem Inf Model 47:1747–1762
42. Fernandez-Lopez M, Gomez-Perez A, Pazos-Sierra J (1999) Building a 43. Chemical ontology using methontology and the ontology development environment. IEEE Intell Syst 14(1):37–46
43. http://www.w3.org/XML. Accessed 1 Oct 2013
44. http://www.w3.org/2007/OWL/wiki/OWLWorkingGroup. Accessed 1 Oct 2013
45. Feldman HJ, Dumontier M, Ling S, Haider N, Hogue CWV (2005) CO: a chemical ontology for identification of functional groups and semantic comparison of small molecules. FEBS Lett 579:4685–4691
46. Sankar P, Krief A, Vijayasarathi D (2013) A conceptual basis to encode and detect organic functional groups in XML. J Mol Graph Model 43:1–10
47. http://www.chemaxon.com/jchem/doc/user/fingerprint.html. Accessed 17 Oct 2013
48. www.moltable.org. Accessed 17 Oct 2013
49. https://scifinder.cas.org/scifinder/view/scifinder/scifinderExplore.jsf. Accessed 17 Oct 2012
50. Jochum C (1994) The Beilstein information system is not a reaction database, or is it? J Chem Inf Comput Sci 34:11–13
51. Blake JE, Dana RC (1990) CASREACT: more than a million reactions. J Chem Inf Comput Sci 30:394–399
52. ChemInform (2010) 41(01)
53. http://chemreact.cambridgesoft.com/chemreact68/index.asp. Accessed 17 Oct 2013
54. http://www.organicworldwide.net/content/reaction-databases. Accessed 17 Oct 2012
55. http://www.molecular-networks.com/databases/biopath. Accessed 17 Oct 2013
56. http://www.chemaxon.com/jchem/doc/user/reactor.html. Accessed 17 Oct 2013
57. Bode JW (2004) Computer software reviews. J Am Chem Soc 126(46):15317–15317
58. http://www.chemaxon.com/jchem/doc/user/chemaxon_reaction_library.html. Accessed 17 Oct 2013
59. Wipke WT, Ouchi GI, Krishnan S (1978) Simulation and evaluation of chemical synthesis: an application of artificial intelligence techniques. Artif Intell 11(12):173–193
60. Cheng H, Scott K (2003) An empirical model approach to gas evolution reactions in a centrifugal field. J Electroanal Chem 544:75–85
61. Friesner RA (2005) Ab initio quantum chemistry methodology and applications. PNAs 102:6648–6653
62. Mulholland A (2007) Chemical accuracy in QM/MM calculations on enzyme-catalyzed reactions. Chem Cent J 1:1–5
63. Warren S, Wyatt P (2008) Organic synthesis: the disconnection approach, 2nd edn. Wiley

64. Thiel W Semiempirical quantum-chemical methods. Wiley interdisciplinary reviews: computational molecular science
65. Siddiqui KA, Tiekink ERT (2013) A supramolecular synthon approach to aid the discovery of architectures sustained by C-H...M hydrogen bonds. Chemical Communications
66. Matyska L, Koca J (1991) MAPOS: a computer program for organic synthesis design based on synthon model of organic chemistry. J Chem Inf Comput Sci 31(3):380–386
67. http://ivan.tubert.org/caos/caos.html. Accessed 17 Oct 2012
68. Gordeeva EV, Lushnikov DE, Zefirov NS (1990) COMPASS program: combination of empirical rules and combinatorial methods for planning of organic synthesis
69. Gillet V, Myatt G, Zsoldos Z, Johnson AP (1995) SPROUT, HIPPO and CAESA: tools for de novo structure generation and estimation of synthetic accessibility. Perspect Drug Discov 3(1):34–50
70. Socorro IM, Goodman JM (2006) The ROBIA program for predicting organic reactivity. J Chem Inf Model 46(2):606–614
71. http://www.reaxys.com. Accessed 17 Oct 2013
72. http://www.molecular-networks.com/products/sylvia. Accessed 17 Oct 2013
73. Nicoletti C, Jain ML, Georgieva P, Azevedo SO (2009) Novel computational methods for modeling and control in chemical and biochemical process systems. In computational intelligence techniques for bioprocess modeling, supervision and control. Springer, Berlin Heidelberg, pp 99–125
74. Serratosa F, Xicart J (1995) Organic chemistry in action, 2nd edn. Elsevier.
75. Siegbahn PE, Borowski T (2006) Modeling enzymatic reactions involving transition metals. Acc Chem Res 39(10):729–738
76. Zhou S, Yung Chan S, Cher Goh B, Chan E, Duan W, Huang M, McLeod HL (2005) Mechanism-based inhibition of cytochrome P450 3A4 by therapeutic drugs. Clin Pharmacokinet 44(3):279–304
77. Cooper GM (2000) The cell: a molecular approach. Sinauer Associate, Sunderland
78. Wright GD (2005) Bacterial resistance to antibiotics: enzymatic degradation and modification. Adv Drug Deliv Rev 57(10):1451–1470
79. Karaman R, Fattash B, Qtait A (2013) The future of prodrugs—design by quantum mechanics methods. Expert Opin Drug Deliv 10(5):713–729

Chapter 7
Predictive Methods for Organic Spectral Data Simulation

Abstract New chemical entities (NCE) with potential bioactivity are synthesized, isolated, and thoroughly characterized for structure elucidation and purity before being subjected to further research. Spectroscopy is one of the most powerful means to deduce the correct structure and configuration of a compound or a fragment. In organic synthesis, the compounds are usually characterized by the spectral techniques such as ultraviolet–visible (UV–Vis), nuclear magnetic resonance (NMR), infrared (IR), mass spectrometry (MS), X-ray, etc. NMR and MS methods are employed in fragment-based drug discovery approaches to identify compounds from a high-throughput screen or a proteomics experiment. However, it is not possible to manually interpret the complex spectral data that require sophisticated computational tools for characterization. These tools aid in spectra analysis, peaks assignment, intensity, etc. and thereby annotate the compound with the appropriate functional group and fragments. The prediction algorithms are developed based on principles of quantum chemistry, machine learning, or simple database/pattern match-based methods. Some of the methods using quantum chemistry are accurate; however, they require more computational time; on the other hand, the machine learning methods such as neural network are faster but require more experimental data for improving their prediction capability. So, there is a trade-off between speed and accuracy, and the user has to decide his/her preference. A number of spectra prediction tools, commercial as well as open source, are discussed in this chapter accompanied with detailed tutorials on the use of some of them. To manage the data, many online servers and spectral databases are available today and a brief introduction to them is also provided. Here, we also describe an in-house-developed carbon and proton NMR chemical shift-based binary fingerprints and their use in virtual screening.

Keywords NMR spectral data · Binary fingerprints · Chemical shift prediction · Classification · Virtual screening

M. Karthikeyan, R. Vyas, *Practical Chemoinformatics*,
DOI 10.1007/978-81-322-1780-0_7, © Springer India 2014

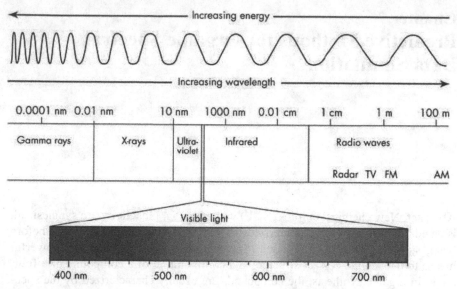

Fig. 7.1 The electromagnetic spectrum. (Source: http://9-4fordham.wikispaces.com/)

7.1 Introduction

The electromagnetic spectrum consists of radio waves, microwaves, infrared (IR) rays, visible light, ultraviolet (UV) light, X-rays, gamma rays, etc. and are classified based on their range of frequencies [1]. Here, we are interested in the frequency regions which provide the diagnostic power to organic chemists for structure determination (Fig. 7.1).

Organic spectroscopy aids chemists immensely to elucidate the structure of complex molecules using a combination of IR, ultraviolet–visible (UV–Vis), nuclear magnetic resonance (NMR), mass spectrometry (MS), and X-ray crystallographic techniques [2]. NMR detects the carbon and hydrogen environment in a molecule [3]. IR spectroscopy helps in detection of functional groups especially the fingerprint regions consisting of hydroxyl and carbonyl groups [4], UV aids in identifying conjugation, if present, between double bonds [5], MS confirms the molecular weight of the molecules along with fragmentation pattern [6], and X-rays give the final crystal structural composition, conformation, and configuration of a molecule [7]. A brief discussion on each of these techniques is given in this section for readers unfamiliar with spectroscopy; however, for the detailed theory, interested readers are encouraged to refer excellent textbooks and reviews on this topic [8–10].

UV spectroscopy is very effective in detecting extended conjugation in molecules like dienes and aromatic dyes. The principle underlying UV–Vis spectroscopy is the Lambert–Beer law which states that absorbance is directly proportional to path length "b" and concentration "c" [11].

The Lambert–Beer law can be stated in the form of Eq. (1) as

$$A = \varepsilon bc \tag{1}$$

where A is the absorbance, ε is the absorbtivity in L mol^{-1} cm^{-1}, b is the path length in cm, and c is the concentration of the compound in solution, in mol/L.

Different molecules absorb at different wavelengths and hence it can be used as a spectroscopic method. The visible region in the spectrum where human eyes can perceive lies in the range of 380–760 nm and the UV region ~300–380 nm. The energy corresponding to this region can promote an electron to a higher-energy orbital. There are a number of electronic excitations possible, such as $n - \pi^*$, $n - \sigma^*$, $\pi - \pi^*$, $\pi - \sigma^*$, and $n - \sigma^*$, each associated with a different energy level. When a sample compound is subjected to light radiation with energy corresponding to any of these transitions, some energy is absorbed. The light-absorbing groups present in a molecule are called chromophores. A UV spectrometer can detect the characteristic wavelength (lambda max, λ_{max}) at which a molecule is absorbed, thereby helping to identify the chromophores. We have predicted the maximum absorption wavelength λ_{max} values for a large set of 374 organic dyes for dye-sensitized solar cells based on extensive structure–property correlation studies [12].

IR spectroscopy is one of the most often used techniques applied in detecting functional groups, and the instrument is known as IR spectrometer. The IR frequency region in the electromagnetic spectrum of interest to organic molecules lies between 11.9×10^{13} and 1.2×10^{14} cm^{-1}. The energy in this region is just sufficient to cause vibrational excitation of covalently bound atoms or groups [13]. The bonds are considered as springs and show bending and stretching movements; there are others like rocking, scissoring, and twisting. The extent of the movement is determined by bond strength and mass of atoms present in the molecular fragment [14]. The absorption spectra show presence of functional groups as they absorb in different regions. For example, the IR spectrum of a molecule with carbonyl functional group shows a distinct sharp peak at 1,720 cm^{-1}. The region from 500 to 1,500 cm^{-1} is termed as the fingerprint region which is characteristic of a compound.

In NMR spectroscopy, the structure of a molecule as well as its purity is determined. A nucleus in a molecule is charged and when it spins, it generates a magnetic field. However, when the spins are not paired in a molecule, it generates a magnetic field dipole [15]. If an external magnetic field is applied, the spin can align with or against the external field creating two energy levels, the difference of which corresponds to the radio frequency region of the electromagnetic spectrum. When the spin returns to the ground level, energy is given out at the same frequency which is then recorded as a signal in the NMR spectrometer [16]. The first step in NMR spectral analysis is the detection of characteristic structural fragments from the chemical shift (δ) values. Chemical shift provides NMR its diagnostic power that reveals conformation and stereochemistry at the functional-group level. It also enables identification of the environment of a proton and the steric, electronic, and spatial arrangement of the neighboring atoms [17]. The chemical shift value of each fragment in a molecule gives rise to a peak in the spectrum as shown in Table 8.2. [18]. The principle behind identification of an atomic environment in carbon 13 and proton ^{1}H are the same, where both nuclei have spin 1/2 but the isotopic abundance of the hydrogen nuclei is 99 % whereas carbon is 1 % [19]. Proton NMR is recorded in the range 0–10 ppm, whereas the range is 20–200 ppm for carbon spectrum. Carbon spectra are proton decoupled to avoid large J couplings between carbon and hydrogen, and couplings between carbons are ignored [20].

Mass Spectroscopy is used to obtain the molecular weight of a sample [21]. When a charged particle passes through a magnetic field, it is deflected along a circular path on a radius which is proportional to the charge to mass ratio (m/e), for example, when an organic molecule is placed in the path of a high-energy beam, then an electron is knocked off to give a radical cation (molecular ion) which can further fragment, and the resulting ions are detected and recorded in a mass spectrometer [22].

7.2　Fragment-Based Drug Discovery

Fragment-based drug discovery (FBDD) methods are gaining precedence in lead identification and optimization phases of drug discovery processes [23]. Virtual drug-like molecules can be generated combinatorially from a fixed number of possible chemical structural fragments, and therefore prescreening fragments instead of fully enumerated libraries seems a more efficient approach. Although fragments sample most of the relevant chemical space, yet they leave scope for ligand optimization in terms of hydrophilicity, hydrophobicity, steric features, etc. to enhance their druglikeness [24]. Apart from structural elucidation of organic molecules, NMR also finds extended application in functional characterization of fragments in biological systems. Each fragment component in a compound makes some contribution to the overall biological activity; specific absorption rate (SAR) by NMR is a prevalent technique in drug discovery to understand ligand interactions with a target using chemical shift mapping to screen low-binding ligands [25]. The fragment libraries are characterized by biophysical analytical techniques like IR, NMR, and MS. The spectral values of common functional groups are given in Table 7.1 for ready reference.

Traditionally, complete structure elucidation of a new organic compound, either synthesized or naturally occurring, is assisted by a combination of elemental analysis, ^{1}H NMR, ^{13}C NMR, and MS techniques [26]. To explain this concept, let us take the example of two molecules 1 and 2 synthesized in our laboratory whose experimentally determined spectra are available [27]; Fig. 7.2.

Compound 1 is the structure containing an aromatic fragment fused with an eight-membered ring related to the class of alkaloids, for example, molecules isolated from autumn crocus [28]. This class of compounds has been studied extensively for their chemical, biological, and medicinal properties. They are effective in the treatment of gout and cancer [29]. Compound 2 shows a spirocyclic structure. Benzo spiroannulation is an important synthetic strategy in organic chemistry. Spirocyclic compounds like 2 either represent an integral part of some biologically active natural product or are utilized as intermediates for the synthesis of some biologically

Table 7.1 Spectral values of commonly occurring functional groups

S.No	Functional group	Molecule	IR(ν cm^{-1})	^{1}H NMR(ppm)	^{13}C NMR(ppm)	Mass(m/Z value)	UV(λ_{max} nm)
1	Alcohol	$H_3C\text{-}\overset{H_2}{C}\text{-}OH$	3200–3600, 3500–3700, 1050–1150	4.7, 3.59 1.18	17.4, 57.9	M$^+$ 46, base peak 29	240
2	Amine	$H_3C\text{-}\overset{NH_2}{\underset{}{C}}H_2$	3300–3500 1080–1360 1600	1.5, 2.69, 1.01	36.9, 19.0		
3	Amide	$H_3C\text{-}\overset{NH_2}{\underset{}{C}}{=}O$	1640–1690 3100–3500 1550–1640	7.0, 2.03	174.3, 22.5		
4	Imine	$\diagdown\diagup{=}NH$		9.36, 0.87	163.7, 16.1		
5	Acid	$H_3C\text{-}\overset{OH}{\underset{}{C}}{=}O$	1700–1725 2500–3300 1210–1320	11.0, 2.10	176.8, 20.8		
6	Alkene	$H_2C{=}CH_2$	3010–3100 675–1000 1620–1680	5.25	123.3		171
7	Aldehyde	$H_3C\text{-}\overset{H}{\underset{}{C}}{=}O$	1740–1720 2820–2850 2720–2750	9.79, 2.20	199.9, 30.7		
8	Ketone	$H_3C\text{-}\overset{O}{\underset{}{C}}\text{-}CH_3$	1670–1820	2.13	30.6, 206.4		290 180
9	Cyanide	$H_3C\text{-}C{\equiv}N$	2210–2260	1.98	117.8, 1.2		200
10	Diene	$\diagup{=}\diagdown\diagup{=}$		6.31, 5.08, 5.19	137.2, 116.1		217
11	Alkyne	$HC{\equiv}CH$	3300, 2100–2260	1.91	71.9		180
12	Alkane	$H_3C\text{-}CH_3$	2850–3000, 1350–1480	0.86	6.5		
13	Halide	$H_3C\text{-}CH_2Cl$	1000–1400 600–800	4.42, 1.49	18.9, 40		205 255
14	Ester	(ethyl ester structure)	1735–1750 1000–1300	4.12, 2.04, 1.26	170.2, 61.0, 20.7, 14.1		280
15	Ether	(diethyl ether structure)	1000–1300	1.21, 3.48	66.3, 15.2		255
16	Aromatic	(benzene ring)	3000–3100 1400–1600	7.26	128.5		295
17	Peroxy	(diethyl peroxide structure)		3.57, 1.10	63.7, 11.5		
18	Azide	$\text{-}N{=}N^+{=}N^-$	2100–2270	1.55, 0.9	43, 12.6		290
19	Sulphide	(diethyl sulphide structure)		2.48, 1.15	25.5, 14.7		
20	Nitro	$\text{-}\overset{O}{\underset{O^-}{N^+}}$	1515–1560 1345–1385				275 200

Fig. 7.2 Structure of the compounds used for spectral data interpretation

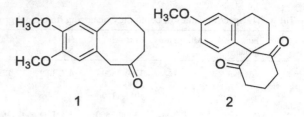

1   2

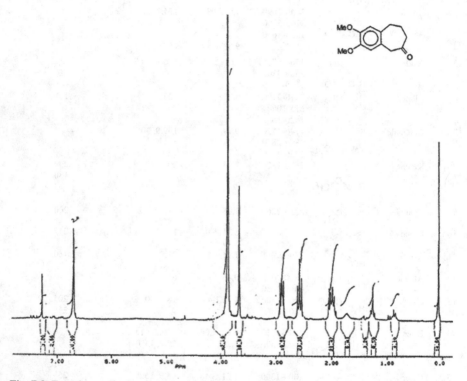

Fig. 7.3 Experimentally determined proton nuclear magnetic resonance (*NMR*) spectrum of compound 1

significant compounds. In this chapter, we will subject them to various spectra prediction tools and compare the results with experimental spectra (Fig. 7.3).

^{1}H NMR spectrum of 1 showed two singlets at δ 6.65 (^{1}H) and 6.70 (^{1}H) for the protons attached to aromatic carbons C1 and C4, respectively. The other two singlets observed at δ 3.90 and 3.85, integrating for three protons each, are assigned to –OMe groups. A sharp singlet appearing at δ 3.70 (2H) corresponds to methylene protons (C5-CH2) confirming the cyclization reaction. Methylene group protons attached to C7 and C10 appeared as triplets at δ 2.35 (*J* = 6.94 Hz) and 2.80 (*J* = 6.94 Hz), respectively. A multiplet at δ 1.80 (4H) corresponds to C8 and C9 methylene protons (Fig. 7.4).

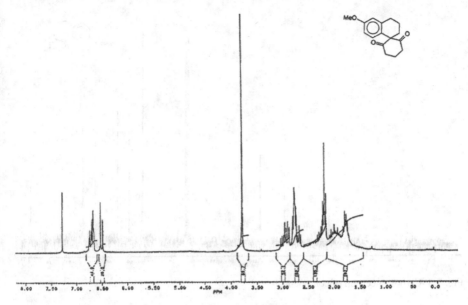

Fig. 7.4 Proton nuclear magnetic resonance (*NMR*) of compound 2

In the ¹H NMR spectrum of 2, C7-H appeared as a double doublet at δ 6.70 (*J1* = 8.78, *J2* = 1.95), C5-H appeared as a broad singlet at δ 6.65, and C8-H aromatic proton appeared as a doublet at δ 6.50 (*J* = 8.78). The OMe group protons appeared at δ 3.80 as a singlet. A multiplet observed at δ 2.97 is assigned to the protons attached to C4 and a multiplet appearing between δ 2.50 and 2.20 (6H) is characterized for the methylene protons attached to C2, C3', and C5', respectively. Another multiplet appearing between δ 1.85 and 1.70 (4H) corresponds to methylene protons attached to C3 and C4 (Fig. 7.5).

The ¹³C NMR spectrum of 1 showed 13 signals and the characterizations of each carbon signal are suggested by the insensitive nuclei enhanced by polarization transfer (INEPT) experiment which are as follows: The signal appearing at δ 211.76 corresponds to C6 keto carbon. The aromatic carbons C2 and C3, bearing –OMe groups, appeared at δ 148.68 and 147.69, respectively. Two aromatic quaternary carbons C4a and C10a, fused with a cyclooctanone moiety, appeared at δ 133.13 and 125.63, respectively. C1 and C4 methine carbon signals appeared at δ 113.38 and 113.15, respectively. Both the methoxy carbons appeared at δ 56.03. The characteristic C5 methylene carbon signal appeared at δ 48.19. Other four methylene carbons (C10, C9, C8, and C7) appeared at δ 32.94, 31.33, 24.71, and 41.12, respectively (Fig. 7.6).

The ¹³C NMR spectrum showed 14 signals. The carbonyl group carbon signals appeared at δ 209.85. The aromatic signal corresponding to C6 appeared at δ 158.38. The other two quaternary carbons C4a and C8a appeared at δ 139.64 and 125.31, respectively. Methine carbon signals for C8, C7, and C5 appeared at δ 131.30, 113.41, and 112.59, respectively. The characteristic quaternary spiro carbon C1 appeared at δ 70.66. The methoxy group carbon appeared at δ 55.04. All other

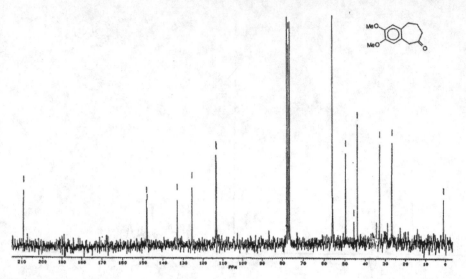

Fig. 7.5 Carbon spectrum of compound 1

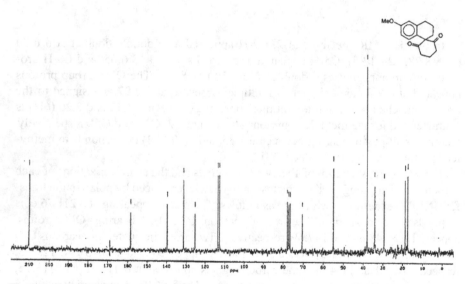

Fig. 7.6 Carbon NMR of compound 2

six methylene carbons such as C3', C5' (2C), C4, C2, C3, and C4' appeared at δ 38.06 (2C), 34.14, 29.47, 18.88, and 17.55, respectively (Fig. 7.7).

Mass spectrum of 1 showed molecular ion peak (m/z) at 234, along with other fragmentation peaks at 206, 191, 175, 165, 121, 107, and 91 (Fig. 7.8).

The mass spectrum of the compound 2 showed molecular ion peak at 258 and base peak at 174.

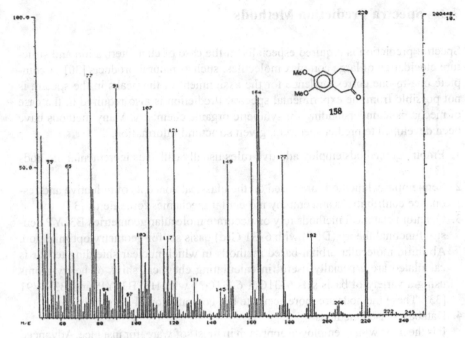

Fig. 7.7 Mass spectrum of compound 1

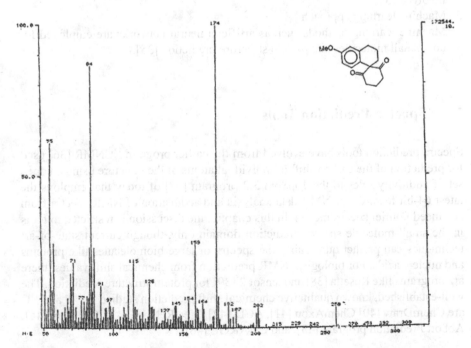

Fig. 7.8 Mass spectrum of compound 2

7.3 Spectra Prediction Methods

Spectral prediction is required especially in the case of characterization and structure elucidation of large complex molecules, such as natural products [30]. A complete one-to-one correspondence for the assignment of the peaks in the spectra is not possible from the experimental spectra. Prediction is also required in the case of mechanistic understanding for synthetic organic chemistry. Many methods have been developed to predict spectrum, given structural information.

1. Empirical methods employ additive rules usually called as incremental methods [31].
2. Semiempirical methods are based on the classical concepts of inductive and resonance contributions and employ molecular mechanics force fields [32].
3. Quantum chemical methods rely on accurate molecular geometries B3LYP density functinal theory (DFT) with 6–31 G(d) basis set for geometry optimization. Ab initio molecular orbital-based methods in which nuclear shielding tensor is calculated are especially useful in calculating chemical shifts of heavy atoms using a variety of basis sets 6–31G*, 6–31G** with HF, B3LYP, and B3PW91 [33]. These methods are more accurate but computationally expensive.
4. Database-based methods
   It is the most widely employed approach in most software, for instance, Advanced Chemistry Development, Inc. (ACD/Labs). It is faster because three-dimensional (3D) geometries are not determined only matching with stored chemical shifts is involved [34].
5. Machine learning approach
   Machine learning methods such as artificial neural networks are employed for both small molecules and protein structure prediction [35].

7.4 Spectra Prediction Tools

Spectra prediction tools have evolved from the earlier program ^{13}CNMR [36] used for prediction of the carbon shift of individual atoms of the structure using an open set of additivity rules to the TopSpin 3.2 program [37] of today that employs the latest 64-bit features for NMR data analysis and acquisition of NMR spectra from advanced Fourier spectrometers. In this chapter, the discussion is restricted to tools in the small-molecule spectra prediction domain only, though current state-of-art techniques can predict quite fairly the spectra of large biomolecules like proteins and nucleic acids. For biological NMR prediction from chemical shift values, there are programs like Rosetta [38] and tensor 2 [39] for protein structure prediction. The well-established, known qualitative chemical shift prediction studied for ^{1}H and ^{13}C are ChemDraw [40] ChemAxon [41], ACD [42], MestReNova [43], Gaussian [44], Abbott Prediction program [45], and CHARGE [46].

7.5 Open-Source Tools

7.5.1 GAMESS

GAMESS is a program for ab initio molecular quantum chemistry which can compute self-consistent field (SCF) wave functions ranging from restricted Hartree–Fock (RHF), ROHF, UHF, GVB, and MCSCF [47]. Computation of the Hessian energy permits prediction of vibrational frequencies with IR or Raman intensities. Solvent effects may be modeled by the discrete Effective Fragment potentials or continuum models such as the polarizable continuum model [48]. Numerous relativistic computations are available, including infinite order two component scalar corrections, with various spin–orbit coupling options [49].

7.6 Proprietary Tools

7.6.1 ACD/NMR Predictors

The program includes predictions for the following nuclei—^1H, ^{13}C, ^{15}N, ^{19}F, and ^{31}P—for 1D spectra, and ^1H and ^{13}C (and ^{15}N) for 2D spectrum prediction [50]. All predictors use both Hierarchical Organisation of Spherical Environments (HOSE) code and neural net algorithms to provide the most accurate chemical shifts in the prediction of spectra also taking into account stereochemistry. The main advantage of the program is that it includes full processing functionality and the ability to train predictions with users' own experimental data [51].

7.6.2 Cambridgesoft Chem3D

This program provides an interface to multiple computational tools like Gaussian, GAMESS, and Jaguar. ChemBio3D provides an interface for Gaussian calculations for computing ^1H, NMR, IR, and Raman spectra [52]. Its 2D drawing tool ChemBioDraw Ultra has provisions to predict NMR spectra [53]. The predicted spectra of compounds 1 and 2 are shown along with the predicted shifts of each atom (Figs. 7.9 and 7.10).

7.6.3 Jaguar

Jaguar is a high-performance ab initio electronic structure package for both gas- and solution-phase simulations, with ability to treat metal-containing systems; Jaguar computes a comprehensive array of molecular properties including NMR, IR, and UV-Vis [54].

ChemNMR ¹H Estimation

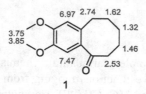

1

Estimation quality is indicated by color: good, medium, rough

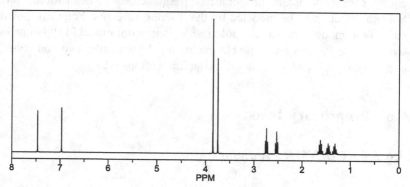

Protocol of the H-1 NMR Prediction (Lib=SU Solvent=DMSO 300 MHz):

Node Shift Base + Inc. Comment (ppm rel. to TMS)

CH 7.47 7.26 1-benzene

CH 6.97 7.26 1-benzen

CH2 2.74 1.37 methylene
CH2 2.53 1.37 methylene
CH2 1.62 1.37 methylene
CH2 1.46 1.37 methylene
CH2 1.32 1.37 methylene
CH3 3.85 0.86 meth
CH3 3.75 0.86 methyl

1H NMR Coupling Constant Prediction

shift atom index coupling partner, constant and vector

7.47	3		
6.97	6		
2.74	12		
	11	7.1	H-CH-CH-H
2.53	8		
	9	7.1	H-CH-CH-H
1.62	11		
	12	7.1	H-CH-CH-H
	10	7.1	H-CH-CH-H

Fig. 7.9 Predicted hydrogen and carbon spectra of compound 1 using ChemBioDraw Ultra

1.46	9		
	8	7.1	H-CH-CH-H
	10	7.1	H-CH-CH-H
1.32	10		
	11	7.1	H-CH-CH-H
	9	7.1	H-CH-CH-H
3.85	17		
3.75	15		

ChemNMR ^{13}C Estimation

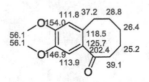

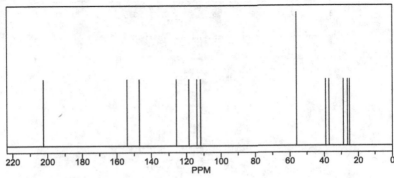

Protocol of the C-13 NMR Prediction: (Lib=S)

Node Shift Base + Inc. Comment (ppm rel. to TMS)

C 146.9 128.5 1-benzene
C 154.0 128.5 1-benzene
C 125.7 128.5 1-benzene
C 118.5 128.5 1-benzene
CH 113.9 128.5 1-benzene
CH 111.8 128.5 1-benzene
C 202.4 193.0 1-carbonyl
CH2 37.2 -2.3 aliphatic
CH2 39.1 -2.3 aliphatic
CH2 28.8 -2.3 aliphatic
CH2 25.2 -2.3 aliphatic
CH2 26.4 -2.3 aliphatic
CH3 56.1 -2.3 aliphatic
CH3 56.1 -2.3 aliphatic

Fig. 7.9 (continued)

ChemNMR ¹H Estimation

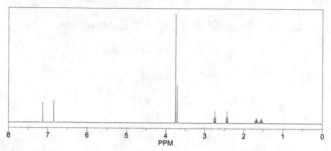

Estimation quality is indicated by color: good, medium, rough

Protocol of the H-1 NMR Prediction (Lib=SU Solvent=DMSO 300 MHz):

```
Node     Shift     Base + Inc.    Comment (ppm rel. to TMS)

CH  7.13             7.26          1-benzene
                    -0.38          1 -O-C
                     0.00          1 -O-C
                     0.16          1 -CC=R
                     0.00          1 -CC
                     0.09          general corrections
CH  6.85             7.26          1-benzene
                     0.00          1 -O-C
                    -0.38          1 -O-C
                     0.00          1 -CC=R
                    -0.08          1 -CC
                     0.05          general corrections
CH2 3.71             1.37          methylene
                     1.22          1 alpha -1:C*C*C*C*C*C*1
                     1.12          1 alpha -C(=O)-C
CH2 2.74             1.37          methylene
                     1.22          1 alpha -1:C*C*C*C*C*C*1
                    -0.06          1 beta -C
                     0.21          general corrections
CH2 2.43             1.37          methylene
                     1.12          1 alpha -C(=O)-C
                    -0.06          1 beta -C
CH2 1.68             1.37          methylene
                     0.29          1 beta -1:C*C*C*C*C*C*1
                    -0.06          1 beta -C
                     0.08          general corrections
CH2 1.55             1.37          methylene
                     0.24          1 beta -C(=O)-C
                    -0.06          1 beta -C
CH3 3.75             0.86          methyl
                     2.87          1 alpha -O-1:C*C*C*C*C*C*1
                     0.02          general corrections
CH3 3.75             0.86          methyl
                     2.87          1 alpha -O-1:C*C*C*C*C*C*1
                     0.02          general corrections
```

1H NMR Coupling Constant Prediction

```
shift    atom index   coupling partner, constant and vector

7.13        3
6.85        6
3.71        7
2.74       12
                       11   7.1   H-CH-CH-H
2.43        9
                       10   7.1   H-CH-CH-H
1.68       11
                       12   7.1   H-CH-CH-H
                       10   7.1   H-CH-CH-H
1.55       10
                        9   7.1   H-CH-CH-H
                       11   7.1   H-CH-CH-H
3.75       17
3.75       15
```

Fig. 7.10 Predicted hydrogen and carbon spectra of compound 2 using ChemBioDraw Ultra

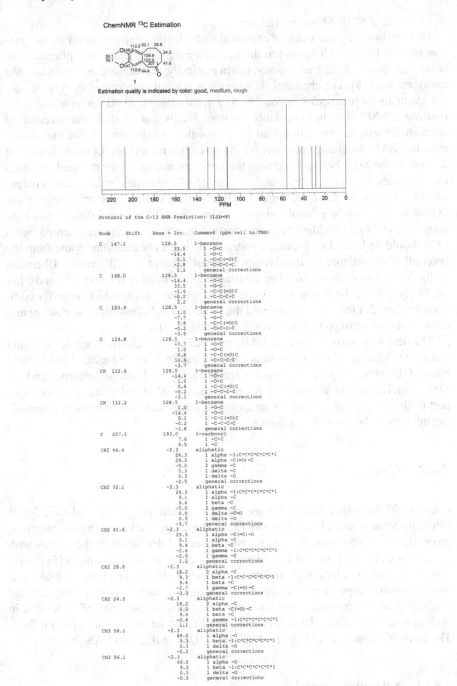

Fig. 7.10 (continued)

Computing NMR Shielding Tensors

Usually in solids, the value of chemical shift changes depending on the orientation of a molecule with respect to the external magnetic fields. This phenomenon is termed chemical shift anisotropy, mathematically represented as chemical shift tensor matrix [55]. The chemical shift tensor is generally described by three diagonal elements or principal components δ_{11}, δ_{22}, and δ_{33} [56]. Gas-phase and solution-phase NMR shielding constants are available for closed-shell and unrestricted open-shell wave functions in Jaguar [57]. To calculate chemical shifts, one should calculate NMR shielding constants for the reference molecules for each element of interest, in the same basis set and with the same method as for the molecule of interest. Shielding constants are returned as atom-level properties in the maestro output file. One can use these values for atom selection, for example, or can display them in labels. Shieldings are calculated for all atoms, including those with effective core potentials (ECPs). Shielding constants for atoms whose core is represented by an ECP should be treated with caution because the main contributions come from the core tail of the valence orbitals, which is largely absent at ECP centers. Chemical shifts derived from these shielding constants might display the correct trends, but are likely to have the wrong magnitude. Here, we have computed tensors for compounds 1 and 2. First, we need to build the structures in Schrodinger workspace and click applications to go to Jaguar (Figs. 7.11 and 7.12).

The next step is to calculate the NMR spectrum of the reference, say tetra methyl silane (TMS) molecule using the same method and basis set in Jaguar. The isotropic parts of the magnetic shielding tensors are extracted for both the reference and the sample molecule. The chemical shift can be calculated by subtracting the isotropic part of the magnetic shielding tensor from the calculated value for the corresponding nucleus in the reference molecule.

7.6.4 Gaussian

Gaussian 98 includes a facility for predicting magnetic properties, including NMR shielding tensors and chemical shifts [58]. These calculations compute magnetic properties from first principles, as the mixed second derivative of the energy with respect to an applied magnetic field and the nuclear magnetic moment [59]. As a result, they can produce high-accuracy results for the entire range of molecular systems studied experimentally via NMR techniques. Gaussian can also be used for predicting IR spectrum as it can compute vibrational frequencies of molecules in their ground and excited states. It can also predict the intensity of the spectral lines. The available methods are Hartree–Fock (HF), DFT, MP2, and CASSF.

7.6.4.1 A Practice Tutorial

Now let us compute the spectra of compound 2 using Gaussian program. The structure is built using the drawing templates provided in the program and energy mini-

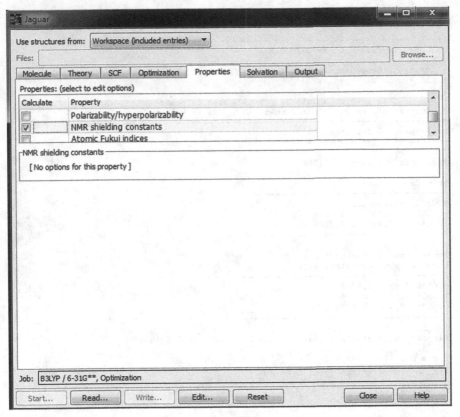

Fig. 7.11 NMR shielding constants calculation for compounds 1 and 2 using the Jaguar module of Schrodinger suite

mized and saved as a Gaussian input file (gif). The NMR option is selected under job type tab and the method chosen is the Gauge Independent Atomic Orbital (GIAO) method. The basis set and method used are specified in the calculation setup screen. We will use 6–31G, a split valence basis set, and HF method to compute the spectra (Figs. 7.13, 7.14, 7.15 and 7.16).

7.6.5 ADF

Amsterdam Density Functional (ADF) is an accurate, parallelized, and powerful computational chemistry program used to understand and predict chemical structure and reactivity with DFT [60]. It is a popular tool to predict and understand magnetic, electric, optical, and vibrational spectra [61]. Heavy elements and transition metals can be modeled with ADF's relativistic zeroth order regular approximation (ZORA) approach and all-electron basis sets for the whole periodic table. It can be used to compute IR frequencies and intensities, vibrational circular dichroism (VCD), mobile block Hes-

```
Job c1cccc_c12_CCCCC__O_C2 started on RenuVyas-VAIO at Sat Jul 27 23:14:14 2013
jobid: RenuVyas-VAIO-0-51f406ec

+---------------------------------------------------------------------+
| Jaguar version 8.0, release 515 |
| |
| Copyright Schrodinger, LLC |
| All Rights Reserved. |
| |
| Use of this program should be acknowledged in publications as: |
| Jaguar, version 8.0, Schrodinger, LLC, New York, NY, 2011. |
+---------------------------------------------------------------------+
NMR Properties for atom C1
==================================

Isotropic shielding: 70.7568

Shielding Tensor
-------------------------------------
87.3542 -3.6096 29.0771
-22.2706 10.5956 -45.1205
32.4488 -46.9852 114.3206
-------------------------------------


Symmetrized Shielding Tensor
Eigenvectors (Principal Axes)
-------------------------------------
0.0136 0.8815 0.4719
0.9359 0.1550 -0.3164
0.3521 -0.4460 0.8229
-------------------------------------


Eigenvalues sigma11, sigma22, sigma33: -6.916 69.516 149.670

sigma_parallel: 149.670
sigma_perpendicular: 31.300

Anisotropy: 118.3704
```

Fig. 7.12 Part of Jaguar output file showing the computed tensors

sian (MBH), Franck–Condon factors and (resonance) Raman, vibrational Raman optical activity (VROA), UV/Vis spectra, etc. NMR spectroscopy parameters like chemical shift, spin–spin coupling, paramagnetic NMR, electron paramagnetic resonance (EPR) g-tensor, hyperfine interaction (A-tensor), and ZFS can also be obtained.

7.6.6   MestreNova

In MestreNova (Mnova), a raw, unprocessed spectrum (free induction decay, FID) can be opened to obtain the fully processed spectrum instantaneously [62]. This involves two fundamental steps—automatic file format recognition and automatic processing of the FID using the concept of real-time frequency domain processing.

Additionally, it provides users with their own choice of processing parameters, changing or adjusting the window function, the Fourier transform (FT), the phasing and baseline correction. Mnova can detect spectra acquired in the arrayed mode (or

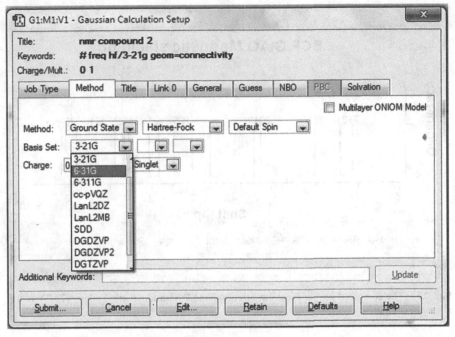

Fig. 7.13 The Gaussian calculation setup screen to select the basis set and the methods for the computation of spectra

Fig. 7.14 The results summary page

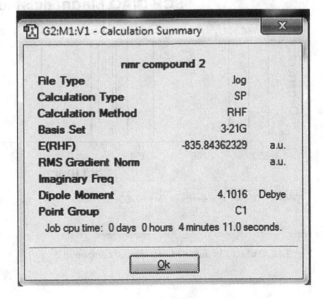

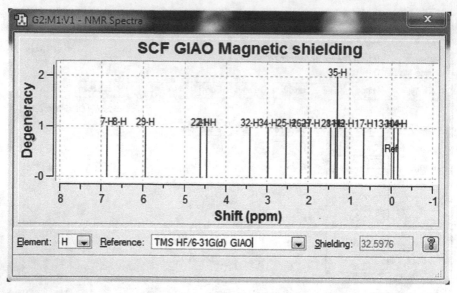

Fig. 7.15 The computed proton spectrum of compound 2 with tetramethyl silane (*TMS*) as the reference compound

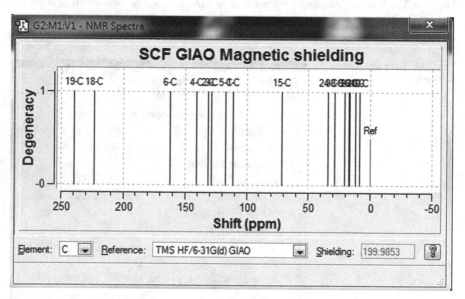

Fig. 7.16 Carbon 13 computed spectrum of compound 2

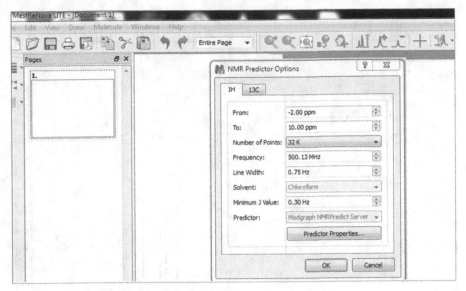

Fig. 7.17 The welcome screen of Mnova Lite

pseudo 2D), typically used in relaxation, kinetics, or diffusion experiments and, by default, will display the spectrum as a stacked plot. Mnova integrates a fast simulation module of ^1H and ^{13}C NMR spectra, called Modgraph NMRPredict Desktop. NMRPredict Desktop uses a neural network system for the prediction of ^{13}C NMR spectra as well as the CHARGE program which offers ^1H NMR prediction based on partial atomic charges and steric interactions. The CHARGE and the Increment algorithms included in NMRPredict Desktop are the same as used in the server-based version to predict ^1H NMR spectra. For the prediction of ^{13}C NMR spectra, it uses a neural network system, but not the HOSE database methodology implemented additionally in the server-based application. The neural network algorithm is much more general and error tolerant than the HOSE code approach (based on a reference spectra database) and is much more accurate at predicting shifts not found in the database [63]. The best algorithm is the combined approach between the Increments and the Conformers algorithm that is capable of producing significantly improved proton NMR predictions [64]. The 4,000,000-assigned chemical shift values of the available 345,000 reference spectra can be predicted with an average deviation between experimental versus calculated of below 2.00 ppm.

Now let us familiarize with the Mnova tool, we shall use it to predict the spectrum of the two compounds 1 and 2. The initial welcome screen of Mnova Lite is shown in Fig. 7.17. With a molecular structure highlighted in the active page of Mnova, we just go to the "Molecule" menu and select "Prediction Options".

One can either import a predrawn structure or draw a molecule here and predict the spectra (Fig. 7.18).

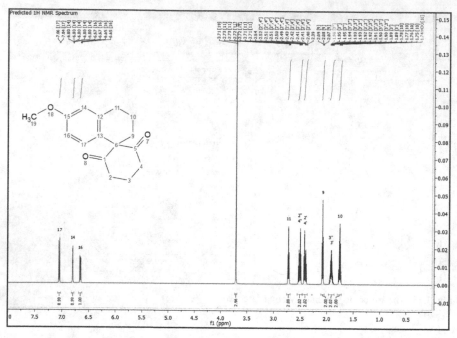

Fig. 7.18 Proton nuclear magnetic resonance (NMR) spectrum of compound 2 predicted using the Mnova program

7.6.7 Spartan

Spartan can compute proton, carbon-13, DEPT and COSY spectra. Additionally it can also be used for large biomolecules like proteins [65].

SPARTA is a database system for empirical prediction of backbone chemical shifts (N, HN, HA, CA, CB, CO) using a combination of backbone phi, psi torsion angles, and side chain chi1 angles from a given protein with known Protein Data Bank (PDB) coordinates [66].

7.6.7.1   A Practice Tutorial

This section will describe how to predict NMR spectrum using the Spartan program. We will predict NMR of the spiro compound 2 using the HF method with 6–31G* basis set. First, the software performs the job of geometry optimization and then calculates the NMR parameters. The structure is initially built using the build option (Fig. 7.19).

An example input file is given here:

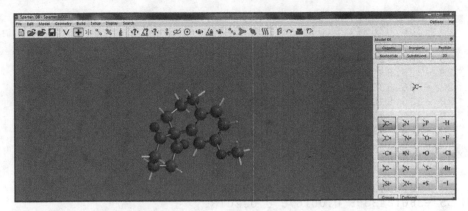

Fig. 7.19 Structure input in Spartan

```
Job type: Geometry optimization.
Method: RHF
Basis set: 6-31G(D)
Number of shells: 68
Number of basis functions: 178
Multiplicity: 1
Parallel Job: 4 threads
SCF model:
A restricted Hartree-Fock SCF calculation will be
performed using Pulay DIIS + Geometric Direct Minimization
Optimization:
Step Energy Max Grad. Max Dist.
1 -386.673151 0.125776 0.117385
2 -386.688700 0.122063 0.107133
3 -386.696587 0.120569 0.142183
4 -386.698430 0.118304 0.150008
5 -386.704271 0.113172 0.098271
6 -386.714717 0.100973 0.103374
7 -386.745129 0.075809 0.152018
8 -386.774432 0.055324 0.164277
9 -386.798198 0.033672 0.151093
10 -386.814140 0.019009 0.137517
11 -386.823137 0.010698 0.124781
12 -386.828546 0.010741 0.123031
13 -386.832040 0.005351 0.093668
14 -386.833871 0.005071 0.099665
15 -386.835186 0.003576 0.083391
16 -386.836045 0.001275 0.088272
17 -386.836623 0.001002 0.160542
18 -386.836981 0.001478 0.081088
19 -386.837315 0.001741 0.182204
20 -386.837579 0.001631 0.093630
21 -386.837832 0.001945 0.167044
```

```
22 -386.838087 0.001819 0.157145
23 -386.838341 0.002085 0.175730
24 -386.838647 0.001906 0.229830
25 -386.838938 0.002454 0.172126
26 -386.839405 0.002223 0.204335
27 -386.839919 0.002714 0.160672
28 -386.840524 0.002916 0.208574
29 -386.841076 0.002545 0.174013
30 -386.841743 0.001362 0.218336
31 -386.842246 0.001513 0.169910
32 -386.842779 0.001132 0.175853
33 -386.843135 0.001326 0.096142
34 -386.843399 0.001687 0.155910
35 -386.843509 0.001670 0.032018
36 -386.843598 0.001049 0.100642
37 -386.843642 0.001629 0.057971
38 -386.843732 0.001170 0.052026
39 -386.843772 0.000849 0.046999
40 -386.843798 0.000369 0.025808
41 -386.843801 0.000068 0.003419
42 -386.843802 0.000027 0.000805
```

<step 2>
Job type: Single point.
Method: RHF
Basis set: 6-31G(D)
SCF total energy: -386.8438016 hartrees
NMR shifts (ppm)
Atom Isotropic Rel. Shift

```
1 H1 25.4238 7.48
2 C1 75.4701 126.25
3 C4 75.0684 126.65
4 C2 62.4297 139.29
5 C6 74.5722 127.15
6 C5 78.1008 123.62
7 C3 74.3070 127.41
8 H6 25.3281 7.57
9 H5 25.4933 7.41
10 H3 25.5653 7.34
11 H4 25.4020 7.50
12 C7 161.8341 39.89
13 H2 30.9525 1.95
14 H7 30.1867 2.72
15 C8 173.9551 27.77
16 H10 31.2758 1.63
17 C9 178.6904 23.03
18 H9 32.0148 0.89
19 H11 31.6720 1.23
20 H12 31.7307 1.17
```

```
21 C10 182.5210 19.20
22 H8 32.3509 0.55
23 H13 32.0721 0.83
24 H14 31.9603 0.94
Reason for exit: Successful completion
Quantum Calculation CPU Time : 8:54.12
Quantum Calculation Wall Time: 17:48.86
SPARTAN '08 Semi-Empirical Program: (PC/x86) Release 132
Semi-empirical Property Calculation
M0001
Guess from Archive
Energy Due to Solvation
Solvation Energy SM5.4/A 2.873
Memory Used: 1.362 Mb
Reason for exit: Successful completion
Semi-Empirical Program CPU Time : .17
Semi-Empirical Program Wall Time: .07
SPARTAN '08 Properties Program: (PC/x86) Release 132
Reason for exit: Successful completion
Properties CPU Time : .83
Properties Wall Time: .77
```

When the structure input file is saved, the calculation is set up and submitted to the program (Fig. 7.20).

Once the job is completed, the display spectra option shows the calculated carbon and proton spectra (Fig. 7.21).

Spartan can also compute advanced NMR spectra like Correlated Spectroscopy (COSY) [67] and Nuclear Overhauser Effect Spectroscopy (NOESY) ([68]; Figs. 7.22 and 7.23).

The IR frequencies can also be calculated using Spartan for the same molecule (Fig. 7.24).

The UV/Vis spectrum is similarly obtained (Fig. 7.25).

7.6.8 Spectral Databases

7.6.8.1 NMRshiftdb2

The NMRshiftdb2 software is open source; the data are published under an open-content license [69].

It has an NMR database (web database) for organic structures and their NMR spectra. It allows for spectrum prediction (^{13}C, ^{1}H, and other nuclei) as well as for searching spectra, structures, and other properties. It also has a collection of peer-reviewed datasets by its users.

Fig. 7.20 Spartan calculation setup

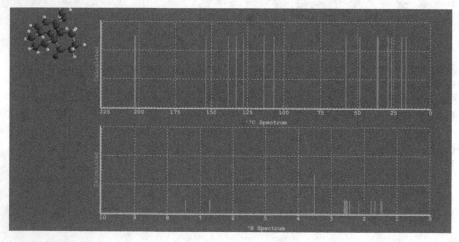

Fig. 7.21 The displayed hydrogen and carbon nuclear magnetic resonance (*NMR*) of compound 2 after computation

7.6.8.2 MassBank

MassBank is the first public repository of mass spectral data for sharing them among the scientific research community [70]. MassBank data are useful for the chemical identification and structure elucidation of chemical compounds detected by MS spectroscopy. The spectra can be searched by exact *m/z* using a browsing interface. One can also perform spectrum, substructure, and peak searches for a given compound. It does substructure searching of chemical compounds. One can retrieve

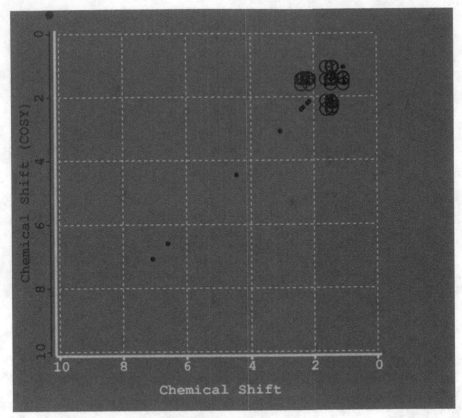

Fig. 7.22 A computed Correlated Spectroscopy (*COSY*) spectrum

spectra similar to the user's spectrum in terms of molecular formulas. This search is helpful to predict the chemical structure of unknown metabolites (Fig. 7.26).

The Spectrum Search feature in MassBank retrieves the chemical compound(s) specified by chemical name or molecular formula and displays its spectra. We gave *spiro* keyword as a query using the quick search option in the browser and retrieved 56 hits, many of which were drug molecules. One can refine results by specifying the instrument and ionization mode (Fig. 7.27).

The MassBank records have one-to-one relation to a specific mass spectrum. Each record has specific information like accession number, record file, license, and author apart from information on the chemical compound regarding its formula, mass, smiles, InChI identifier etc. The analytical information available is the instrument type and make, Msn type data. A typical MassBank record is shown here (Fig. 7.28).

One can also get chemical structures of unknown metabolites from the query compound (Fig. 7.29).

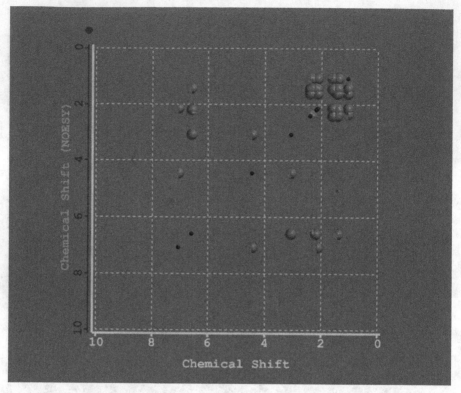

Fig. 7.23 Nuclear Overhauser Effect Spectroscopy (*NOESY*) spectrum of compound 2

7.6.8.3   SWGDRUG Mass Spectral Library

SWGDRUG has compiled a mass spectral library from a variety of sources containing drugs and drug-related compounds [71]. All spectra were collected using electron-impact MS systems. This library is available for download from its website.

7.6.8.4   SDBS

Spectral Database for Organic Compounds (SDBS) is an integrated spectral database system for organic compounds, which includes six different types of spectra, an electron-impact mass spectrum (EI-MS), a Fourier transform infrared spectrum (FT-IR), a ^{1}H NMR spectrum, a ^{13}C NMR spectrum, a laser Raman spectrum, and an electron spin resonance (ESR) spectrum [72]. SDBS is maintained by the National Metrology Institute of Japan (NMIJ) under the National Institute of Advanced Industrial Science and technology (AIST). Currently, EI-MS spectrum, ^{1}H NMR spectrum, ^{13}C NMR spectrum, FT-IR spectrum, and the compound dictionary are

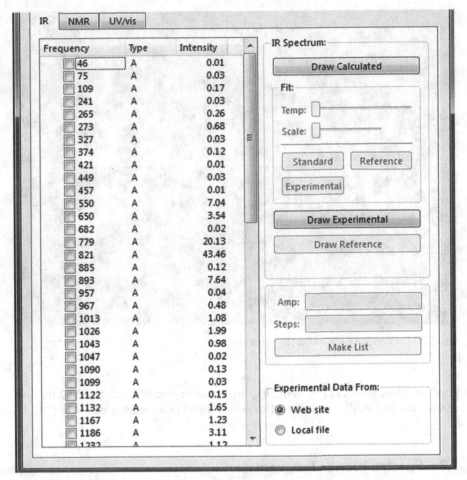

Fig. 7.24 The infrared (*IR*) frequencies and the displayed spectrum of compound 2

active for correcting and maintenance of the data. Since 1997, SDBS is free to the public through Tsukuba Advanced Computing Center (TACC) as Research Information Data Base (RIO-DB). A compound name search using *spiro* keyword gave 183 hits of NMR spectrum one of which is shown here. ^{1}H NMR was measured with a JEOL FX-90Q (89.56 MHz), a JEOL GX-400 (399.65 MHz), or a JEOL AL-400 (399.65 MHz) (Fig. 7.30).

7.6.8.5 Spectral Libraries

Sigma Aldrich libraries having text- and data-searching capabilities for FT-NMR, FT-IR, and attenuated total reflectance-infrared (ATR-IR) spectra are available as

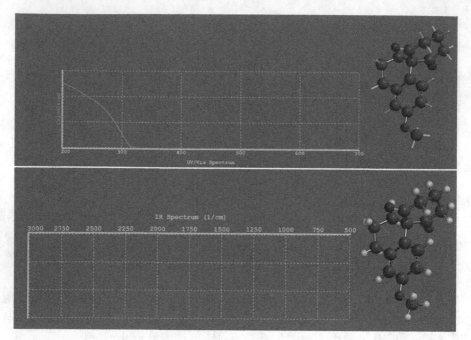

Fig. 7.25 The computed ultraviolet (*UV*) spectrum

good reference sources [73]. Scifinder, the search engine of CAS, has 59.2 million carbon and 59.1 million proton NMR spectra stored in its registry database [74].

7.7   Spectra Viewer Programs

JSpecView is a viewer for spectral data in the JCAMP-DX and AnIML/CML format [75]. The program was initially developed at the Department of Chemistry of the University of the West Indies, Mona, Jamaica, West Indies and is available via sourceforge net under the GNU Lesser General Public License. It is an open-source viewer and converter for multiple spectra (Fig. 7.31).

7.8   In-House Tools for Spectra Prediction

40,000 compounds stored in NMRshiftDB and an in-house NMR data archive for computing binary fingerprints from chemical shift data were processed [76]. From this dataset of original NMR spectra, we used reported chemical shift val-

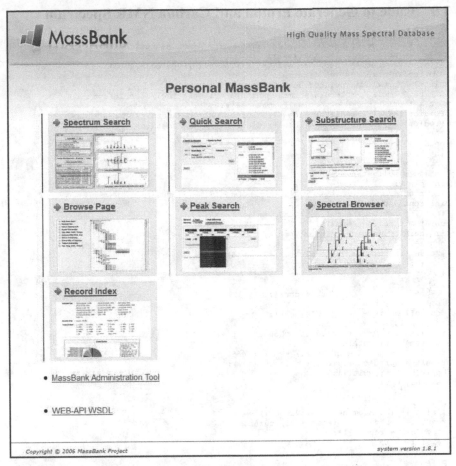

Fig. 7.26 MassBank home page displaying the various search options in the database

ues to generate the binary fingerprints. Conventionally, the area of the peak at specific positions represents the number of atoms with similar environment. In our approach, if there was a peak in the region, the bit was allocated to the highest peak; peak intensity analysis was performed via atom count with the same chemical shifts (Fig. 7.32). Next, we statistically analyzed the bins based on frequency of occurrence of particular peaks in the NMR spectra. We were able to calculate NMR shift-based binary fingerprints of entire PubChem [77], ChEMBL [78], and HMDB [79] database molecules using high-performance computing (HPC) tools.

7.9   Code to Generate Proton and Carbon NMR Spectrum

```
public String getBothHandCNMRdata(String smi) throws Exception {
 Molecule mol = getMolFromSmi(smi);
 AtomContainerManipulator.convertImplicitToExplicitHydrogens(mol);
 IMolecularFormula moleculeFormula =
MolecularFormulaManipulator.getMolecularFormula(mol);
 String formula = MolecularFormulaManipulator.getString(moleculeFormula);
 DecimalFormat df = new DecimalFormat("####.00");
 PredictionTool cpredictor = new PredictionTool("cNMRdata.csv");
 PredictionTool hpredictor = new PredictionTool("hNMRdata.csv");
 IAtom curAtom;
 double[] result;
 int h1 = 0;
 int c1 = 0;
 int ac = mol.getAtomCount();
 int p = 0;
 int hcnt = 0;
 int ccnt = 0;
 for (int i = 0; i < mol.getAtomCount(); i++) {
 curAtom = mol.getAtom(i);
 if (curAtom.getAtomicNumber() == 1) {
 hcnt++;
 } else if (curAtom.getAtomicNumber() == 6) {
 ccnt++;
 }
 }
 double[] cppmv = new double[ccnt];
 int[] caid = new int[ccnt];
 double[] hppmv = new double[hcnt];
 int[] haid = new int[hcnt];
 for (int i = 0; i < mol.getAtomCount(); i++) {
 curAtom = mol.getAtom(i);
 if (curAtom.getAtomicNumber() == 1) { //6
 result = hpredictor.predict(mol, curAtom);
 hppmv[h1] = result[1];
 haid[h1] = i;
 h1++;
 p++;
 } else if (curAtom.getAtomicNumber() == 6) {
 result = cpredictor.predict(mol, curAtom);
 cppmv[c1] = result[1];
 caid[c1] = i;
 c1++;
 p++;
 }
 }

 String[][] dbppm = new String[p][3];
 String[] hppms = new String[h1];
 String[] cppms = new String[c1];
 for (int c0 = 0; c0 < c1; c0++) {
 dbppm[c0][0] = caid[c0] + 1 + "";
 dbppm[c0][1] = "C";
 dbppm[c0][2] = df.format(cppmv[c0]);

 }

 for (int h = 0; h < h1; h++) {
 int h0 = h + c1;
 dbppm[h0][0] = haid[h] + 1 + "";
 dbppm[h0][1] = "H";
 dbppm[h0][2] = df.format(hppmv[h]);

 }

 String alldata = "";
 for (int i = 0; i < dbppm.length; i++) {
 for (int j = 0; j < dbppm[0].length; j++) {
 alldata += dbppm[i][j] + " ";
 }
 alldata += "\n";
 }
 return alldata;
}
```

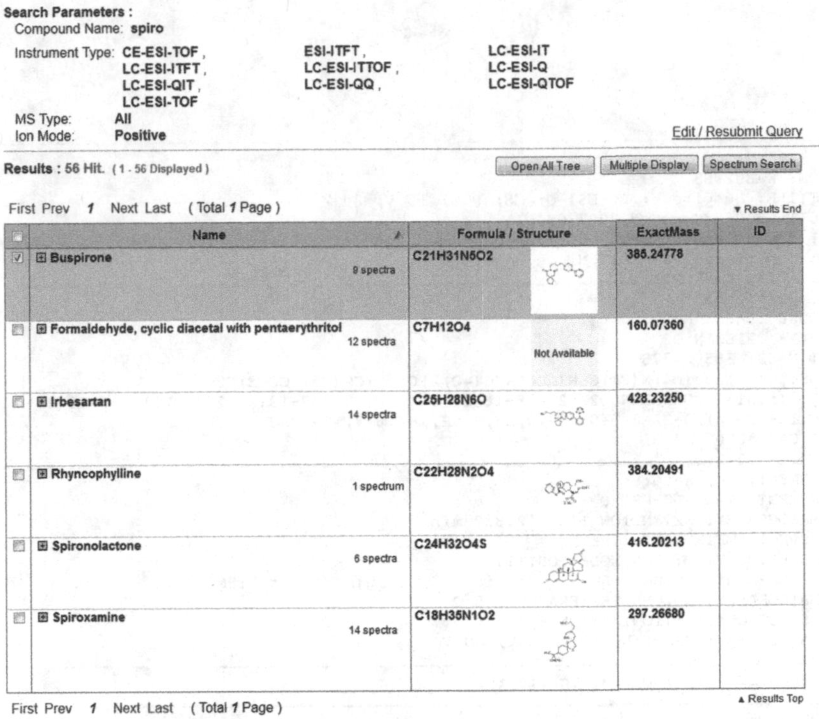

Search Parameters :
Compound Name: **spiro**

Instrument Type: **CE-ESI-TOF** ,	**ESI-ITFT** ,	**LC-ESI-IT**
LC-ESI-ITFT ,	**LC-ESI-ITTOF** ,	**LC-ESI-Q**
LC-ESI-QIT ,	**LC-ESI-QQ** ,	**LC-ESI-QTOF**
LC-ESI-TOF		

MS Type: **All**
Ion Mode: **Positive**

Edit / Resubmit Query

Results : 56 Hit. (1 - 56 Displayed)

Open All Tree   Multiple Display   Spectrum Search

First Prev *1* Next Last   (Total *1* Page)

▼ Results End

	Name	Formula / Structure	ExactMass	ID
☑	⊞ Buspirone *9 spectra*	C21H31N5O2	385.24778	
☐	⊞ Formaldehyde, cyclic diacetal with pentaerythritol *12 spectra*	C7H12O4 Not Available	160.07360	
☐	⊞ Irbesartan *14 spectra*	C25H28N6O	428.23250	
☐	⊞ Rhyncophylline *1 spectrum*	C22H28N2O4	384.20491	
☐	⊞ Spironolactone *6 spectra*	C24H32O4S	416.20213	
☐	⊞ Spiroxamine *14 spectra*	C18H35N1O2	297.26680	

First Prev *1* Next Last   (Total *1* Page)

▲ Results Top

Fig. 7.27 Hits for the keyword query *spiro* in MassBank

The chemical shift-based binary fingerprints were applied to map the entire drug space. "Cumulative" NMR spectra of proton and carbon nuclei of 1,200 compounds deposited in the Food and Drug Administration (FDA) database were generated [80] (Fig. 7.33). Statistically significant regions of corresponding fingerprints of these reference spectra were used for virtual screening library of compounds. A molecule whose predicted NMR matched either with other molecules in the dataset or with the cumulative NMR model qualified as a hit.

The binary fingerprints were used to discriminate between various bioactivity classes for the purpose of virtual screening of huge libraries. Here, one of the examples of cumulative carbon NMR of an antifungal class of molecules is shown. The three representative molecules highlighted are the ones having the maximum bit occupancy for certain preferred fragments. The bit position at 192 corresponding to 48 ppm on the carbon NMR scale encodes for the methyl carbon attached to oxygen, bit position 498 corresponding to 125 δ in carbon NMR encodes for the naphthyl region in second representative compound, and bit position at 568 (142 ppm) possesses the fragment with a triazole ring system (Fig. 7.34).

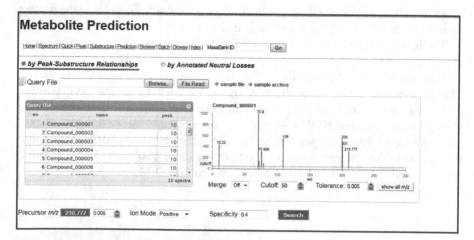

```
ACCESSION: WA002983
RECORD_TITLE: Buspirone; LC-ESI-Q; MS; POS; 15 V, 30 V
DATE: 2011.05.06 (Created 2007.08.01)
AUTHORS: Nihon Waters K.K.
LICENSE: Copyright 2007-2011 Nihon Waters K.K.

CH$NAME: Buspirone
CH$COMPOUND_CLASS: N/A
CH$FORMULA: C21H31N5O2
CH$EXACT_MASS: 385.24778
CH$SMILES: c(c4)cnc(n4)N(C1)CCN(CCCCN(C(=O)2)C(=O)CC(C3)(CCC3)C2)C1
CH$IUPAC: InChI=1S/C21H31N5O2/c27-18-16-21(6-1-2-7-21)17-19(28)26(18)11-4-3-10-24-
12-14-25(15-13-24)20-22-8-5-9-23-20/h5,8-9H,1-4,6-7,10-17H2
CH$LINK: CAS 36505-84-7

AC$INSTRUMENT: ZQ, Waters
AC$INSTRUMENT_TYPE: LC-ESI-Q
AC$CHROMATOGRAPHY: RETENTION_TIME 12.830 min
AC$MASS_SPECTROMETRY: MS_TYPE MS
AC$MASS_SPECTROMETRY: ION_MODE POSITIVE
AC$CHROMATOGRAPHY: COLUMN_NAME 2.1 mm id - 3. 5{mu}m XTerra C18MS
AC$CHROMATOGRAPHY: COLUMN_TEMPERATURE 35 C
AC$MASS_SPECTROMETRY: IONIZATION ESI
AC$CHROMATOGRAPHY: SAMPLING_CONE 15 V, 30 V

MS$DATA_PROCESSING: FIND_PEAK ignore rel.int. < 5

PK$NUM_PEAK: 5
PK$PEAK: m/z int. rel.int.
 150 6 6
 219 35 35
 386 999 999
 387 212 212
 388 24 24
//
```

Fig. 7.28 A MassBank record of a spiro compound

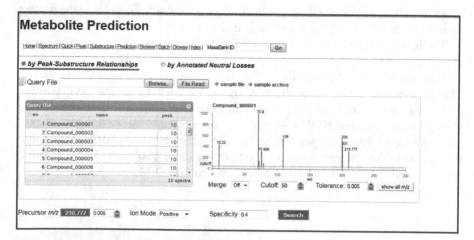

Fig. 7.29 Metabolite prediction option in MassBank

7.10   Thumb Rules for Spectral Data Handling and Prediction

- Choose the spectra prediction program tailored to your needs. There is always a trade-off between speed and accuracy. Quantum chemistry programs based on first principles show less deviations in their predicted chemical shifts from the experimental values but are computationally expensive and time consuming

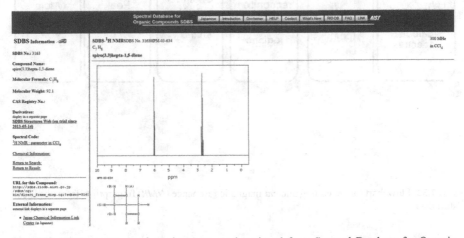

Fig. 7.30 NMR spectrum of a spiro compound retrieved from Spectral Database for Organic Compounds (SDBS) by chemical name search

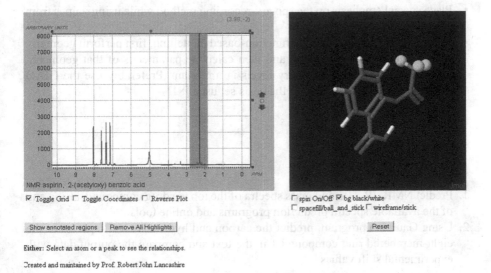

Fig. 7.31 NMR spectrum of aspirin molecule as visualized in JSpecView program

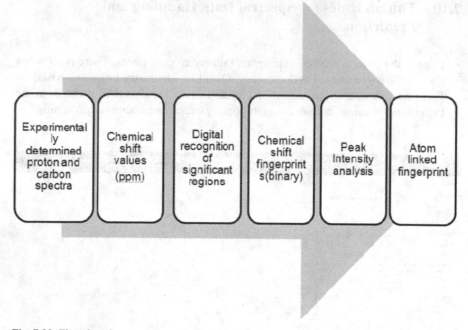

Fig. 7.32 Flowchart for generating nuclear magnetic resonance (*NMR*) fingerprints from chemical shift data

- Recheck and make sure the experimental values of the NMR parameters viz. chemical shifts, shielding tensors, coupling constants, etc. used in modelling studies are correct
- Place special emphasis on the spectra-recording method while using values from a database
- In case of ab initio and density function-based modelling, first perform geometry optimization of the compound and then calculate parameters of that geometry. The right combination of theory levels is important. Preferably use the GIAO method as it is less sensitive to the basis set used [81]

7.11 Do it Yourself

1. Predict NMR, IR, UV, and mass spectra of the top ten drug compounds using any of the available spectra prediction programs and online tools
2. Using Gaussian program, predict the carbon and hydrogen NMR spectra of the eight-membered ring compound 1 in the text and compare the output data with experimental shift values

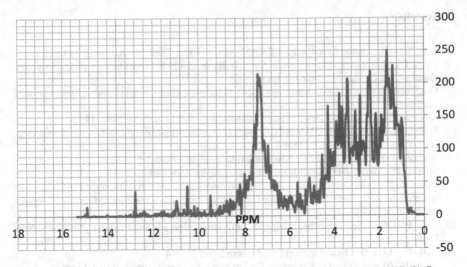

Fig. 7.33 Cumulative spectrum generated using the nuclear magnetic resonance (*NMR*) finger-prints of Food and Drug Administration (FDA) drugs

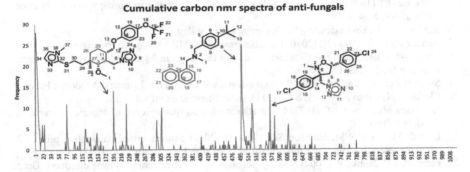

Fig. 7.34 The cumulative specturm of 30 antifungal compounds

7.12 Questions

1. Write a brief essay on known spectra prediction methods. Highlight the advantages and disadvantages of each method.
2. How are vibrational frequencies computed? Explain with the help of an example.
3. Write a short note on databases used in spectra prediction programs.
4. Give a stepwise account of how NMR tensor values can be computed in Guassian3W program.
5. Using Mnova program, predict the carbon NMR spectra of example compounds 1 and 2 discussed in the chapter.

References

1. http://missionscience.nasa.gov/ems/TourOfEMS_Booklet_Web.pdf. Accessed 31 Oct 2013
2. Pavia DL, Lampman GM, Kriz GS, Vyvyan JR (eds) (2009) Introduction to spectroscopy. Brooks/Cole Cengage Learning, USA
3. Gunther H (1995) NMR spectroscopy: basic principles, concepts and applications in organic chemistry. Wiley
4. McDonald RS (1986) Review: infrared spectrometry. Anal Chem 58:1906–1925
5. Schoonheydt RA (2010) UV-VIS-NIR spectroscopy and microscopy of heterogeneous catalysts. Chem Soc Rev 39:5051–5066
6. Watson JT, Sparkman D (2007) Introduction to mass spectrometry. Wiley
7. Smyth MS, Martin JHJ (2000) X Ray crystallography. Mol Path 53:8–14
8. Dyer JR (1965) Applications of organic spectroscopy of compounds. Prentice Hall
9. Silverstein RRM, Webster FK, Kiemle DJ (2005) The spectrometric identification of organic compounds. Wiley
10. Kalsi PS (2004) Spectroscopy of organic compounds. New age international publishers, New Delhi
11. Calloway D (1997) Beer-Lambert law. J Chem Educ 74:744
12. Stuart B (2004) Infrared spectroscopy fundamentals and applications. Wiley, England
13. Karthikeyan M, Imran (unpublished results)
14. Hamm P, Zani M (2011) Concepts and methods of 2D Infra red spectroscopy. Cambridge University press, New York
15. Callaghan PT (1991) Principles of nuclear magnetic resonance spectroscopy. Oxford Science Publications, New York
16. Keeler J (2010) Understanding NMR spectroscopy. Wiley
17. Richards SA, Hollerton JC (2010) Essential practical NMR for organic chemistry. Wiley
18. Campos-Olivas R (2011) NMR screening and hit validation in fragment based drug discovery. Curr Top Med Chem 11(1):43–67
19. Stothers J (1972) Carbon 13 NMR spectroscopy, vol 24 organic chemistry. Academic Press
20. Clayden J, Greeves N, Warren S (2012) Organic chemistry. Oxford
21. Hamming MC, Foster NC (1972) Interpretation of mass spectroscopy of organic compounds. Academic Press
22. Berardi MJ, Shih WM, Harrison SC, Chou JJ (2011) Mitochondrial uncoupling protein 2 structure determined by NMR molecular fragment searching. Nature 476:109–113
23. Fernandez C, Jahnke W (2004) New Approaches for NMR screening in drug discovery. Drug Discov Today Technol 1(3):277–283
24. Liu P, Lu M, Zheng Q et al (2013) Recent advances of electrochemical mass spectrometry. Analyst
25. Kang EH, Lee EY, Lee YJ et al (2008) Clinical features and risk factors of postsurgical gout. Ann Rheum Dis 67:1271–1275
26. Taber DF (2007) Organic Spectroscopic structure determination: a problem based learning approach. Oxford University Press
27. http://moltable.ncl.res.in/c/document_library/get_file?p_l_id=12401&folderId=12410&name=DLFE-1102.pdf
28. Buchnicek J (1950) Colchicine in ripening seeds of the wild saffron (Colchicum autumnale L). Pharm Acta Helv 25:389–401
29. Shuker SB, Hajduk PJ, Meadows RP, Fesik SW (1996) Discovering high affinity ligands for proteins SAR by NMR. Science 274(5292):1531–1534
30. Rynchnovsky SD (2006) Predicting NMR spectra by computational Methods: structure revision of hexacyclinol. Org Lett 8:2995–2898
31. Elyashberg M, Blinov K, Smurnyy Y, Churanova T, Williams A (2010) Empirical and DFT GIAO quantum mechanical methods of 13C chemical shifts prediction competitors or collaborators. Magnet Reson Chem 48(3):209–229

32. Hu Y, Li Y, Lam H (2011) A semiemprirical approach for predicting unobserved peptide MS MS spectra from spectral libraries. Proteomics 11(4702):4711

33. Charpenier T (2011) The PAW/GIPAW approach for computing NMR parameters: a new dimension added to NMR study of solids. Solid State Nucl Magn Reson 40(1):1–20

34. Will M, Joachim R (1997) Spec-Solv an innovation at work. J Chem Inf Comput Sci 37: 403–404

35. Blinov KA, Smurnyy YD, Elyashberg ME, Churanova TS, Kvasha M, Steinbeck C, Lefebvre BA, Williams AJ (2008) Performance validation of neural network of 13C NMR prediction using a publicly available data source. J Chem Inf Model 48:550–555

36. Pretsch E, Furst A, Bodertscher M, Burgin R (1992) C13Shift: A computer program for the prediction of 13CNMR spectra based on an open set of additivity rules. J Chem Inf Model 32:291–295

37. http://www.bruker.com/products/mr/NMR/NMR-software/software/topspin/overview.html

38. Shen Y, Lange O, Delaglio F, Rossi P, Aramini JA, Liu G, Eletsky A, Wu Y, Singarapu KK, Lemak A, Ignatchenko A, Arrowsmith CH, Szyperski T, Montelione GT, Baker D, Bax A (2008) Consistent blind protein structure generation from NMR chemical shift data. Proc Natl Acad Sci USA 105(12):4685–4690

39. http://www.bioNMR.com/forum/NMR-dynamics-21/tensor-2-analysis-overall-internal-dynamics-54/. Accessed 31 Oct 2013

40. http://www.cambridgesoft.com/Ensemble_for_Chemistry/ChemDraw/. Accessed 31 Oct 2013

41. http://www.chemaxon.com/marvin/help/calculations/NMRpredict.html. Accessed 31 Oct 2013

42. http://www.acdlabs.com/products/adh/NMR/NMR_pred/. Accessed 31 Oct 2013

43. http://mestrelab.com/software/mnova-NMRpredict-desktop/. Accessed 31 Oct 2013

44. Wiberg KB, Hammer JD, Zilm KW, Cheeseman JR (1999) NMR chemical shifts. 3. A comparison of acetylene, allene, and the higher cumulenes. J Org Chem 64:6394

45. Spanton SG, Whittern D (2009) The development of an NMR chemical shift prediction application with the accuracy necessary to grade proton NMR spectra for identity. Magn Reson Chem 47(12):1055–1061

46. Raymond AJ, Mehdi M (2004) The prediction of 1H NMR chemical shifts in organic compounds. Spectrosc Eur 16(4):20–22

47. http://www.msg.ameslab.gov/gamess/. Accessed 31 Oct 2013

48. Roesky HW, Walawalkar MG, Ramaswamy M (2001) Is water a friend or foe in organometallic chemistry? The case of group 13 organometallic compounds. Acc Chem Res 34(3): 201–211

49. Gordon MS, Schmidt MW (2005) Advances in electronic structure theory: GAMESS a decade later. In: Dykstra CE, Frenking G, Kim KS, Scuseria GE (eds) Theory and applications of computational chemistry: the first forty years, pp. 1167–1189. Elsevier, Amsterdam

50. Pagenkopf B (2005) ACD/HNMR Predictor and ACD/CNMR Predictor. J Am Chem Soc 127(9):3232

51. Chemicke L (2008) Drasar, Pavel Bulletin presents. Comparison of advantages and disadvantages of ACD/1D NMR Assistant, ACD/1D NMR Processor, and ACD/Labs NMR Predictor software. 102(4):299–300

52. http://insideinformatics.cambridgesoft.com/VideosAndDemos/Default.aspx?ID=52

53. Wang H (2005) Application of chemdraw nmr tool: correlation of program-generated 13c chemical shifts and pKa values of para-substituted benzoic acids. J Chem Educ 82(9):1340

54. http://www.schrodinger.com/productpage/14/7/. Accessed on 31 Oct 2013

55. Saitoa H, Andob I, Ramamoorthy A (2010) Tensors the heart of NMR. Prog Nucl Magn Reson Spectrosc 57(20):181–228

56. Mason J (1993) Solid State Nucl Magn Reson 2:285

57. Facelli JC (2011) Chemical shift tensors: theory and application to molecular structure problems. Prog Nucl Magn Reson Spectrosc 58(3–4):176–201

58. http://www.gaussian.com/g_whitepap/NMRcomp.htm. Accessed 31 Oct 2013

59. Nikolic G, Shimazaki T, Yoshihiro A (eds) (2011) Fourier transforms, Gaussian and Fourier Transform (GFT) method and screened Hartree-Fock exchange potential for first-principles band structure calculations. 15–36
60. http://www.scm.com/Products/Capabilities/SpectroscopicProperties.html. Accessed 31 Oct 2013
61. Francesco RD, Stener M, Fronzoni G (2012) Theoretical study of near-edge X-ray absorption fine structure spectra of metal phthalocyanines at C and N K-edges. J Phys Chem A 116:2285–22894
62. Cobas C, Seoane F, Dominiguez S, Sykora S, Davies AN (2011) A new approach to improving automated analysis of proton NMR spectra through global spectral deconvolution(GSD) 23(1)
63. Jens M, Maier W, Martin W, Reinhard M (2002) Using neural networks for 13C NMR chemical shift prediction-comparison with traditional methods. J Magn Reson 157(2):242–252
64. Pearlman DA (1996) Fingar: a new genetic algorithm based method for fitting NMR data. J Biomol NMR 8(1):49–66
65. http://www.wavefun.com/products/spartan.html. Accessed 31 October 2013
66. Yang S, Bax A (2010) SPARTA+A modest improvement in empirical NMR chemical shift prediction by an artificial neural network. J Biomol NMR 48(1):13–22
67. Plainchont B, Nuzillard JM (2013) Structure verification though computer assisted spectral assignment of NMR spectra. Magn Reson Chem 51(1):54–59
68. Bertini I, Felli IC, Kuemmerle R, Moskau D, Pierattelli R (2004) 13C-13C NOESY: an attractive alternative for studying large macromolecules. J Am Chem Soc 126(2):464–465
69. Kuhn S, Schlorer NE (2012) NMR structure determination in synthetic chemistry. Nachrichten aus der Cemie 60(11):1106–1107
70. http://www.massbank.jp/?lang=en. Accessed 31 Oct 2013
71. http://www.swgdrug.org. Accessed 31 Oct 2013
72. Kazutoshi T, Hayamizu K, Shuitiro O (1991) Analytical sciences, spectral database system on PC with CD-ROM 7 (Suppl., Proc. Int. Congr. Anal. Sci., Pt. 1), 711–712
73. http://www.sigmaaldrich.com/labware/labware-products.html?TablePage=19816610
74. Nitsche C (1996) SciFinder 2.0: Preserving the partnership between chemistry and the information professional. Database (Oxford) 19:51
75. http://jspecview.sourceforge.net/. Accessed 31 Oct 2013
76. unpublished results
77. http://pubchem.ncbi.nlm.nih.gov/. Accessed 31 Oct 2013
78. https://www.ebi.ac.uk/chembl/. Accessed 31 Oct 2013
79. http://www.hmdb.ca/. Accessed 31 Oct 2013
80. http://www.accessdata.fda.gov/scripts/cder/drugsatfda/. Accessed 31 Oct 2013
81. Toukach FV, Ananikov VP (2013) Recent advances in computational prediction of NMR parameters for the structural elucidation of carbohydrates: methods and limitations. Chem Soc Rev 42:8376

Chapter 8
Chemical Text Mining for Lead Discovery

Abstract With the growth of the Internet, the information disseminated and available in public resources has expanded enormously. There is a need for the development of new tools to navigate through each and every document automatically, word by word to extract useful patterns, concepts, knowledge, or discover something which is not explicitly mentioned in a document to derive useful conclusions. Recently, computational linguistics developers and scientists have devised several text-mining tools and techniques for converting the natural language and processing the information content into facts and data for interpretation, analysis, and predictions. Text mining comprises data mining, information retrieval, natural language processing (NLP), and machine learning (ML) methods. Text mining provides researchers with metadata to ascertain meaningful associations of terms prevalent in their respective domains. Thus, it aids in finding meaning, context, semantics, identifying hidden concepts, trends, and discovering hitherto unknown relationships and correlations from heaps of largely fragmented, unstructured, and scattered information lying in public realm. In this chapter, we highlight the general concept of text mining followed by its features and tools especially for handling biomedical and chemical literature data for drug/lead discovery available in over 22.9 million abstracts in PubMed. The emphasis is on building and using simple text-mining tools in a practical way by harnessing the power of open source and commercially available tools and comprehending the overall strategic challenges in this field. An open-source-based tool for text mining literature with chemical significance that can be effectively used for solving chemoinformatics problems related to lead discovery has been developed. MegaMiner can directly predict lead molecules for a target disease of interest by submitting a text-based query in a distributed computing platform.

Keywords Text-mining · Clustering · Stemming · Chemoinformatics · Lead discovery · MegaMiner · Open-source tools

M. Karthikeyan, R. Vyas, *Practical Chemoinformatics,*
DOI 10.1007/978-81-322-1780-0_8, © Springer India 2014

8.1 What is Text Mining?

Information is widely dispersed across numerous articles, publications, patents, books, blogs, discussion forums, and scientific literature databases [1]. Just plain textual data as such is a large resource of information. The most important accessible resource for this freely available information is the Internet [2]. But one major problem with such large data is that the information is mostly unstructured and not available in ready-to-query databases. Hence, computer-based processing and analysis of such information is a tedious task. Such data need to be explored for keywords to discover knowledge and only a small portion will actually be of use to a given user. Judicious selection of this bit of information can be performed by a text-mining protocol. Text mining deals with scanning text data for patterns, connections, profiles, and trends. In fact, text mining automates finding, reading, storing, understanding, and consolidating data [3]. The researcher has to only make sense out of it and derive his/her own inferences. For instance, if one is interested in studying gene–protein, protein–protein, or target–ligand interactions involved in a biological pathway, one can collect all available textual data and use appropriate text-mining tools. The tool will facilitate annotating terms and look for co-occurring entities. Further, terms can be visualized with their relations in a network to derive information to validate a hypothesis. Another text-mining application is for automated biocuration [4]. Manual curation for data straight into databases is very helpful but very time consuming. A mixed approach where manual curation is used with automated text mining is beneficial. With the emergence of the first publicly funded text-mining center in the world, NacTem, newer tools and techniques have gained more importance and acceptance [5].

In essence, text mining can be defined as extracting high-quality information from plain text. It is a derived discipline which takes help from information retrieval, data mining, web mining, statistical modelling, computational linguistics, and natural language processing (NLP). Text mining has been defined vividly but the most apt definition is "the process of recognizing pattern from a wealth of information hidden latent in unstructured text and deducing explicit relationship among data entities by using data mining tools." [6]. It is a highly data-intensive process which enables a user to find meaning from heaps of largely fragmented, unstructured, and scattered information available in a public domain using a suite of text analysis tools. This field provides methods and techniques to find patterns and trends across textual data, sort, and rank documents according to importance and relevance and compare documents.

8.1.1 Text Mining vis-a-vis Data Mining

Text mining is akin to data-mining systems in that it shares similar architectural features as well as robust browsing capabilities to draw logical inferences [7]. Text mining can be considered as a subset of data mining which is a process of extracting useful information, as per the user requirements, from large amounts of datasets [8]. Both are provided with visualization tools for facilitating user interactivity to identify patterns in the data but differ in the presence of feature extraction and feature selection steps in the former. Data mining deals with databases containing tables linked through certain relationships (relational database management system; RDBMS) which is straightforward as the data are represented in proper formats (int, float, text, char, blob, binary, etc.), and applying mathematical and statistical tools to identify the trends and patterns [9]. For example, when we search the Web or any database with a query, we get the results in seconds. This is not so in text mining where it is not possible to dynamically search for the keyword and display results in a fraction of a second. The main challenge in text mining is that as the number of words in any text increases, its dimensionality increases [10]. Moreover, all the relevant information that we are looking for is a complex combination of words and phrases and so there is always a possibility of word ambiguity or semantic ambiguity. For example, there can be two words with the same meaning or one sentence can have multiple meanings. Added to this, there is a presence of noisy data which can be spelling mistakes, stop words, abbreviations, etc. Next, the most important challenge in text mining is to identify relevant data and classify them properly as numeric or text [11]. This becomes even more difficult while handling scientific documents where there are several mathematical expressions and scientific terms which are not usually classified by conventional NLP programs [12, 13].

Text mining is thus different from data mining and is carried out in a series of steps. A typical text-mining work flow involves text categorization, text clustering, concept/entity extraction, production of granular taxonomies, sentiment analysis, document summarization, and entity relation modelling [14]. The core text-mining operations that focus on query creation algorithms are distributions, frequent and near-frequent sets and associations which enable the user to explore the data in collected volumes [15]. Named entity recognition (NER) and NLP, ML are the text analytical tools [16]. The input for text-mining procedure is raw text, i.e., text without label/classification which usually comes from a data source like PubMed hosted by the National Library of Medicine which is expanding daily and contains the most important published biomedical research literature. PubMed is a large repository of citation entries of scientific articles. It is the most commonly used source for biomedical information [17]. It is a service provided by the National Library of Medicine and National Institutes of Health. Presently, it contains approximately 23 million citations. It hosts articles and reviews from ~36,000 journals. It includes links to full-text articles and other related sources. The search interface for National Center for Biotechnology Information (NCBI) is Entrez. Entrez is an integrated, text-based search-and-retrieval system at NCBI used for all the major databases [18]. Annotating PubMed data is a huge task. It requires a lot of computing power. But annotation of PubMed will help researchers to find trends and patterns in similar fields of research. The information from the abstracts can be converted to knowledge. One can manually search PubMed by directing the browser to the following link: http://www.ncbi.nlm.nih.gov/sites/entrez?db=pubmed. There are filters to search by authors, journals, dates, languages, and article type. The results can be retrieved as Extensible Markup Language (XML), citations, abstracts, summary, etc. and can be accessed as text or downloaded as files. To do the same, programmatically, NCBI provides Entrez Programming Utilities or E-utilities. Entrez Pro-

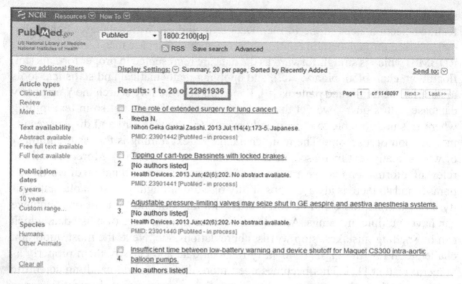

Fig. 8.1 A PubMed search results page displaying the current abstract entries

gramming Utilities are tools that provide access to Entrez data outside of the regular Web query interface and may be helpful in retrieving search results for future use in another environment ([19]; Fig. 8.1).

For example, to search for pmids related to kinases:

http://eutils.ncbi.nlm.nih.gov/entrez/eutils/esearch.fcgi?db=pubmed&term=kinases

To retrieve only seven results, the uniform resource locator (URL) would be http://eutils.ncbi.nlm.nih.gov/entrez/eutils/efetch.fcgi?db=pubmed&retmax=7& id=19232228

The results can be returned as Abstract/Citation/Medline/Full:

http://eutils.ncbi.nlm.nih.gov/entrez/eutils/efetch.fcgi?db=pubmed&rettype=ab stract&id=19232228

To retrieve results as XML/Text/HTML:

http://eutils.ncbi.nlm.nih.gov/entrez/eutils/efetch.fcgi?db=pubmed&retmode=x ml&id=19232228

For example, to fetch results for swine flu on PubMed, the URL would be http://eutils.ncbi.nlm.nih.gov/entrez/eutils/esearch.fcgi?db=pubmed&term= swine+flu

8.1.2   A Snippet of Java Code Using the Above URL

```
URL url =
"http://eutils.ncbi.nlm.nih.gov/entrez/eutils/esearch.fcgi?db=pubmed&term=swine+flu&re
tmax=1&retmode=xml&rettype=abstract"
URLConnection con = url.openConnection();
InputStream in = url.openStream();
BufferedReader br = new BufferedReader (
 new InputStreamReader (con.getInputStream()));
```

The PMID returned is: 19232228. This PMID is used to fetch the entry

```
URL url =
"http://eutils.ncbi.nlm.nih.gov/entrez/eutils/efetch.fcgi?db=pubmed&retmode=xml&rettyp
e=abstract&id=19232228"
```

The search result for the above entry is

```
<PubmedArticle>
<MedlineCitation Owner="NLM" Status="In-Process">
<PMID>19232228</PMID>
<DateCreated>
<Year>2009</Year>
<Month>02</Month>
<Day>23</Day>
</DateCreated>
<Article PubModel="Electronic">
<Journal>
<ISSN IssnType="Electronic">1560-7917</ISSN>
<JournalIssue CitedMedium="Internet">
<Volume>14</Volume>
<Issue>7</Issue>
<PubDate>
<Year>2009</Year>
</PubDate>
</JournalIssue>
<Title>Euro surveillance: bulletin européen sur les maladies transmissibles = European
communicable disease bulletin</Title>

<ISOAbbreviation>Euro Surveill.</ISOAbbreviation>

</Journal>

        <ArticleTitle>Human case of swine influenza A (H1N1), Aragon, Spain, November
2008.</ArticleTitle>

<ELocationID EIdType="pii" ValidYN="Y">19120</ELocationID>
<Abstract>
<AbstractText>A human case of swine influenza A (H1N1) in a 50-year-old woman from a
village near Teruel (Aragon, in the north-east of Spain), with a population of about
200 inhabitants, has been reported in November 2008.</AbstractText>
</Abstract>
<Affiliation>Direccion General de Salud Publica (Directorate General of Public
Health), Zaragoza, Spain. mbadiego@aragon.es</Affiliation>
```

All the keywords are indexed. They are linked to relevant web pages and databases. The keywords are ranked according to the number of times they are searched. Statistical data are generated for the number of occurrences of the terms and also for their occurrences with other terms. All this facilitates faster searches.

8.2   What are the Components of Text Mining?

Usually, the tasks involved are text preprocessing or tokenization, part-of-speech (POS) tagging, stemming, text transformation, attribute generation and attribute selection, NER, data mining, or pattern discovery and evaluation [20]. There are other steps involved in information retrieval like linguistic preprocessing, removing stop words, stemming and finding synonyms. Let us elaborate on some of them in detail (Fig. 8.2).

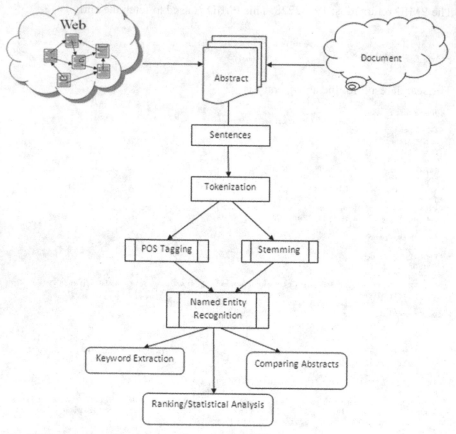

Fig. 8.2 Flowchart depicting steps for a general text-mining process

Tokenization: Text preprocessing is important because data may contain special characters like punctuations and stop words which are not usually scientific entities. Data may contain other symbols, formats like number, date, e-mail, etc. Token is an unclassified word from the text. Tokenization is a compilation of all words in a given document or dataset with the help of a parser [21]. Relevant text is identified from the abstracts which contain largely unstructured free textual data and subjected to tokenization to break up the text into constituent sentences and words. The raw text is divided into sentences and the sentences are further divided into tokens. Processing tokens is easier than considering the whole text every time. Stop words are generally prepositions, articles, pronouns, and other user-defined keywords which are often eliminated for better system performance and the further menace of irrelevant data handling.

Stemming: It is a process to find the root of a word to achieve reduction in the word space. For example, the root for keywords connection, connections, connective, connected, and connecting all relate to "connect." Stemming allows reduction in data dimension and data overload [22].

POS tagging: The process of assigning the best part of speech to a word in proper context is POS tagging [23]. Noun, adjective, verb, adverb, etc. are POS tags. This step takes a stream of words as inputs and the output is the best POS tag for every word. Thus, POS tagging assigns a POS-like noun, verb, pronoun, preposition, adverb, adjective, or other lexical class markers to each word in a sentence. The input to a tagging algorithm is a string of words of a natural language sentence and a specified tag set. The output is a single best POS tag for each word. POS tagging is harder than just having a list of words and their parts of speech, because some words can represent more than one part of speech at different times. Stop words can be identified from this step. Stop words are most unlikely to help text mining [24]. The words that appear in documents often have many morphological variants. Therefore, each word that is not a stop word is reduced to its corresponding stem word (term), i.e., the words are stemmed to obtain their *root form* by removing common prefixes and suffixes [25]. In this way, we can identify groups of corresponding words where the words in the group are syntactical variants of each other and can collect only one word per group. This reduces the dimensionality in the data. For instance, the words *disease, diseases,* and *diseased* share a common stem term *disease,* and can be treated as different occurrences of this word. POS tagging may be rule based, most often grammatical rules, or based on statistical models like different word order probabilities or simply corpus based, for instance, Brown corpus which is a compilation of one million pre-tagged English words [26].

NER: The next phase is the NER phase, an information extraction step wherein the text-mining engine identifies all mentions of proper names, dates, and time in the text [27]. NER is the recognition of the entities relevant to the domain. It is information linking where a term is assigned to a predefined category, e.g., protein, gene, disease. In the biomedical domain, these entities would be genes, proteins, diseases, chemicals, and so on. Other domains will naturally have different entities, for example, the typical entities in the financial news domain are companies, persons, products, and so forth. The last steps are categorization and clustering which are standard supervised and unsupervised learning techniques, respectively [28]. The interested reader is referred to excellent reviews and books on the text-mining processes for an in-depth understanding of all the basic processes especially in the context of biology [29–31].

8.3 Text-mining Methods

Text-mining methods employ algorithms that use similarity-based functions in order to obtain k nearest neighbors for novel query objects [32]. Term weighting is performed to measure the importance of a term in representing the information contained in the document [33]. For mining literature, the two most common approaches are ML-based and the rule-based approaches, though in practice a combination of approaches works best [34].

8.3.1   Statistics/ML-based Approach

In this approach, systems work by building classifiers that may operate on any level, from labeling POS to choosing syntactic parse trees to classifying full sentences or documents [35]. Statistical systems typically require large amounts of expensive-to-get labeled training data. Generally, in this approach dictionaries are used. Some of the few popular ones are GENIA corpus from the GENIA project [36], BioCreative corpus [37]. These are dictionaries containing labeled and structured data. They can be used to extract biological keywords from text. Generally, binary versions of these dictionaries are compiled and these binary files are used, with the help of specific taggers to label tokens in plain text. If one does not want to use these dictionaries, one can create their own dictionaries with the help of statistical NLP tools [38]. Building a nonredundant dictionary is a difficult task. The initial task to building a dictionary is gathering the data to be added to the dictionary. The data should be in the form of tokens and can be chemical terms, IDs, registration numbers, etc. and every token should be labeled [39]. However, the labeling can be in any format; just the code should be modified accordingly. For example,

```
Acetaminophen|Chemical,
p53|Gene,
Influenza|Disease,
1-Benzyl-5-Methoxy-2-Methyl-1h-Indol-3-Yl)-Acetic Acid|Chemical
```

A complete list with all the entries should be created. A snippet of creating a dictionary is given below:

```
MapDictionary dictionary = new MapDictionary();
dictionary.addEntry (new DictionaryEntry
(token, label, CHUNK_SCORE));
AbstractExternalizable.compileTo (dictionary, <filename>);
```

The tokens and the labels are represented as feature vectors, n-dimensional vectors of numerical features. It is the statistical representation of the input text [40]. The dictionary file can be compiled to binary or hexadecimal formats. This makes it difficult to interpret the file without proper readers. Such compiled files facilitate faster tagging of text. For reading a dictionary and using it to tag text, here is a snippet of code:

```
MapDictionary dictionary = (MapDictionary)
AbstractExternalizable.readObject (modelFile);
ExactDictionaryChunker dictionaryChunker =
      new ExactDictionaryChunker
(dictionary, IndoEuropeanTokenizerFactory.FACTORY, true, false);

String[][] result = chunk (dictionaryChunker, <testData>);
```

One can obtain the labels and offsets of every term from the text. There are various statistical models that can be used in this process. Hidden Markov model (HMM) is the simplest of dynamic Bayesian model. HMM is a finite set of states, each of which is associated with a (generally multidimensional) probability distribution [41]. HMMs are a form of generative models that define a joint probability distribution p(X, Y) where X and Y are random variables, respectively, ranging over observation sequences and their corresponding label sequences [42]. In contrast to HMMs, in which the current observation only depends on the current state, the current observation in a maximum entropy Markov model may also depend on the previous state [43]. Conditional random fields (CRFs) are a probabilistic framework for labeling and segmenting sequential data, based on the conditional approach [44]. A CRF is a form of undirected statistical graphical model that defines a single log-linear distribution over label sequences given a particular observation sequence. They have demonstrated state-of-the-art accuracy on a wide variety of sequence-labeling tasks.

8.3.2   Rule-based Approach

Rule-based methods are based on rules written by human developers that capture syntactical, lexical, and semantic knowledge required for identifying the entities and the relationships, e.g., Java Annotation Pattern Engine (JAPE) [45, 46]. Rule-based systems make use of some sort of knowledge. The knowledge can be related to general language structure or domain-specific literature [47]. It is worth noting that useful systems have been built using technologies at both ends of the spectrum, and at many points in between [48]. The rules are set by the developers depending upon the data. The idea is to look for patterns in text. For example,

```
1. Chemical: 1,2,3,4-tetrahydroisoquinoline
   Pattern: ^[1-9]{1}[,][1-9]{1}[,][1-9]{1}[,][1-9]{1}[-][A- Z]{1,10}
2. Chemical: 1-(Isopropylthio)-Beta-Galactopyranside
   Pattern: ^[1-9]{1}[-][(][A-Z]{1,15}
3. CASRN: 2889-31-8
   Pattern: ^[1-9]{1,5}[-][1-9]{1,3}[-][1-9]{1,3}
4. String: Adinazolam is a benzodiazepine derivative
   Pattern: <Drug>......<Drug property>
```

If following is the test data(PMID: 172459),

"Mouse 3T3, Simian virus 40 transformed 3T3 cells (SV3T3) and two SV3T3 lines showing reversion of their transformed phenotype (Rev 3 and Rev 5) have been studied with respect to electrophoretic mobilities and colloidal iron hydroxide (CIH) binding density visible by electron microscopy, before and after incubation with neuraminidase or ribonuclease. The results show that, in general, the marked changes in both sets of surface parameters associated with transformation are largely reversed in the Rev 5 revertant, and only partially reversed in the Rev 3 line. It was also observed that, in common with Ehrlich ascites tumor (EAT) cells examined previously, the densities of CIH-particles bound over the microvilli of all the cell types was 1.5–2.7 times higher than those bound to the spaces between them. In contrast to the EAT cells, the higher density of CIH particles bound over the microvilli was not due to neuraminidase-sensitive binding sites."

Results will be, depending upon how good the dictionary is,

"**Mouse|ORGANISM** 3T3|O **Simian virus|ORGANISM** 40|O transformed|O
3T3|O cells|O SV3T3|O and|O two|O SV3T3|O lines|O showing|O reversion|O
of|O their|O transformed|O phenotype|O Rev|O 3|O and|O Rev|O 5|O have|O
been|O studied|O with|O respect|O to|O electrophoretic|O mobilities|O
and|O colloidal|O **iron hydroxide|CHEMICAL** CIH|O binding|O density|O
visible|O by|O electron|O microscopy|O before|O and|O after|O incubation|O
with|O **neuraminidase|PROTEIN** or|O ribonuclease|O The|O results|O
show|O that|O in|O general|O the|O marked|O changes|O in|O both|O sets|O
of|O surface|O parameters|O associated|O with|O transformation|O are|O
largely|O reversed|O in|O the|O Rev|O 5|O revertant|O and|O only|O partially|O
reversed|O in|O the|O Rev|O 3|O line|O It|O was|O also|O observed|O that|O
in|O common|O with|O **Ehrlich ascites tumor|DISEASE** EAT|O cells|O
examined|O previously|O the|O densities|O of|O CIH|O particles|O bound|O
over|O the|O microvilli|O of|O all|O the|O cell|O types|O was|O 15|O to|O
27|O times|O higher|O than|O those|O bound|O to|O the|O spaces|O between|O
them|O In|O contrast|O to|Othe|O EAT|O cells|O the|O higher|O density|O
of|O CIH|O particles|O bound|O over|O the|O microvilli|O was|O not|O due|O
to|O **neuraminidase|PROTEIN** sensitive|O binding|O sites|O"

8.4   Why Text Mining

Almost 80 % of the biochemical data are available in text format, excluding audio,
images, and videos which is a lot of information to be handled manually. Generally, while looking for information, we normally use search engines. It returns
ranked hits which are just URLs. The daunting task is how to look for information
from millions of hits returned if a user is looking for certain patterns, such as reported side effects of a certain drug from clinical outcome data or hits. This is when
automated text mining becomes very important. The applications of text-mining
techniques are wide from extracting protein–protein interaction (PPIs) networks,
drug repurposing, side effect profiling, and bridging hidden information through
a network of biological entities [49]. Finding new uses for existing drugs is more
feasible from an academic perspective and thus more promising. Text mining gives
more insight into digging out novel uses for existing drugs while profiling side
effects on the systems considered for study.

8.5   General Text-mining Tools

There are a number of general text-mining tools available to choose from. Only a
brief introduction is given here. Mallet is a collection of tools in Java for statistical
NLP, text classification, and clustering [50]. GATE is a toolkit for text mining and

information extraction provided with a graphical user interface (GUI) [51]. The natural language toolkit (NLTK) is a tool for teaching and researching classification, clustering tagging, and speech parsing [52]. LingPipe [53] and OpenNLP [54] are among the important open-source NLP tools. The OpenNLP site hosts a variety of Java-based NLP tools which perform sentence detection, tokenization, POS tagging, chunking and parsing, named-entity detection, and co-reference analysis using the Maxent ML package [55].

Stanford Parser is a Java package for sentence parsing from the Stanford NLP group. It has implementations of probabilistic natural language parsers, both highly optimized probabilistic context-free grammar (PCFG) and lexicalized dependency parsers, and a lexicalized PCFG parser [56]. OpenEphyra is a full-featured, end-to-end system for QA written in Java and developed at Carnegie Mellon University's (CMU's) Language Technologies Institute (LTI) department [57]. Other tools worth mentioning are GENIA Tagger [58], MetaMap [59], and Yamcha [60]. Comparative studies have been done to highlight their chunking capability. In one such study, OpenNLP outperformed all of the above-mentioned tools to give F score values of 89.7 and 95.7% for nounphrase chunking and verb-phrase chunking, respectively [61].

Carrot2 is another open-source search result clustering software written in Java [62]. There are some string-similarity-matching tools like Simmetrics maintained by Sheffield University [63]. Weka is a collection of ML algorithms for data mining. It is probably the most widely used text classification framework [64]. It has implemented a wide variety of algorithms including Naive Bayes and Support Vector Machine (SVM). Alias-I's LingPipe is a Java tool for information extraction and data mining including entity extraction, speech tagging, clustering, classification, string similarity, etc. It is one of the most mature and widely used open-source Internet Explorer (IE) toolkits in industry. LingPipe, *royalty version*, is a freely available text-mining tool from Alias-i that has been used for classification [65]. The GENIA corpus for biomedical data, which is a part of the LingPipe package, has been used to cluster textual data [66].

8.5.1 A Practice Tutorial with an Open-source Tool

LingPipe is a toolkit for processing text using computational linguistics. LingPipe is used to do tasks like finding the names of people, organizations, or locations in news, and automatically classifying search results into categories and suggesting correct spellings of queries [67]. LingPipe's architecture is designed to be efficient, scalable, reusable, and robust with features like Java application programming interface (API) with source code and unit tests.

In the following section, we will explore a practical way to text mining biomedical literature from MEDLINE. For demonstration, we used LingPipe, a free tool available for downloading from the Internet. In order to use LingPipe effectively, the interested readers are encouraged to visit the website and download the "jar" file or "zip" file containing all the detailed instructions. Here, we will describe how to integrate the power of LingPipe with custom-designed programs to achieve chemically intelligent text mining. The first step is to design a database containing tables to hold the plain text data retrieved from PubMed or any other Internet resource. After loading the data to the database, the LingPipe will retrieve the data from database or XML files to annotate each and every term into any one of the 36 classes from GENIA corpus. In order to recognize a chemical term, the user has to build a dictionary containing a list of chemical terms. The same strategy is applicable for building protein, species, gene, bioactivity, diseases of interest, or any other class of terms. Once the dictionary is built for the selected classes of interest, it is necessary to compile them to make them compatible for any text-mining tool to seek the terms and annotate with the name of the class. Once the terms are correctly recognized, the next step is to identify the frequency of occurrence of those terms with class details to build the network of information connecting molecule to disease or molecule to species, etc. The stepwise procedure and code snippets are discussed below:

Step 1: Fetch URL/PMID based on query to public databases (Internet/PubMed)
Step 2: Retrieve the document or abstracts (URL/PMID)
Step 3: Load the document to database (remove redundancy)
Step 4: Retrieve document contents (plain text) sequentially/distributed way from database
Step 5: Apply text-mining tools (LingPipe, OSCAR, Abner, etc.) to annotate the text to class (chemical, protein, disease, gene)
Step 6: Write the output annotation to comma-separated value (CSV) or database tables
Step 7: Frequency analysis (to identify relevant terms) to prioritize the contents
Step 8: Build the network based on relationships (ML applied to remove false positives*). Optimize the ML tools to build the models to automatically alert the relationships between terms (molecule–disease, molecule–target, molecule–activity) with confidence score
Step 9: Network analysis and interpretation
Step 10: Extract Scaffolds from chemicals (ring compounds) and functional groups, Linkers
Step 11: Build virtual library enumeration (to get new molecules that are not used for training)
Step 12: Random selection or complete scanning of the virtual library (VL) to select molecules of interest
Step 13: Compute molecular descriptors for screening by evaluating the scores DrugLike, Lead Like, Progressive DrugLike, Progressive LeadLike (DL, LL, PDL, PLL), toxicophore-based scores, pharmacophore-based scores, etc. as filters
Step 14: Compile the new hits and convert them in a three-dimensional (3D) format for further studies (docking, pharmacophore search)

To implement the security features (encoding, decoding, compression, un-compression) are required for efficient text mining in a distributed computing environment.

Encoding:

```
sun.misc.BASE64Encoder encoder = new sun.misc.BASE64Encoder();
String encodedUserPwd
=encoder.encode("<proxyUsername>:<proxyPassword>".getBytes());
con.setRequestProperty("Proxy-Authorization", "Basic " + encodedUserPwd);
```

Parse XML:

```
DocumentBuilderFactory docBuilderFactory = DocumentBuilderFactory.newInstance();
DocumentBuilder docBuilder = docBuilderFactory.newDocumentBuilder();
Document doc = docBuilder.parse(file);
Normalizing text representation and pulling out pubmed id (pmid) from xml
structure.
doc.getDocumentElement ().normalize ();
```

Passing pmid (str1) to fetch XML:

```
String
pmidString="http://eutils.ncbi.nlm.nih.gov/entrez/eutils/efetch.fcgi?db=pubmed&id="
+str1+"&retmode=xml&rettype=abstract";
fetchXml(pmidString);
```

LoadMedLineDb:

```
LoadMedlineDb.MedlineDbLoader dbLoader = new
LoadMedlineDb.MedlineDbLoader("db.properties");
dbLoader.openDb();
loadXML(dbLoader, new File(fileNme));
 dbLoader.closeDb();
```

AnnotateMedlineDb:

```
AnnotateMedlineDb amd = new AnnotateMedlineDb("db.properties");
Integer[] ids = amd.getCitationIds();
amd.annotateCitation(ids[i].intValue());
```

AnnotateMedlineDb Class
Instantiate chunkers

```
tokenizerFactory = IndoEuropeanTokenizerFactory.INSTANCE;
sentenceModel = new IndoEuropeanSentenceModel();
sentenceChunker = new SentenceChunker(tokenizerFactory,sentenceModel);
genomicsModelfile = new File("ne-en-bio-genia.TokenShapeChunker");
neChunker = (Chunker)AbstractExternalizable .readObject(genomicsModelfile);
```

*..getCitationIds
*..annotateCitation (Title and Abstracts, Full text if available)

```
annotateSentences(citationId,"Title",title);
annotateSentences(citationId,"Abstr",abstr);
annotateSentences(citationId,"FullText",fulltext);

 Chunking chunking = sentenceChunker.chunk(text.toCharArray(),0,text.length());
 for (Chunk sentence : chunking.chunkSet()) {
 int start = sentence.start();
 int end = sentence.end();
 int sentenceId = storeSentence(citationId,start,end-start,type);
 annotateMentions(sentenceId,text.substring(start,end));
 }
```

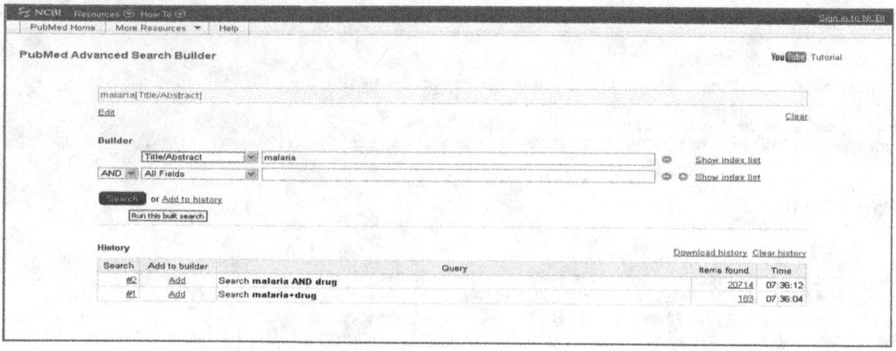

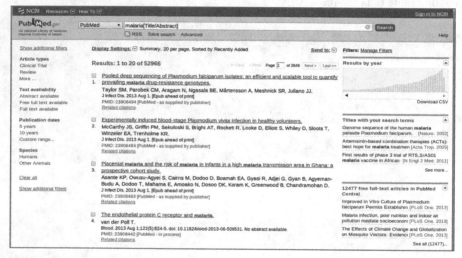

Fig. 8.3 PubMed advanced query builder

Fig. 8.4 Search results

Method annotateMentions (Linking "Word" to "class" based on GENIA)

```
Chunking chunking = neChunker.chunk(text.toCharArray(),0,text.length());
 for (Chunk mention : chunking.chunkSet()) {
int start = mention.start();
int end = mention.end();
storeMention(sentenceId,start,mention.type(),text.substring(start,end));
 }
```

Once we have understood the functioning of the code with a snippet-by-snippet explanation, let us use malaria dataset downloaded from PubMed to perform text-mining operations. The steps for downloading the data are provided here (Figs. 8.3, 8.4, and 8.5).

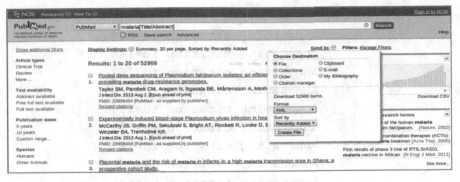

Fig. 8.5 Add all results to a XML file for download

Use Core 3 functions, once you have PubMed XML file with all records searched for malaria keyword (pubmed.xml) to load data into "citation" table of "medline" database programmatically.

The output in JAVA Netbeans output console is shown below:

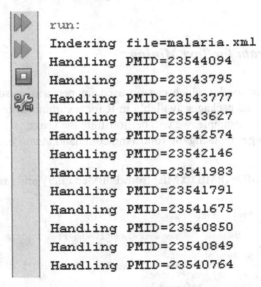

```
run:
Indexing file=malaria.xml
Handling PMID=23544094
Handling PMID=23543795
Handling PMID=23543777
Handling PMID=23543627
Handling PMID=23542574
Handling PMID=23542146
Handling PMID=23541983
Handling PMID=23541791
Handling PMID=23541675
Handling PMID=23540850
Handling PMID=23540849
Handling PMID=23540764
```

After the citation table has been populated, run use Core 4 functions to annotate all the abstracts in 36 classes and populate "sentence" and "mention" tables.

The output from Netbeans output console would look like this:

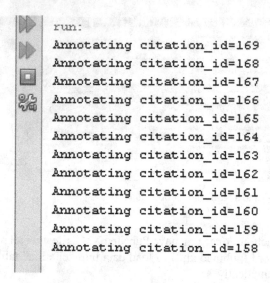

```
run:
Annotating citation_id=169
Annotating citation_id=168
Annotating citation_id=167
Annotating citation_id=166
Annotating citation_id=165
Annotating citation_id=164
Annotating citation_id=163
Annotating citation_id=162
Annotating citation_id=161
Annotating citation_id=160
Annotating citation_id=159
Annotating citation_id=158
```

8.5.2 R Program for Text Mining

R is an open-source toolkit which can be used for performing some of the text-mining tasks [68]. The packages available in R for text mining are tm RCurl XML SnowballC. Using an example, we will demonstrate the usage.

Install the packages by using the following command *install.packages* ("*packageName*").

Step 1. Retrieve PMIDs (PMID XML) from PubMed query and save them to a file

```
library(XML)
query='Tuberculosis[Title/Abstract]'
query=gsub('\\s+','+',query)
url = "http://eutils.ncbi.nlm.nih.gov/entrez/eutils/esearch.fcgi?retmax=5000"
url = paste(url, "&db=pubmed&term=", query,sep = "")
datafile = tempfile(pattern = "pub")
try(download.file(url, destfile = datafile, method = "internal", mode =
"wb", quiet = TRUE), silent = TRUE)
xml <- xmlTreeParse(datafile, asTree = TRUE)
nid = xmlValue(xmlElementsByTagName(xmlRoot(xml), "Count")[[1]])
lid = xmlElementsByTagName(xmlRoot(xml), "IdList", recursive = TRUE)[[1]]
write.table(as.data.frame(unlist(lapply(xmlElementsByTagName(lid, "Id"),
xmlValue))), quote=FALSE, file = "pmid.txt")
```

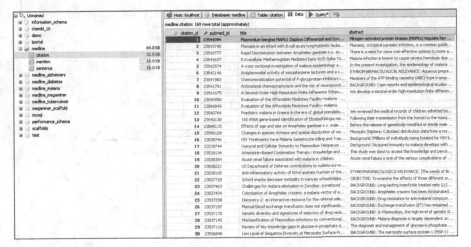

Fig. 8.6 Citation table data

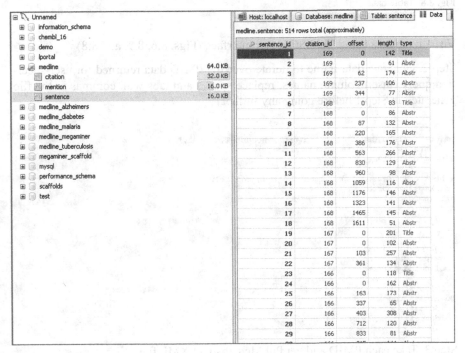

Fig. 8.7 Sentence table data

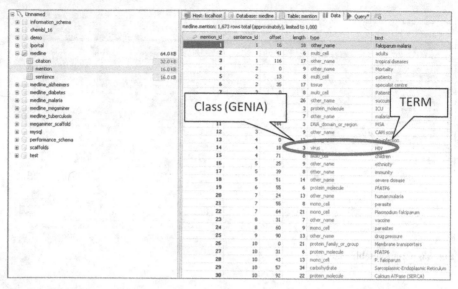

Fig. 8.8 Table data

Visualizing data using HeidiSQL user interface (Figs. 8.6, 8.7, and 8.8)

Step 2. Generate data frame to iterate over the PMID data returned for user-speci-fied query (insert column names, replace spaces and tabs with commas in the file generated in Step 1 before going any further)

```
pmiddata<-read.csv(file='pmid.txt', header=T, sep=",")
pmiddataframe<-as.data.frame(pmiddata)
dim(pmiddataframe)

#Check col names, count number of rows and columns
colnames(pmiddata)
nrow(pmiddata)
ncol(pmiddata)

or

colnames(pmiddataframe)
nrow(pmiddataframe)
ncol(pmiddataframe)

# PMIDs stored in second column of data frame named "pmiddataframe"
To access them, type
pmiddataframe[1,2]
pmiddataframe[2,2] and so on..
```

Step 3. Take each PMID and get PubMed abstract XML file

```
# Iterate over pmids in data frame and fetching abstracts from corresponding Pubmed
XML data.
# Fetching xml data from each pmid
# Parse xml and store abstracts in a file
require(RCurl)
get_pubmed <- function(query){
url="http://eutils.ncbi.nlm.nih.gov/entrez/eutils/efetch.fcgi?db=pubmed"
url=paste(url,"&id=",query,"&retmode=xml&rettype=abstract",sep="")
datafile = tempfile(pattern = "pub")
try(download.file(url, destfile = datafile, method ="internal", mode="wb", quiet =
TRUE), silent = TRUE)
xml <- xmlTreeParse(datafile, asTree = TRUE)
lid = xmlElementsByTagName(xmlRoot(xml), "Abstract", recursive = TRUE)[[1]]
write.table(as.data.frame(unlist(lapply(xmlElementsByTagName(lid, "AbstractText"),
xmlValue))), quote=FALSE, file = "xml_from_pmid.txt", append=TRUE)
}

for (i in 1:5000){
 df<-data.frame()
 df<-get_pubmed(pmiddataframe[i,2])
 }
```

Step 4. Build corpus from the data

```
library(tm)
a <- Corpus(DirSource("pubmedR"), readerControl = list(language="lat"))
summary(a)
```

Step 5. Remove numbers and punctuation from corpus data

```
a <- tm_map(a, removeNumbers)
a <- tm_map(a, removePunctuation)
```

Step 6. Remove white spaces and stop words from corpus data

```
a <- tm_map(a , stripWhitespace)
a <- tm_map(a, tolower)
a <- tm_map(a, removeWords, stopwords("english"))# this stopword file is at
C:\Users\[username]\Documents\R\win-library\2.13\tm\stopwords
```

Step 7. Stem words

```
a <- tm_map(a, stemDocument, language = "english")
```

Step 8. Build a document term matrix using refined corpus

```
adtm <-DocumentTermMatrix(a)
adtm <- removeSparseTerms(adtm, 0.75)
```

#Inspect the matrix
```
inspect(adtm[1,1:10]) # first document in directory "pubmedR" with 10 frequent
terms
```

Step 9. Find frequent terms

```
findFreqTerms(adtm, lowfreq=20)
```

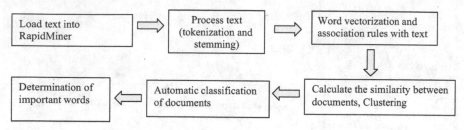

Fig. 8.9 Steps for text mining in rapid miner program

Further one can do classification, clustering, associations, word cloud with these data in R.

8.6   Free Tools for Text Mining

RapidMiner, formerly known as YALE (Yet Another Learning Environment), is an environment for ML and data-mining experiments [69]. It allows experiments to be made up of a large number of arbitrarily nestable operators, described in XML files which are created with RapidMiner's graphical user interface. For text mining, one needs to install the text-mining extensions available at their site [70]. The broad general steps are briefly outlined as follows; for further details, there are a number of video tutorials elucidating each and every component of the text-mining process (Fig. 8.9).

8.7   Biomedical Text Mining

Biomedical text mining deals with mining biologically or chemically relevant entities from an unstructured source of literature data. The trouble with the ever-growing literature data is the increasing complexity and ambiguity if the same data need to be browsed for entity or relation mining. It has been reported that more than 80 % of biomedical data are embedded in plain text form which accounts for the high degree of "unstructuredness" of biomedical data [71]. The very first step in text mining will be to convert these data to semi-structured formats like XML or more structured forms like relational databases. Most of the publicly available biomedical data are in the form of abstracts and are semi-structured, i.e., neither structured nor unstructured [72]. There are structured fields like authors, references, keywords, title, date, and also some unstructured fields like abstract, text, or concepts. Also, an important problem with the biomedical data is that a single term is linked to structures, sub-structures, Ids, or pathways. Because of the ambiguity in gene and protein nomenclature, it is often difficult to predict the use of hyphen, period, and triplet contextually. Thus, there is a need for specialized parsers in tokenization. Moreover, the biomedical literature is a complex set of information which makes use of heavy domain-specific terminologies. Word sense disambiguation is another issue one faces, as the meanings are not singled out. Also, the important data that occur are sparse as the words have very low frequency. New terms and names are created. One can also witness typographical variants and different writing styles depending upon the origin of the information. The best solution to this problem would be to build a standard protocol that can be easily followed and which will be easy for the computers to interpret. But this will not solve the problem for already published information. And extraction of useful information while maintaining all the links and relevance is quite a challenge. The real challenge is to overcome ambiguity of context in the biological science literature. But the field of text mining has evolved to solve such problems. Text-mining techniques have been extensively applied in annotating the biomedical literature to reveal interesting patterns and relationships between organisms, proteins, genes, disease, metabolism, therapeutic categories, etc. A number of biomedical text-mining tools have been reported; however, a few significant ones are briefly highlighted here. FACTA+ mines associations between biomedical entities such as drug, diseases, symptoms, enzymes, etc. [73]. KLEIO has many methods for acronym recognition and disambiguation, gene/protein name recognition [74]. @Note built on top of AI bench, a Java application development framework, provides a work bench environment to process abstracts, full-text information retrieval, tokenization, stop word removal etc [75]. A collaborative text annotation tool Bionotate [76] was developed for disease-centered relation extraction from biomedical text. BioRAT [77] covers the full journal articles for text mining instead of just PubMed abstracts. Another program MedKit [78] solved many of the downloading and parsing limitations encountered while mining PubMed literature. GIFT [79] was specially developed to find gene interactions in text and applied to a fly database. IdMap [80] was created to infer relationships between targets and chemicals using text mining and chemical structure information. PathTexts [81] consisting of a pathway visualizer, text-mining algorithms, and annotation tools is available for systems biologists. Other important text-mining tools specifically built for medical informatics are Biocontrast [82] and BioText Quest [83]. AbNER [84], an open-source software tool for biomedical text mining, provides a GUI for tagging genes, proteins, and other entity names in the given text.

8.8 Chemically Intelligent Text-mining Tools

In this section, we will discuss the manipulation of text-mining tools for chemoinformatics. Text mining of chemical synthesis literature is fraught with many problems, the foremost being the number of synonyms possible for a compound. A chemical can be present in the text as International Union of Pure and Applied

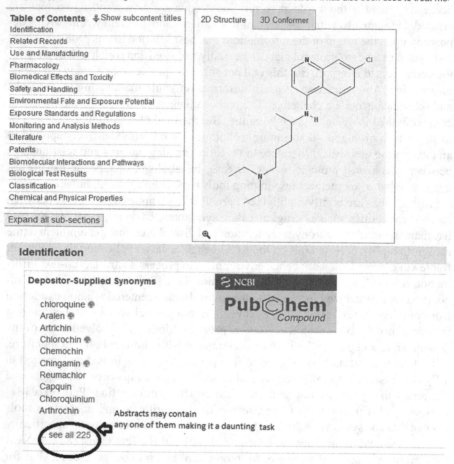

Compound Summary for: CID 2719

Chloroquine

Also known as: Aralen, Artrichin, Chlorochin, Chemochin, Chingamin, Reumachlor, Capquin, Chloroquinium, Arthrochin
Molecular Formula: $C_{18}H_{26}ClN_3$ **Molecular Weight:** 319.87214   **InChIKey:** WHTVZRBIWZFKQO-UHFFFAOYSA-N
The prototypical antimalarial agent with a mechanism that is not well understood. It has also been used to treat rhe...

Table of Contents   ⇩ Show subcontent titles
Identification
Related Records
Use and Manufacturing
Pharmacology
Biomedical Effects and Toxicity
Safety and Handling
Environmental Fate and Exposure Potential
Exposure Standards and Regulations
Monitoring and Analysis Methods
Literature
Patents
Biomolecular Interactions and Pathways
Biological Test Results
Classification
Chemical and Physical Properties

Expand all sub-sections

2D Structure   3D Conformer

Identification

Depositor-Supplied Synonyms

≋ NCBI

Pub◯hem
Compound

chloroquine
Aralen
Artrichin
Chlorochin
Chemochin
Chingamin
Reumachlor
Capquin
Chloroquinium
Arthrochin

Abstracts may contain
any one of them making it a daunting task

see all 225

Fig. 8.10 A PubChem search results for chloroquine retrieves 222 synonyms

Chemistry (IUPAC) name, common name, Chemical Abstracts Service (CAS) number, or corporate ID (Pfizer, Bayer), etc. Apparently the lack of a global standard like gene id or protein id in bioinformatics impedes the efficient application of text-mining tools (Fig. 8.10).

Much effort has been devoted toward efficient chemical text mining. A chemically intelligent tool is OSCAR4 (Open Source Chemical Analysis Routine), designed for chemistry-specific NLP [85]. It performs chemical NLP, chemical entity recognition (CER), chemical name recognition by direct lookup or ML. Its parsers can identify words and phrases representing chemical concepts. For example, acetyl salicylic acid is a single word and should not be interpreted as acetyl or salicylic.

It integrates name to structure parsing using OPSIN [86] and ChEBI (Chemical Entities of Biological Interest) identifiers [87]. OPSIN converts an IUPAC name to SMILES or INChI to structure. OSCAR performs all the important tasks such as identifying chemical names, reaction name, and small compound and enzyme prefix, suffix, and adjectives. It comes with an extensive API for developing extensions with other tools such as Taverna [88] Mendeley [89] and U-Compare [90].

To get started with OSCAR4, we use the following code to search for NER from the given "text":

```
Oscar oscar = new Oscar();
List < NamedEntity >namedEntities
= oscar.findNamedEntities(text);
```

Return hits only if named entity is resolved to a structure

```
Oscar oscar = newOscar();
List < ResolvedNamedEntity >entities
= oscar.findAndResolveNamedEntities(s);
for (ResolvedNamedEntity entity : entities) {
ChemicalStructure structure = entity.get-
FirstChemicalStructure (FormatType.INCHI));
}
```

Make the system use different classifiers

```
ChemicalEntityRecogniser myRecogniser = new-
PatternRecogniser()
Oscar oscar = newOscar();
oscar.setRecogniser(myRecogniser);
oscar.setDictionaryRegistry
(myDictionaryRegistry);
List < ResolvedNamedEntity >entities = oscar.
findResolvableEntities(s);
```

A combination of rule-based chemical text and formal grammar parser has been developed known as ChemicalTAgger [91]. It is a freely available open-source Java-based software which uses both OSCAR and open NLP programs (Fig. 8.11).

8.9   In-house Tools for Text-mining Applications for Chemoinformatics

When it comes to processing millions of biomedical-related documents, one computer may not be sufficient. This is where distributed platforms for text mining can be applied. Distributed text mining is text mining in a distributed computing

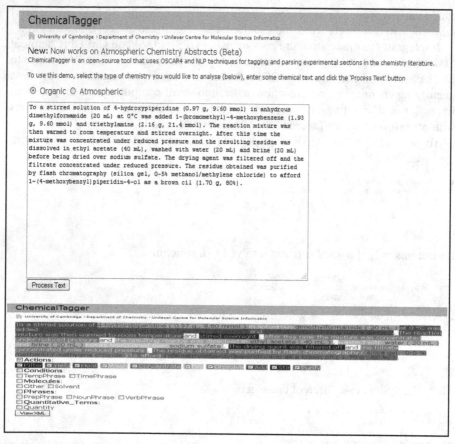

Fig. 8.11 ChemicalTagger used for parsing experimental chemical literature

environment [92]. The technology has been demonstrated for chemical computing in ChemStar [93] for property prediction. A similar architecture can be applied for text mining. A text-mining portal termed MegaMiner has been developed for applying text-mining techniques to solve chemoinformatics problems [94]. MegaMiner supports many data input types and file formats from the user via the portlet. There is a provision to input data or upload files, and the number of distributed nodes can be specified. MegaMiner segregates massive data in parallel to build entity network in a biomedical context after intensive treatment with text-mining algorithms like NER in a cloud environment. The search engine, upon receiving a query, generates an equivalent XML which is parsed using LingPipe and MegaMiner libraries. The obtained datasets are ranked according to the frequency, co-occurrence, and uniqueness to filter out the most relevant one. A PubMed search for malaria with "All Fields" retrieved more than 60,000 articles denoting the significant amount of research work in this field, and 31,403 of these articles can be safely assumed to

Table 8.1 Text-mining results for query term "malaria"

Protein 1	Protein 2	PMID	Relevant text
IFN gamma	IL2	26226	In some rodent malaria models, Th1 cells producing primarily "IL2 and IFN gamma" give rise to protection in early infection while Th2 cells producing IL4 are essential for parasite clearance in late infection

Table 8.2 A list of top-ranking protein names

Protein	Frequency
IFN-gamma	819
MSP-1	599
CD36	489
MSP1	441
TNF-alpha	435
IL-10	424
TNF	421

Table 8.3 List of top-occurring terms found by text mining

S. No.	Term	Frequency of occurrence
1	Plasmodium falciparum	25,926
2	Chloroquine	5,995
3	Plasmodium vivax	3,747
4	Plasmodium berghei	2,436
5	Drug resistance	1,120
6	Mefloquine	1,049
7	Quinine	1,031

be the most relevant hits as the keyword *malaria* appeared in the title of the article. A query keyword *malaria* in the portlet retrieved the text-mining results shown in Table 8.1.

From these data, a list of most frequently occurring proteins, terms, and drugs could be extracted (Tables 8.2, 8.3, and 8.4).

The relational database was queried to separate out the unique terms and find out the number of times they occurred. A total of 4.3 million biomedical terms were identified and put in an internal dictionary. These terms were classified into five major groups of proteins, genes, chemicals, diseases, and organisms. The co-occurrence, in the abstracts, of each of the terms with the others was calculated. This co-occurrence is just considering the keywords, the noise words will help to understand whether the co-occurrence is positive or negative.

As the system handles an array of complex tasks like those mentioned above, it has to be designed for crash handling and be a self-sufficient secure system. This has been specifically addressed by the use of a portal system, which uses industry-standard encryption technologies including DES, MD5, and RSA, load balancing, and portlet and code performance monitoring [95].

Table 8.4 Co-occurring proteins and drugs/organic compounds related to malaria

S. No.	Proteins	Drugs/organic compounds
1	Cytochrome chain	Atovaquone
2	Human serum albumin	Cationomycin
3	Mouse TNF receptor R75	Liposome encapsulated
4	NOS	Aminoguanidine
5	NOS inhibitor	Aminoguanidine
6	Pf155/RESA	PD
7	PfTrxR	Natural substrates
8	Purine salvage enzyme	Hypoxanthine–guanine–xanthine phosphoribosyltransferase
9	Purine salvage enzyme	HGXPRT
10	rhTNF-alpha	Liposome encapsulated
11	*Staphylococcus aureus* protein A	PD
12	*Staphylococcus aureus* protein A	SpA
13	Stereospecific transporter	Cytochalasin B
14	Thioredoxin reductase	5,5'-Dithiobis(2-nitrobenzamides)
15	Trypanothione reductase	5,5'-Dithiobis(2-nitrobenzamides)
16	Xanthine	5-Phospho-alpha-D-ribosyl-1-pyrophosphate
17	Xanthine	Naturally occurring 6-oxopurine
18	Xanthine	Allopurinol

The data security and system stability issues are also taken care of by using MySQL Cluster, enabled clustering of in-memory databases in a shared-nothing system [96]. The architecture is built such that one can use inexpensive hardware with minimum specific requirements of both software and hardware. It is designed not to have a single point of failure and integrated standard MySQL server with an in-memory clustered storage engine called Network DataBase (NDB) [97]. A typical MySQL cluster consists of a set of computers called as nodes, MySQL servers for access to NDB data, data nodes for storage of the data, and management nodes for managing and monitoring (Fig. 8.12).

The in-house-built tool MegaMiner was used to find antihypertensive lead molecules and apicoplast inhibitors (antimalarials) from simple text queries. In a fully automated system, the text-mining module extracted 50 organic compounds and drugs from PubMed abstracts followed by conversion of textual chemical names to Simplified Molecular Input Line Entry Specification (SMILES) format after which Scaffold, Linkers, and Building-blocks were generated. Text-mining application in the MegaMiner server was used to obtain lead molecules for hypertension. The PubMed records were queried using the text *hypertension* to retrieve 346,017 hits. The XML file containing 500 top citations related to hypertension was downloaded. LingPipe was used to load the titles and abstracts of the citations in a database in MySQL. 36 classes were obtained by annotating the data and classifying the text. Top 500 proteins and genes were identified. The table selected_terms_20 (which contained the text terms which have occurred with a frequency greater than 100 in the citations and their corresponding class) from the hypertension database was exported as a text file. This text file was imported into Cytoscape [98] and a con-

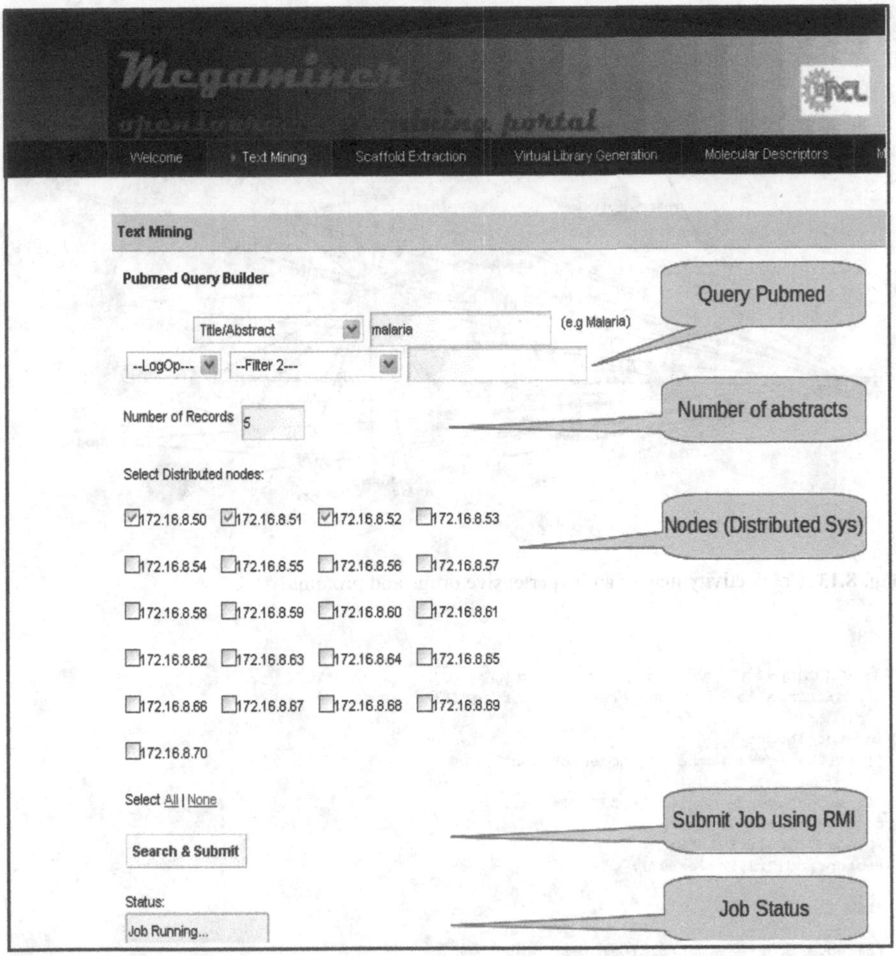

Fig. 8.12 MegaMiner homepage

nectivity map for hypertension was obtained. This connectivity map revealed various relationships such as organism–drugs, organism–proteins, drugs–proteins, etc. which were obtained using a text-mining approach. The results generated through text mining were validated using DrugBank database ([99]; Fig. 8.13).

8.9.1 Java Code Snippet for Data Distribution

Array list named ip and lookup stores the node selection made by the user

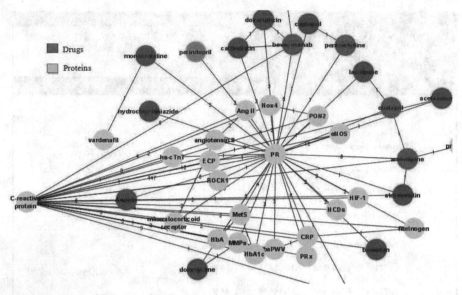

Fig. 8.13 Connectivity map of antihypertensive drugs and proteins

```
List<String> ip = new ArrayList<String>();
List<String> lookup = new ArrayList<String>();

int totalNodes=0;
//get the total number of nodes chosen

if(request.getParameter("ip50") != null) {
totalNodes++;
ip.add("172.16.8.50");
lookup.add("rmiServer50");
}
else {
}
if(request.getParameter("ip51") != null) {
totalNodes++;
ip.add("172.16.8.51");
lookup.add("rmiServer51");
}
```

and so on till the last node.

Variable "ip" stores the ip address and "lookup" stores which client to look up for getting work done in Java Remote Method Invocation (RMI) server/client architecture.

1. Array "text" is used for storing total nodes selected by the user and its size is initialized accordingly.

```
String[] text=new String[ip.size()];
```

2. Array text is populated with values read from a file which stores smiles input from the user.

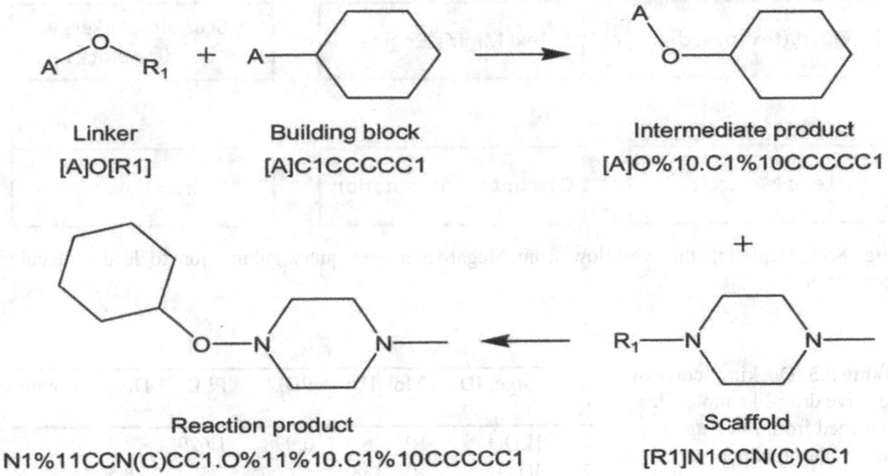

Fig. 8.14 MegaMiner virtual library synthesis

```
text=distributeData(fname); //fname is the file location

public static String[] distributeData(String fname) throws FileNotFoundException,
IOException{
  FileReader fileReader = new FileReader(fname);
  BufferedReader bufferedReader = new BufferedReader(fileReader);
  List<String> lines = new ArrayList<String>();
  String line = null;
  while ((line = bufferedReader.readLine()) != null) {
  lines.add(line);
  }
  bufferedReader.close();
  return lines.toArray(new String[lines.size()]);
  }
```

3. Data are distributed evenly between available clients using Rmiclient.

```
RmiClient rc= new RmiClient();
for (int i = 0; i < ip.size(); i++) {
  System.out.println(ip.get(i)+"->"+lookup.get(i));
  rc.rmiClient(ip.get(i), text[i], lookup.get(i));
  }
```

Complete source code and Javadocs are available at http://172.16.8.69:8080/web/guest/text-mining.

MegaMiner Virtual Library (MVL) is generated from extended scaffolds, linkers, and building blocks using the previously discussed ChemScreener program [100]. Progressive Druglike and leadlike scores are calculated for every compound in the MVL to rank them in order of priority ([101]; Fig. 8.14) .

To demonstrate text to lead prototype, a PubMed query was constructed with keywords "malaria" in conjunction with "drugs" filtered by title category in the MegaMiner portal. The limit on abstracts to be returned as hits was set to 15. This

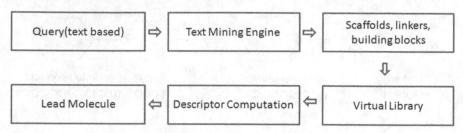

Fig. 8.15 Depicting the workflow from MegaMiner text query submission to lead molecule generation

Table 8.5 Docking scores of top five drug-like molecules screened from MVL against four malarial targets

S. No.	Target ID	Mol_ID	PDL[a]	PLL[b]	Docking score[c]
1	1LDG	RV_16	0.998	1.629	−7.5
2	3Q43	RV_138			−8.5
3	3UG9	RV_635			−8.5
4	4B1B	RV_989			−7.5
5	1LDG	RV_16	1	1	−5.8
6	3Q43	RV_138			−6.9
7	3UG9	RV_635			−6.5
8	4B1B	RV_989			−6.7
9	1LDG	RV_16	1	1	−5.2
10	3Q43	RV_138			−5.9
11	3UG9	RV_635			−7.2
12	4B1B	RV_989			−5.3
13	1LDG	RV_16	1.055	1.729	−6.6
14	3Q43	RV_138			−7.8
15	3UG9	RV_635			−6.5
16	4B1B	RV_989			−6.7
17	1LDG	RV_16	1.222	1.982	−59
18	3Q43	RV_138			−7.3
19	3UG9	RV_635			−6.8
20	4B1B	RV_989			−6.5

[a] Progressive drug-like score
[b] Progressive lead-like score
[c] Using Autodock Vina

returned top ten ranking proteins, genes, organic compounds/drugs, general terms, and co-occurring proteins based on frequency count. Finally, MegaMiner mined five leads from ChEMBL and MVL with their PDL and PLL scores and docked with validated malaria targets 1LDG, 3Q43, 3UJ9, and 4B1B to further validate it (Fig. 8.15; Table 8.5).

8.10 Thumb Rules While Performing and Using Text-mining Results

- Do not use the early biomedical text-mining systems which give co-occurrence as output, as mere co-occurrence of two terms cannot be indicative of a definitive relationship between any two entities, say a gene and a disease.
- Clearly define the inputs and outputs, whether they are terms or identifiers or database entries or plain text prior to building your text-mining system
- Chemical name ambiguity has to be dealt with carefully while text mining synthesis literature.s

8.11 Do it Yourself

1. Give a text query *kinase* in PubMed and search for relevant genes and proteins using a text-mining program of your choice
2. Using text-mining methods, retrieve the drugs associated with cancer

8.12 Questions

1. Highlight the major steps required in a general text-mining process.
2. What are the open-source tools available for text-mining process? Highlight any one.
3. What are the applications of text mining in chemoinformatics?
4. Briefly discuss the visualization programs used for biomedical text-mining results.

References

1. http://www.datanami.com/datanami/2012-12-24/how_object_storage_and_information_dispersal_address_big_data_challenges.html. Accessed 31 Oct 2013
2. Karthikeyan M, Krishnan S, Pandey AK, Bender A (2006) Harvesting chemical information from the Internet using a distributed approach: ChemXtreme. J Chem Inf Model 46:452–461
3. Cohen KB, Hunter L (2008) Getting started in text mining. Plos Comput Biol 4(1)
4. Wei CH, Kao HY, Lu Z (2013) PubTutor: a web based text mining tool for assisting biocuration. Nucleic Acids Res 41(Web Server issue):W518–22
5. http://www.nactem.ac.uk/research.php. Accessed 31 Oct 2013
6. Aguiar-Pulido V, Seoane JA, Gestal M, Dorado J (2013) Exploring patterns of epigenetic information with data mining techniques. Curr Pharm Des 19(4):779–789

7. Yang Y, Adelstein SJ, Kassis AI (2012) Target discovery from data mining approaches. Drug Discov Today 17(Suppl), S16–S23

8. Guha R, Gilbert K, Fox G, Pierce M, Wild D, Yuan H (2010) Advances in chemoinformatics methodologies and infrastructure to support the data mining of large heterogeneous chemical datasets. Curr Comput Aided Drug Des 6(1):50–67

9. http://www.kdnuggets.com/gpspubs/aimag-kdd-overview-1996-Fayyad.pdf. Accessed 31 Oct 2013

10. http://dound.com/wp/wp-content/uploads/2007-Exploring_Dimensionality_Reduction_for_Text_Mining.pdf. Accessed 31 Oct 2013

11. Macskassy SA, Hirsh H, Banerjee A, Dayanik AA (2003) Converting numerical classification into text classification. Artif Intell 143:51–77

12. Manning CD, Schutze H (1999) Foundations of statistical natural language processing. MIT Press

13. Indurkhya N, Damerau F (2010) Handbook of natural language processing. Boca Raton

14. Miner G, Elder J, Hill T, Nisbe R, Delen D, Fast A (2012) Practical text mining and statistical analysis for non-structured text data applications. Elsevier Academic Press

15. Feldman R, Sanger J (2006) The text mining handbook advanced approaches in analyzing unstructured data. Hebrew University of Jerusalem, ABS Ventures, Boston

16. Cunningham H, Tablan V, Angus RB, Kalina B (2013) Getting more out of biomedical documents with GATE's full lifecycle open source text analytics. PLoS Comput Biol 9(2):e1002854

17. http://www.ncbi.nlm.nih.gov/pubmed. Accessed 31 Oct 2013

18. http://en.wikipedia.org/wiki/Entrez. Accessed 31 Oct 2013

19. http://www.ncbi.nlm.nih.gov/books/NBK25497/

20. Fayyad U, Piatetsky-Shapiro G, Smyth P (1996) From data mining to knowledge discovery: an overview. In: Fayyad U, Piatetsky-Shapiro G, Smyth P, Uthurusamy R (eds) Advances in knowledge discovery and data mining. MIT Press, Cambridge, pp 1–36

21. Webster JJ, Kit C (1992) Tokenization as the initial phase in NLP, vol 4. University of Trier, pp 1106–1110

22. Popovic M, Willett P (1992) The effectiveness of stemming for natural-language access to slovene textual data. J Am Soc Inform Sci 43(5):384–390

23. DeRose SJ (1988) Grammatical category disambiguation by statistical optimization. Comput Linguist 14(1):31–39

24. http://norm.al/2009/04/14/list-of-english-stop-words/. Accessed 31 Oct 2013

25. Papanikolaou N, Pafilis E, Nikolaou S, Ouzounis CA, Iliopoulos I, Promponas VJ (2011) BioTextQuest: a web-based biomedical text mining suite for concept discovery. Bioinformatics 27(23):3327–3328

26. Francis WN, Kucera H (1964) A standard corpus of present-day edited American english, for use with digital computers. Department of Linguistics, Brown University, Providence

27. Ananiadou S, Sullivan D, Black W, Levow Gi-A, Gillespie JJ, Mao C, Pyysalo S, Kolluru B, Tsujii J, Sobral B (2011) Named entity recognition for bacterial Type IV secretion systems. PLoS One 6(3):e14780

28. Berry MW, Castellanos M (eds) (2007) Survey of text mining: clustering, classification, and retrieval. Springer

29. Baker NC, Hemminger BM (2010) Mining connections between chemicals, proteins, and diseases extracted from Medline annotations. J Biomed Inform 43(4):510–519

30. Korhonen A, Seaghdha DO, Silins I, Sun L, Hoegberg J, Stenius U (2012) Text mining for literature review and knowledge discovery in cancer risk assessment and research. PLoS One 7(4):e33427

31. Berry MW, Jacob KJ (eds) (2010) Text mining: applications and theory. Wiley

32. Zhou Y (2009) An improved KNN text classification algorithm based on clustering. J Comput 4(3)

33. Lan M, Tan C, Low H, Sungy S (2005) A comprehensive comparative study on term weighting schemes for text categorization with support vector machines. In: Proceedings of the 14th international conference on World Wide Web, pp 1032–1033

34. Wu X, Zhang L, Chen Y, Rhodes J, Griffin TD, Boyer SK, Alba A, Cai K (2010) Chem-Browser: a flexible framework for mining chemical documents. Adv Exp Med Biol 680:57–64 (Advances in Computational Biology)

35. Khan A, Baharudin B, Lee LH, Khan KA (2010) Review of machine learning algorithms for text-documents classification. J Adv Inf Technol 1(1)

36. http://www.nactem.ac.uk/GENIA/tagger/. Accessed 31 Oct 2013

37. http://biocreative.sourceforge.net/bio_corpora_links.html. Accessed 31 Oct 2013

38. http://www-nlp.stanford.edu/links/statnlp.html. Accessed 31 Oct 2013

39. Witten IH, Frank E, Hall MA (2011) Data mining: practical machine learning tools and techniques. Elsevier, MK (The Morgan Kaufmann series in data management systems)

40. Joachims T (1998) Text categorization with support vector machines: learning with many relevant features. Lect Notes Comput Sci Springer 1398:137–142

41. Baum LE, Petrie T (1966) Statistical inference for probabilistic functions of finite state Markov chains

42. Jang H, Song S, Myaeng S (2006) Text mining for medical documents using a hidden Markov model. In: Ng H, Leong M-K, Kan M-Y, Ji D (eds) Information retrieval technology, vol 4182. pp 553–559

43. Mccallum A, Freitag D (2000) Maximum entropy Markov models for information extraction and segmentation

44. Lafferty J, McCallum A, Pereira F (2001) Conditional random fields: probabilistic models for segmenting and labeling sequence data. ICML

45. Cohen KB, Hunter L (2008) Getting started in text mining. PLoS Comput Biol 4(1):e20

46. http://gate.ac.uk/sale/tao/splitch8.html#chap:jape. Accessed 31 Oct 2013

47. Nahm UY, Mooney RJ (2001) Mining soft-matching rules from textual data. In: Proceedings of the seventeenth International Joint Conference on Artificial Intelligence(IJCAI-01), pp 979–984, Seattle, WA

48. Miwa M, Ohta T, Rak R, Rowley A, Douglas BK, Pyysalo S, Ananiadou S (2013) A method for integrating and ranking the evidence for biochemical pathways by mining reactions from text. Bioinformatics 29(13):i44–i52

49. Srivastava A, Sahami M (2009) Text mining: classification, clustering, and applications. CRC Press, Boca Raton

50. http://mallet.cs.umass.edu/. Accessed 31 Oct 2013

51. http://gate.ac.uk/. Accessed 31 Oct 2013

52. http://nltk.org/book/ch07.html. Accessed 31 Oct 2013

53. http://alias-i.com/lingpipe/demos/tutorial/db/read-me.html. Accessed 31 Oct 2013

54. http://opennlp.apache.org/. Accessed 31 Oct 2013

55. Nigam K, Leffarty J, Maccallum A (1999) Using maximum entropy for text classification IJCAI-99 workshop on machine learning

56. http://nlp.stanford.edu/software/lex-parser.shtml. Accessed 31 Oct 2013

57. http://sourceforge.net/projects/openephyra/. Accessed 31 Oct 2013

58. Ning K, van Mulligen EM, Kors JA (2011) Comparing and combining chunkers of biomedical text. J Biomed Inform 44(2):354–360

59. http://metamap.nlm.nih.gov/. Accessed 31 Oct 2013

60. http://compbio.ucdenver.edu/corpora/bcresources.html. Accessed 31 Oct 2013

61. Yonghui W, Joshua DC, Trent RS, Miller RA, Giuse DA, Xu H (2012) A comparative study of current clinical natural language processing systems on handling abbreviations in discharge summaries annual symposium proceedings AMIA Symposium, 997–1003

62. http://project.carrot2.org/. Accessed 31 Oct 2013

63. http://sourceforge.net/projects/simmetrics/. Accessed 31 Oct 2013

64. http://www.cs.waikato.ac.nz/ml/weka/. Accessed 31 Oct 2013

65. http://alias-i.com/lingpipe/web/licensing.html. Accessed 31 Oct 2013

66. http://www.alias-i.com:8080/lingpipe-demos/ne_en_bio_genia/textInput. Accessed 31 Oct 2013

67. Wellner B, Huyck M, Mardis S, Aberdeen J, Morgan A, Peshkin L, Yeh A, Hitzeman J, Hirschman L (2007) Rapidly retargetable approaches to de-identification in medical records. J Am Med Inform Assoc 14(5):564–567
68. http://cran.r-project.org/web/packages/tm/vignettes/tm.pdf. Accessed 31 Oct 2013
69. Mierswa I, Wurst M, Klinkenberg R, Scholz M, Euler T (2006) YALE: rapid prototyping for complex data mining tasks. In: Proceedings of the 12th ACM SIGKDD international conference on Knowledge Discovery and Data Mining (KDD-06)
70. http://rapid-i.com/content/view/55/85/. Accessed 31 Oct 2013
71. Feng D, Burns G, Hovy E (2007) Extracting data records from unstructured biomedical full text proceedings of the 2007 joint conference on empirical methods in natural language processing and computational natural language learning, Prague, Association for Computational Linguistics, pp. 837–846
72. Rodriguez-Esteban R (2009) Biomedical text mining and its applications. PLoS Comput Biol 5(12):e1000597
73. http://refine1-nactem.mc.man.ac.uk/facta/. Accessed 31 Oct 2013
74. http://www.nactem.ac.uk/Kleio/. Accessed 31 Oct 2013
75. Lourenco A, Carreira R, Carneiro S, Maia P, Glez-Pena D, Fdez-Riverola F, Ferreira EC, Rocha I, Rocha M (2009) @Note: a workbench for biomedical text mining. J Biomed Inf 42:710–720
76. Kano C, Monaghan T, Blance A, Wall DP, Peshkin L (2009) Collaborative text annotation resource for disease centered relation extraction from biomedical text. J Biomed Inform 42(5):967–977
77. Corney DPA, Buxton BF, Langdon WB, Jones DT (2004) BioRAT: extracting biological information from full-length papers. Bioinformatics 20:3206–3213
78. Ding J, Berleant D (2005) MedKit: a helper toolkit for automatic mining of MEDLINE/PubMed citations. Bioinformatics 21:694–695
79. Domedel-Puig N, Wernisch L (2005) Applying GIFT, a gene interactions finder in text, to fly literature. Bioinformatics 21:3582–3583
80. Kim J-J, Zhang Z, Park JC, Ng S-K (2006) BioContrasts: extracting and exploiting protein-protein contrastive relations from biomedical literature. Bioinformatics 22:597–605
81. Papanikolaou N, Pafilis E, Nikolaou S, Ouzounis CA, Iliopoulos I, Promponas VJ (2011) BioTextQuest: a web-based biomedical text mining suite for concept discovery. Bioinformatics 27:3327–3328
82. Settles B (2005) ABNER: an open source tool for automatically tagging genes, proteins and other entity names in text. Bioinformatics 21(14):3191–3192
83. Jessop DM, Adams SE, Willighagen EL, Hawizy L, Murray-Rust P (2011) OSCAR4: a flexible architecture for chemical text-mining. J Cheminform 3:41 (and references cited therein)
84. Ha S, Seo YJ, Kwon M-S, Chang BH, Han C-K, Yoon J-H (2008) IDMap: facilitating the detection of potential leads with therapeutic targets. Bioinformatics 24:1413–1415
85. Kemper B, Matsuzaki T, Matsuoka Y, Tsuruoka Y, Kitano H, Ananiadou S, Tsuji J (2010) PathText: a text mining integrator for biological pathway visualizations. Bioinformatics 26:i374–i381
86. http://opsin.ch.cam.ac.uk/. Accessed 31 Oct 2013
87. http://www.ebi.ac.uk/chebi/. Accessed 31 Oct 2013
88. http://www.taverna.org.uk/. Accessed 31 Oct 2013
89. http://www.mendeley.com/. Accessed 31 Oct 2013
90. http://u-compare.org/. Accessed 31 Oct 2013
91. Hawizy L, Jessop DM, Adams N, Murray-Rust P (2011) ChemicalTagger: a tool for semantic text mining in chemistry. J Chemoinform 3:17
92. Attiya H, Welch J (2004) Distributed computing: fundamentals, simulations and advanced topics. Wiley-Interscience
93. Karthikeyan M, Krishnan S, Pandey AK (2008) Distributed chemical computing using ChemStar: an open source java remote method invocation architecture applied to large scale molecular data from PubChem. J Chem Inf Model 48:691–703

94. Unpublished results
95. http://www.liferay.com/products/liferay-portal/overview. Accessed 31 Oct 2013
96. http://www.mysql.com/. Accessed 31 Oct 2013
97. http://dev.mysql.com/doc/refman/5.5/en/mysql-cluster.html. Accessed 31 Oct 2013
98. Shannon P, Markiel A, Ozier O, Baliga NS, Wang JT, Ramage D, Amin N, Schnikowski B, Idekar T (2003) Cytoscape: a software environment for integrated models of biomolecular interaction network. Genome Res 13:2498–2504
99. http://www.drugbank.ca/. Accessed 31 Oct 2013
100. Karthikeyan M, Pandit D, Bhavasar A, Bender A, Vyas R (2013) ChemScreener: a distributed computing tool for scaffold based virtual screening. Comb Chem High T Scr:xx
101. Monge A, Arrault A, Marot C, Morin-Allory L (2006) Managing, profiling and analyzing a library of 2.6 million compounds gathered from 32 chemical providers. Mol Diversity 10:389–403

Chapter 9
Integration of Automated Workflow in Chemoinformatics for Drug Discovery

Abstract The ever-increasing data and restricted execution time require automated computational workflow systems to handle it. Several tools are emerging to support this activity. Automated workflow systems require scripting to define the repetitive tasks on new data to generate desired output. They help in focussing on what a particular virtual experiment will achieve rather than how the process is executed. The theme of this chapter is identification of the repetitive tasks which can be automated to employ workflows for streamlining a series of computational tasks efficiently. A brief introduction to workflows and their components is followed by in-depth tutorials using today's state-of-art workflow-based applications in the field of chemoinformatics for drug discovery research. An in-house-developed stand-alone application for chemo-bioinformatics workflow for performing protein–ligand networks J-ProLINE is also presented.

Keywords Workflow · Chemoinformatics · Drug design · Pipeline

9.1 What is a Workflow?

A workflow consists of a sequence of connected steps or modules (as nodes) where each step follows without delay or gap and ends just before the subsequent step may begin [1]. It is a depiction of a sequence of operations and an abstract virtual representation of actual work. It is related to many fields like artificial intelligence and operations research [2]. Workflows indicate any systematic pattern of any activity wherein different components interact to provide a function or a service. A workflow is composed of essentially three parameters, the input, the algorithms and the output description. It is described using flow diagramming techniques and in mathematical form using Petri nets. There are huge scientific workflow systems for instance in domains of earth science and astronomy [3]. In chemoinformatics context the workflow can be illustrated as a set of three steps involving three different modules for evaluation of pre-built quantitative structure–activity relationship (QSAR) models (Fig. 9.1).

M. Karthikeyan, R. Vyas, *Practical Chemoinformatics*,
DOI 10.1007/978-81-322-1780-0_9, © Springer India 2014

Fig. 9.1 A basic workflow
with three modules

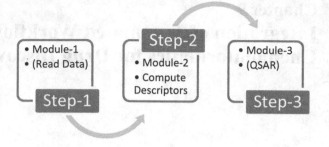

In the first step, the molecules are read by module-1 (recognizing different file formats like Simplified Molecular-Input Line-Entry System, SMILES; MOL; structure data file, SDF etc.) which verifies the entities for errors if any. Once all the molecules are processed, they are automatically passed to module-2 in the second step to generate a selected set of molecular descriptors (two-dimensional (2D) or 3D options as desired) and subjected to QSAR model evaluation in step 3 using module-3. In the final step, the model will evaluate the molecule as a hit or no-hit. Once the user completes the cycle with one data set, he can again reuse the modules in sequence with other data sets with minimum or no-manual intervention. Here, each step consists of only one input and one output components and in complex workflow system, each node (or module/step) might contain several input and output components. It is at the discretion of the user to choose the modules in sequence to accomplish the desired tasks by selecting appropriate steps, modules and methods available in the workflow-enabled tools for chemoinformatics. Konstanz Information Miner (KNIME) and Pipeline Pilot programs are the ones used for lead identification and lead optimization process in the drug discovery research.

9.2 Need for Workflows

The workflow management software systems are required to automate redundant tasks and make sure the task is completed before moving to the next one. Workflow-based engines provide rational, adaptive and responsive environment to the users clearly pointing out the dependencies for each task [4]. They can import data, perform statistical tests and generate reports efficiently. Workflows have become indispensable in scientific disciplines of biology and chemistry where there is a dire need for multiple interconnected tools and multiple data formats. They help users to perform executable processing without knowledge of programming with the assistance of a visual front end. One of the greatest advantages of workflows is analysis and control of dataflow [5]. Before proceeding to the next step, they check whether the output of one process is the correct input of next one, thus requiring no manual intervention. The user need not be concerned with the intricacies of the data processing which goes on in some remote component. Workflows allow for independent development and easy modification of each of its constituent components. KNIME allows development of individual components by defining the tasks and compiling them for distribution.

9.3   General Workflows in Bioinformatics

The bioinformatics-based workflow applications can deal with heterogeneous data types and provide a number of in-built functionalities, and some of them even allow users to add new components into a process. There are many popular tools like Galaxy [6] initially developed for genomics but can be used for any integrated workflow in bioinformatics. BioBIKE is an open-source cloud-based program that uses artificial intelligence for biocomputing [7]. Chipster is an advanced bioinformatics platform for performing next-generation sequencing (NGS) analysis and handling proteomics and microarray data [8]. Anduril is another open-source workflow-based framework which can be used for single-nucleotide polymorphism (SNP), next generation sequencing (NGS) flow cytometry and cell imaging analysis [9]. VisTrails combines workflow and visualization tasks mainly developed for exploratory computational tasks [10].

9.4   General Workflows in Chemistry Domain

Workflow is the connection of sequential steps for data management and analysis in chemistry. There are several tools for creating a workflow or pipelines: Accelrys ipeline Pilot [11], IDBS Chemsense (Inforsense suite) [12], chemistry development kit (CDK) Taverna [13], KNIME [14] etc.

9.4.1   Accelrys Pipeline Pilot

It is a scientific visual and dataflow programming language, used in various scientific domains, such as cheminformatics and QSAR, NGS, image analysis, text analytics, etc.

The graphical user interface (GUI), called the Pipeline Pilot Professional Client, allows users to drag and drop components, connect them together in pipelines and save the application developed as a protocol [15].

There are several nodes that have specific tasks on the data. Predefined components can be chosen from the library, configured, redesigned or even created from scratch and documented. When a new component is made by collapsing a few components together, it is called subprotocol. Many custom script components are available in Pipeline Pilot that allows to include the code directly into the pipelines and maintain a library of components based on a preferred language, such as Perl, Java, VBScript, .NET, JavaScript, Python, Matlab etc.

Figure 9.2 shows a typical workflow for importing CAP sample in Pipeline Pilot.

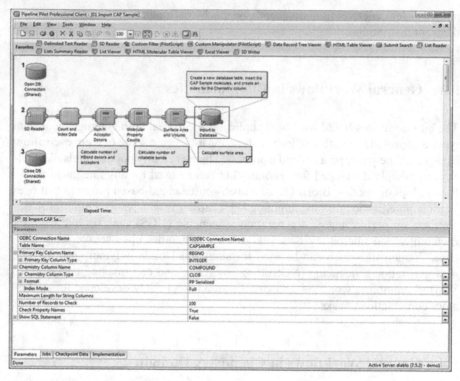

Fig. 9.2 A screenshot of Pipeline Pilot program

9.4.2   IDBS Chemsense (Inforsense Suite)

IDBS Chemsense in the Inforsense suite can be used to build chemical workflows as it adds a chemistry domain to Inforsense. It can be used for importing and exporting to chemical formats like SMILES, MOL, Chemical Markup Language (CML), reaction file (RXN), reaction data file (RDF), SDF, IUPAC International Chemical Identifier (InChI), etc. Common chemical structure-drawing tools like Accelrys, Perkin Elmer and ChemAxon can be used in Chemsense to render chemical structures and reactions. It can be used to interact with Oracle chemistry data cartridges to search and insert chemical structures and reactions. It has provision to integrate chemoinformatics functionalities from ChemAxon [16]. It is also used to build database solutions to hold chemical information (chemical reagent database), automate and publish complex cheminformatics workflows, integrate data from multiple sources and visualize chemical data from the Web. Figure 9.3 shows a workflow on Chemsense (Markush).

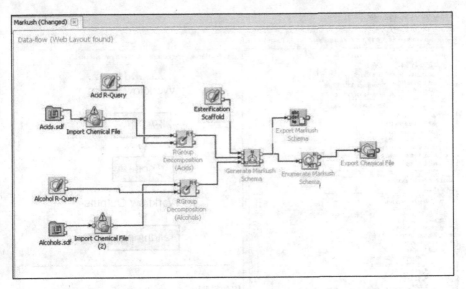

Fig. 9.3 IDBS Chemsense

9.4.3   CDK Taverna

CDK Taverna is an open-source tool. CDK Taverna can be used to create chemical workflows. Recurring tasks can be automated using CDK Taverna. This can be applied for chemical data filtering, transformation, curation, migrating workflows, chemical documentation and information retrieval-related workflows (structures, reactions, pharmacophores, object relational data etc.), data analysis workflows (statistics and clustering/machine learning for QSAR, diversity analysis etc.) [17] (Fig. 9.4).

9.4.4   KNIME

KNIME stands for the Konstanz Information Miner and is a visualization platform for creating and editing data evaluation pipelines and workflows using certain features called as 'Node Repository'. It is an open-source tool for creating chemical workflows and was developed by Prof. Michael Berthold [18]. KNIME is downloadable from www.knime.org. CDK chemistry project was incorporated in KNIME and was written in Java. It can work in integration with chemoinformatics software.

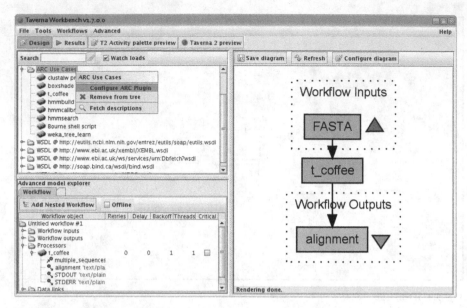

Fig. 9.4 The Taverna workbench

9.4.4.1   KNIME A practice tutorial

Downloading and Installation Instructions for KNIME

1. Go to www.knime.org
2. Go to the 'Download KNIME' option under 'Getting Started' tab
3. Select 'KNIME Desktop' and you can select one of the two options (with registration or without registration)
4. Choose the version for your platform
5. Accept the terms and conditions before downloading
6. After the download is complete, extract the zip file into a desired destination.
7. Open the KNIME executable file
8. Select the destination for the workflow and click 'OK'
9. Go to the 'File' menu in the KNIME interface
10. Click on INSTALL 'KNIME EXTENSION'

Chemical Workflow Development Pipeline of analysis process is 'Reading Data', 'Cleaning Data', 'Filtering Data' and 'Training a Model'. KNIME implements its workflow graphically. Each step of the data analysis is executed by a box called 'node'. A node is a single processing unit of a workflow. It takes data as input, processes it and makes it available on the output port, where another node of the corresponding output is attached. The 'processing' action of a node ranges from modelling, like an artificial neural network learner node, to data manipulation, like transposing.

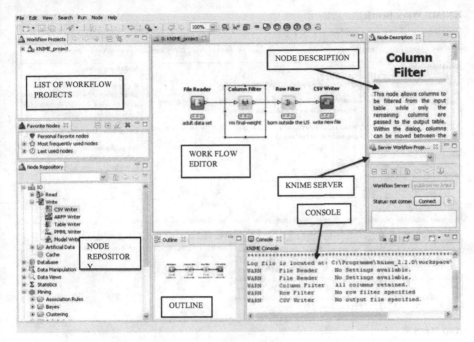

Fig. 9.5 KNIME application GUI

Every node in KNIME has three stages.

1. Inactive and not yet configured (red traffic light)
2. Configured but not yet executed (yellow traffic light)
3. Executed successfully (green traffic light)

If the node is executed with errors (unsuccessfully), its status stays at the yellow traffic light. Nodes containing other nodes are called meta nodes.

The KNIME Workbench After accepting the path of the workspace, KNIME opens the KNIME workbench. The KNIME workbench includes a workflow editor where the user can create the workflows. The KNIME workbench was developed as an Eclipse Plug-in and many of its features are inherited from the Eclipse environment. The 'KNIME Workbench' consists of a top menu, a tool bar and a few panels. Panels can be closed and moved around (Figs. 9.5 and 9.6).

Import/Export KNIME Workflow 'File' → 'Import KNIME workflow' is a link function for workflows. It links a workflow.

Import workflows from another workspace to the local workspace. It also works from zipped files. If flag 'Copy projects into workspace' is enabled, the workflow files are copied as well and not only linked into the local workspace. Changing the linked workflows changes the original workflows.

KNIME Work Bench

Top Menu: File, Edit, View, Search, Run, Node, Help		
Tool Bar: Create, Save, Run, Open Report (if reporting was installed), Open the "Add Meta node" Dialog, Buttons to reset and/or run selected or all nodes		
Workflow Projects This panel shows the list of workflow projects in the selected workspace.	**Workflow Editor** The central area consists of the "Workflow Editor" itself. A node can be selected from the "Node Repository" panel and dragged and dropped here, in the "Workflow Editor" panel.	**Node Description** If a node is selected in the "Workflow Editor" or in the "Node Repository", this panel displays a summary description of the selected node's functionalities.
Favorite Nodes This panel helps you find the nodes that are used most often or most recently or that for some other reason you want to keep at hand.	Nodes can then be connected by clicking the exit of one node and releasing the mouse at the entrance of the next node.	**Server Workflow Projects** This panel is dedicated to work on the KNIME Server, which is not part of the KNIME Desktop open source product.
Node Repository This panel contains all the nodes that you can use. It is something similar to a palette of tools when working in a report or with web designer software. There we use graphical tools, while in KNIME we use data analysis tools.	**Outline** The "Outline" panel contains a small overview of the contents of the "Workflow Editor". The "Outline" panel might not be of so much interest for small workflows. However, as soon as the workflows reach a considerable size, all the workflow's nodes may no longer be visible in the "Workflow Editor" without scrolling. The "Outline" panel can help you locate newly created nodes faster.	**Console** The "Console" panel displays error and warning messages to the user. This panel also shows the location of the log file, which might be of interest when the console does not show all messages.

Top Menu

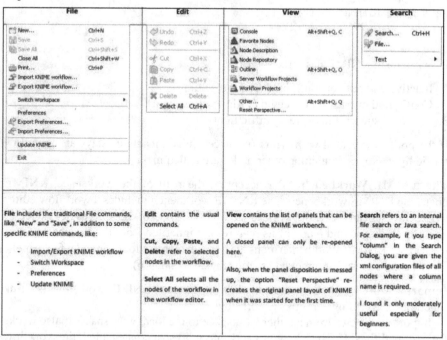

File		Edit		View		Search	
New...	Ctrl+N	Undo	Ctrl+Z	Console	Alt+Shift+Q, C	Search...	Ctrl+H
Save	Ctrl+S	Redo	Ctrl+Y	Favorite Nodes		File...	
Save All	Ctrl+Shift+S			Node Description			
Close All	Ctrl+Shift+W	Cut	Ctrl+X	Node Repository		Text	▶
Print...	Ctrl+P	Copy	Ctrl+C	Outline	Alt+Shift+Q, O		
Import KNIME workflow...		Paste	Ctrl+V	Server Workflow Projects			
Export KNIME workflow...				Workflow Projects			
Switch Workspace	▶	Delete	Delete	Other...	Alt+Shift+Q, Q		
		Select All	Ctrl+A	Reset Perspective...			
Preferences							
Export Preferences...							
Import Preferences...							
Update KNIME...							
Exit							
File includes the traditional File commands, like "New" and "Save", in addition to some specific KNIME commands, like: - Import/Export KNIME workflow - Switch Workspace - Preferences - Update KNIME		**Edit** contains the usual commands. **Cut, Copy, Paste,** and **Delete** refer to selected nodes in the workflow. **Select All** selects all the nodes of the workflow in the workflow editor.		**View** contains the list of panels that can be opened on the KNIME workbench. A closed panel can only be re-opened here. Also, when the panel disposition is messed up, the option "Reset Perspective" re-creates the original panel layout of KNIME when it was started for the first time.		**Search** refers to an internal file search or Java search. For example, if you type "column" in the Search Dialog, you are given the xml configuration files of all nodes where a column name is required. I found it only moderately useful especially for beginners.	

Fig. 9.6 The KNIME workbench

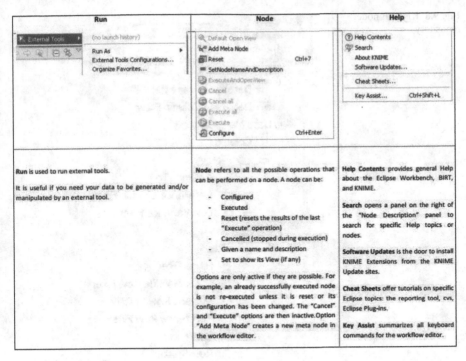

Run	Node	Help

External Tools ▶ (no launch history)
 	
Run As
External Tools Configurations...
Organize Favorites...
	🔍 Default Open View
	🔷 Add Meta Node
	📋 Reset Ctrl+7
	▬ SetNodeNameAndDescription
	ExecuteAndOpenView
	⊘ Cancel
	⊘ Cancel all
	⊘ Execute all
	⊘ Execute
	📋 Configure Ctrl+Enter
		❓ Help Contents
		🔍 Search
		  About KNIME
		  Software Updates...
		
		  Cheat Sheets...
		
		  Key Assist... Ctrl+Shift+L

Run is used to run external tools. It is useful if you need your data to be generated and/or manipulated by an external tool.	**Node** refers to all the possible operations that can be performed on a node. A node can be: - Configured - Executed - Reset (resets the results of the last "Execute" operation) - Cancelled (stopped during execution) - Given a name and description - Set to show its View (if any) Options are only active if they are possible. For example, an already successfully executed node is not re-executed unless it is reset or its configuration has been changed. The "Cancel" and "Execute" options are then inactive. Option "Add Meta Node" creates a new meta node in the workflow editor.	**Help Contents** provides general Help about the Eclipse Workbench, BIRT, and KNIME. **Search** opens a panel on the right of the "Node Description" panel to search for specific Help topics or nodes. **Software Updates** is the door to install KNIME Extensions from the KNIME Update sites. **Cheat Sheets** offer tutorials on specific Eclipse topics: the reporting tool, cvs, Eclipse Plug-ins. **Key Assist** summarizes all keyboard commands for the workflow editor.

Fig. 9.6 (continued)

Fig. 9.7 Workflow import selection in Knime

Fig. 9.8 KNIME node
repository

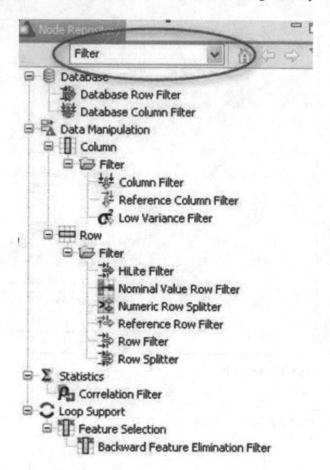

'File' → 'Export KNIME workflow' writes the selected local workflow to a zipped file. The option 'Exclude data from export' enables the export of only the nodes without the intermediate data. This generates considerably smaller export files (Figs. 9.7 and 9.8).

In the 'Node Repository' panel, there is a search box. If you type a keyword in the search box and then hit 'Enter', a list of nodes with that keyword in the name is obtained. Press the 'Esc' key to view all nodes again. For example, all the nodes with the keyword 'Filter' in their name are searched.

Workflow Operations

Creating a new workflow Right-click Local (Local Workspace) in the KNIME Explorer panel and click 'New KNIME Workflow'. Type a name for the workflow and click 'Finish'. (Destination can be changed if desired) (Fig. 9.9).

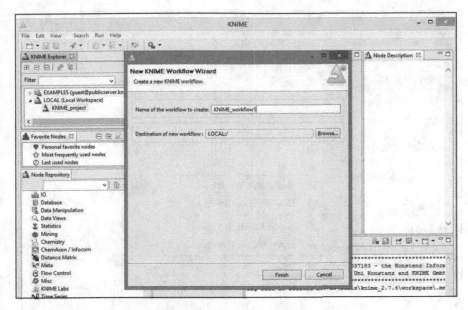

Fig. 9.9 A new workflow creation in KNIME

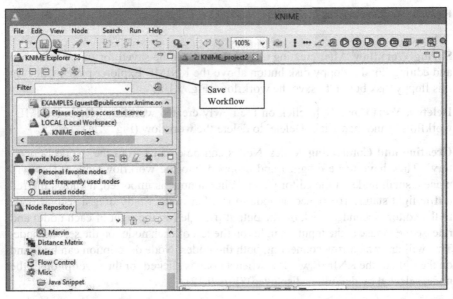

Fig. 9.10 Saving a workflow

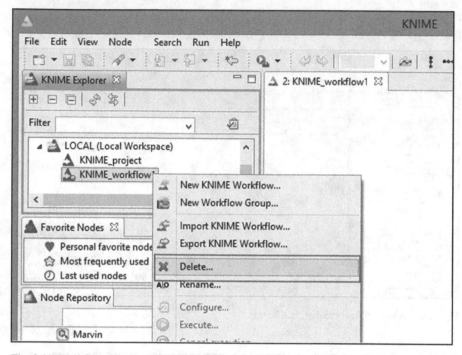

Fig. 9.11 Workflow deletion in KNIME

Saving Workflow After creating a workflow, it can be saved for future reference and editing. Find a floppy disk button above the KNIME Explorer panel. Click on this floppy disk button to save the workflow (Fig. 9.10).

Delete a Workflow Right-click on the newly created workflow (Here: KNIME_workflow1) and then click 'Delete' to delete the workflow (Fig. 9.11).

Creating and Connecting Nodes Nodes can be created from the 'Node Repository'. They have to be dragged and dropped into the workflow editor. This will place a small node on the editor panel. When a node is imported, it shows the red traffic light status. To connect a node to another node, firstly drag a second node to the editor. Secondly, click on the output triangle (on the right of each node) and release the mouse at the input triangle (on the left of each node) on the second node. This will draw an arrow connecting both the nodes. Node description can be found on the right of the KNIME window when a node is clicked (or the description of the node, selected by default, is displayed) (Fig. 9.12).

Configuring a Node This step is performed to load entities to the node or to accept input from a previous node. Double-click on the node to open the menu or right-click the node to open the menu. Click on the configure menu. Every node has different configure dialog box and can be used to fill the configuration settings. When the configuration is successful, the node turns to the yellow traffic signal (means, it is ready to be run).

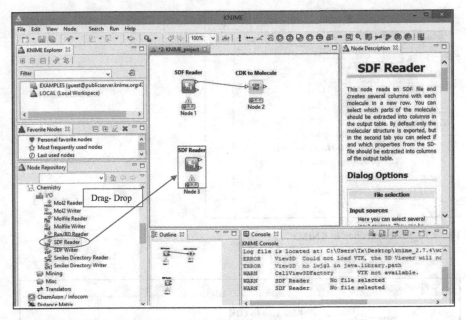

Fig. 9.12 Creating and connecting nodes in KNIME

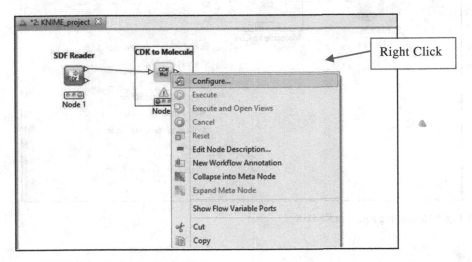

Fig. 9.13 Node configuration in KNIME

Note: Input ports of the node must be connected to a previous node (which is at green traffic signal) (Fig. 9.13).

Executing a Node When a node is configured (yellow traffic signal), right-click on the node and select 'Execute' on the menu. This will run the function of the node and it turns to green traffic light (if successful). Sometimes, the process can

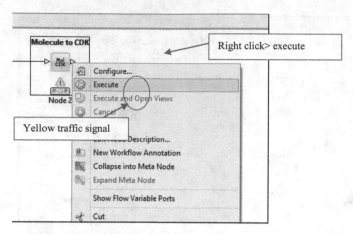

Fig. 9.14 Node execution in KNIME

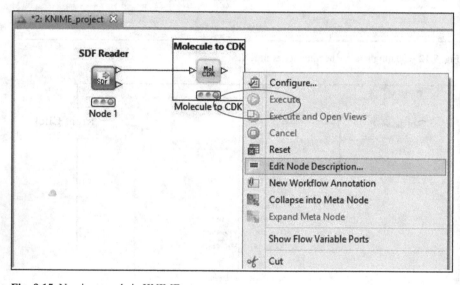

Fig. 9.15 Naming a node in KNIME

be lengthy and will happen only after the queue is complete, which will show the status (Fig. 9.14).

Node Name and Description Node can be given a name and a description. Rename the node by double-clicking on node name ('node 1'). Right-click on the node and then select 'Edit description'. In the dialog box, enter a desired description.

Note the difference between the node names as shown in Figs. 9.14 and 9.15 (node name is changed).

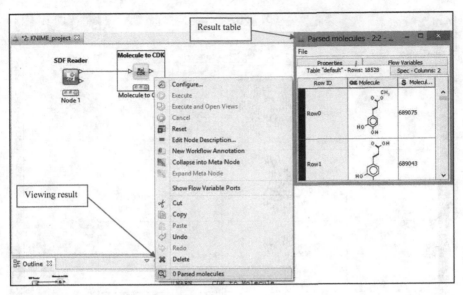

Fig. 9.16 Result visualization in KNIME

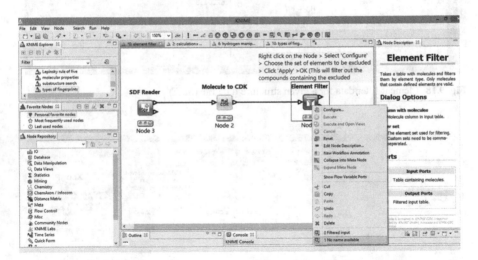

Fig. 9.17 Element filter data in KNIME

View Processed Data (Result) If the execution was successful, the traffic signal turns to green. To view the processed data, right-click on the node and select the final menu. This gives the result table (Fig. 9.16).

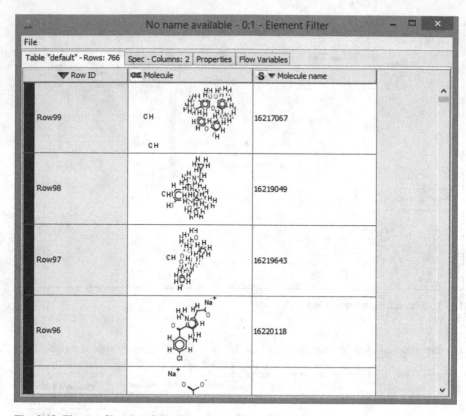

Fig. 9.18 Element filter data for a given structure data file (*SDF*)

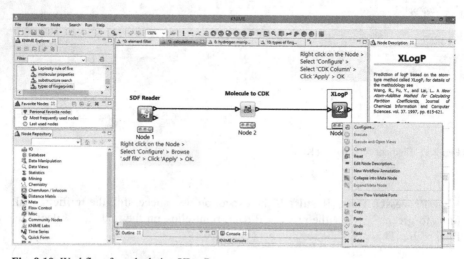

Fig. 9.19 Workflow for calculating XLogP

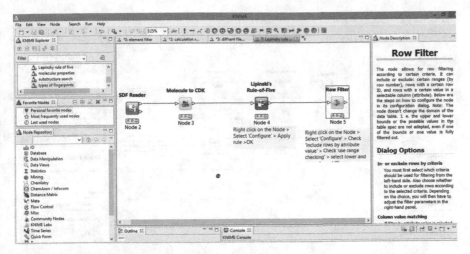

Fig. 9.20 Workflow for filtering molecules

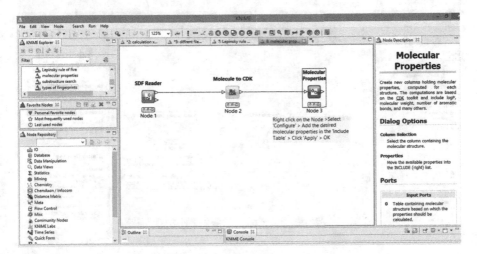

Fig. 9.21 Workflow for calculation of molecular property

9.4.5 Workflow Examples

1. Workflow for filtering compounds using element filter (Figs. 9.17 and 9.18)
2. Workflow for calculating XLogP (Fig. 9.19)
3. Workflow for filtering molecule (Lipinski's rule of five) (Fig. 9.20)
4. Workflow for calculating molecular property (Fig. 9.21)
5. Workflow for calculating fingerprint similarity (Fig. 9.22)
6. Workflow for substructure search (Fig. 9.23)
NOTE: Sample workflows can be downloaded from the following links

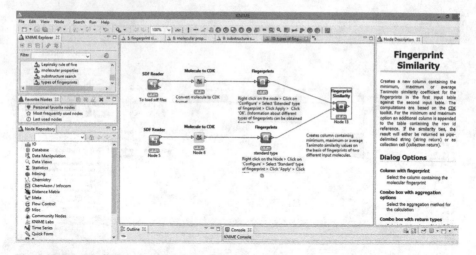

Fig. 9.22 Workflow for computing fingerprint similarity

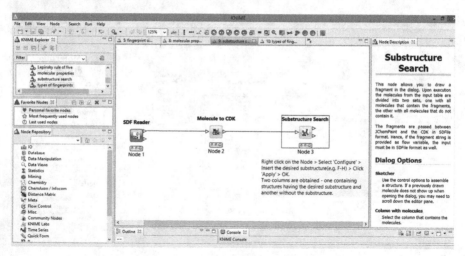

Fig. 9.23 Workflow for substructure searching of molecules in KNIME

https://docs.google.com/file/d/0B2heHCCmonQQTWNibXFJdC1yNDg/
edit?usp=sharing

https://docs.google.com/file/d/0B2heHCCmonQQbTFfckVjbFl4Vlk/
edit?usp=sharing

https://docs.google.com/file/d/0B2heHCCmonQQZjFRMjlBU19RWU0/
edit?usp=sharing

https://docs.google.com/file/d/0B2heHCCmonQQZ3VZaW9jejJOVjg/
edit?usp=sharing

https://docs.google.com/file/d/0B2heHCCmonQQSFpHNm5NazJENlk/
edit?usp=sharing

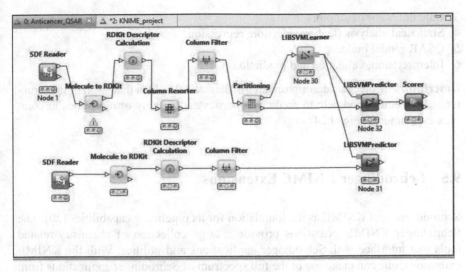

Fig. 9.24 Workflow for QSAR model building

https://docs.google.com/file/d/0B2heHCCmonQQcFI2V1Fkd09MQkE/
edit?usp=sharing

https://docs.google.com/file/d/0B2heHCCmonQQOEtsVnRIS2JHdEk/
edit?usp=sharing

https://docs.google.com/file/d/0B2heHCCmonQQVnBIaFdYc0xRM0k/
edit?usp=sharing

https://docs.google.com/file/d/0B2heHCCmonQQVVpObXQ0emlBX3c/
edit?usp=sharing

https://docs.google.com/file/d/0B2heHCCmonQQMlFGVFR5QVhTajQ/
edit?usp=sharing

9.4.6   Workflow for QSAR (Anti-cancer)

Quantitative structure–activity relationships (QSARs)

QSAR is an important technique in ligand–structure-based drug design [19]. Potency or toxicity of a set of similar drugs is correlated with a variety of molecular descriptors with the help of QSAR. Empirical formula is used to rapidly calculate multiple descriptors based on the structure and the connectivity of atoms in the molecule. For example, descriptors such as the molecular weight and the number of H-bond acceptors are easily concluded. Some descriptors, such as logP and molecular polarizability, can be approximated from atomic or group contributions.

Steps for a QSAR Model Generation:

1. Preparation of input data (structures, known biological activities)
2. 3D Geometry optimization (conformation generation, alignment)

3. Calculation of descriptors
4. Statistical analysis (feature selection, regression)
5. QSAR model building
6. Interpretation, validation and prediction

Descriptors Molecular descriptors are mathematical values that explain the structure of molecules and help to predict properties and activity of molecules in complex experiments (Fig. 9.24).

9.5   Schrodinger KNIME Extensions

Schrodinger uses KNIME as the foundation for its pipelining capabilities [20]. The Schrodinger KNIME extensions provide a large collection of chemistry-related tools that interface with Schrodinger applications and utilities. With the KNIME extensions, one can make use of the full spectrum of Schrodinger applications from within KNIME workflows. The version of KNIME that the Schrodinger extensions are built on is a freely available core KNIME distribution. One can of course develop their own extensions that make use of Schrödinger software. To develop custom nodes, at least a basic understanding of Java and the KNIME application programming interface (API) is required.

When one installs KNIME and the Schrödinger KNIME extensions from the Schrödinger distribution, they are installed into $SCHRODINGER/knime-v*version*, and a script is installed with which KNIME can be run. To start KNIME, use this command: %SCHRODINGER\knime.bat [*options*]

Some of the important features that are available through the KNIME extensions are:

- Ability to assemble, edit and execute workflows using a graphical tool
- Access to most of Schrödinger's modelling and cheminformatics tools
- Ability to integrate existing command-line tools and scripts
- Interoperability with third-party applications
- Web services integration
- Support for distributed and high-throughput computing and compute-intensive modelling tasks
- Ability to visualize and interact with data at every step of a workflow
- Ability to share workflows

The Schrödinger KNIME extensions can be downloaded or updated from the Schrödinger website, through the KNIME interface. Readers are referred to the Schrodinger manual for details. A collection of entire workflows is also available for download from the Schrödinger website, at http://www.schrodinger.com/knime-workflows.

Select Workflows available at the site.

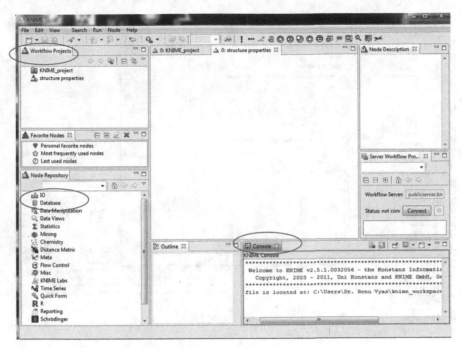

Fig. 9.25 Initial screen of KNIME Schrodinger

Listed below are example KNIME workflows that utilize many of the Schröding-er KNIME extensions (nodes) as well as many other built-in tools.

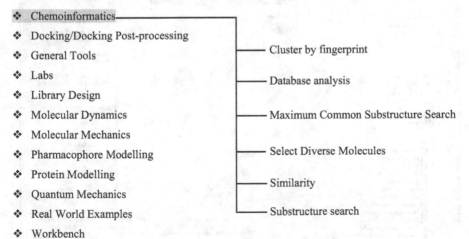

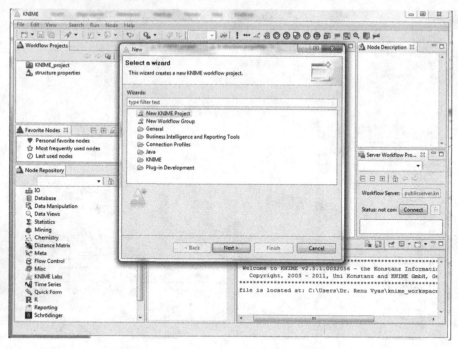

Fig. 9.26 Creating a new KNIME project

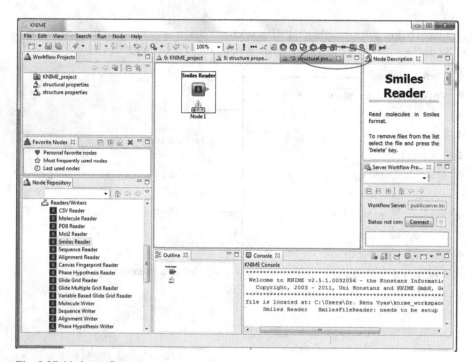

Fig. 9.27 Node configuration

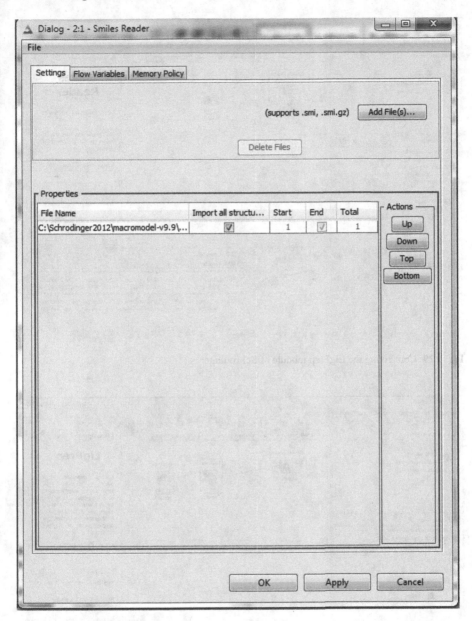

Fig. 9.28 Reading SMILES of a molecular data set

9.5.1 A Practice Tutorial

In this tutorial, we will learn how to use LigPrep and QikProp modules of Schrodinger to calculate properties of molecules in the KNIME workbench.

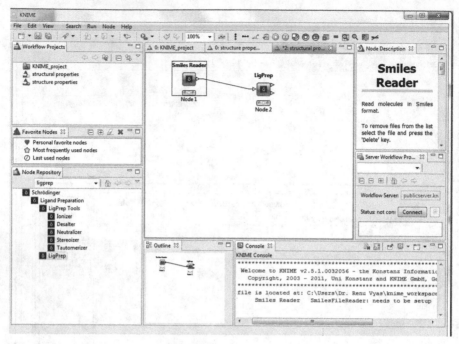

Fig. 9.29 Connecting the LigPrep module of Schrodinger

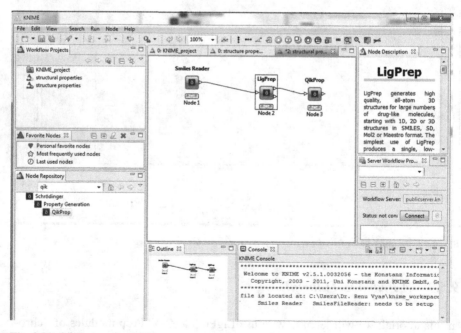

Fig. 9.30 QikProp node addition

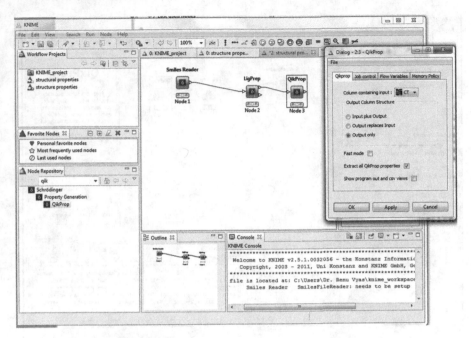

Fig. 9.31 The QikProp dialog box

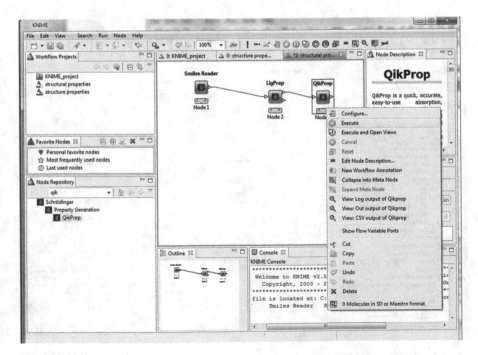

Fig. 9.32 Node execution

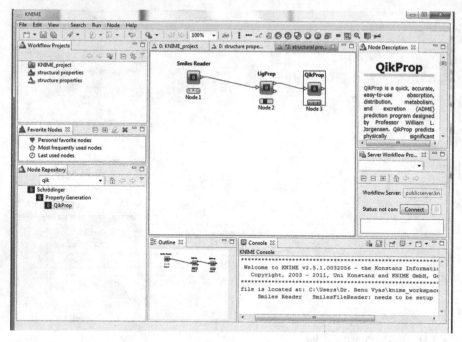

Fig. 9.33 Workflow creation

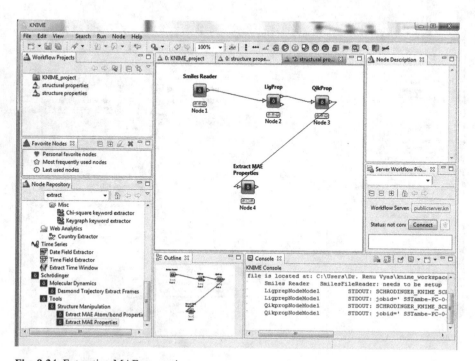

Fig. 9.34 Extracting MAE properties

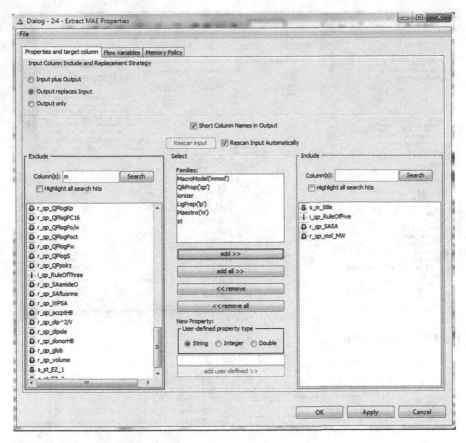

Fig. 9.35 Extract MAE dialog box

Launch KNIME in windows by double-clicking the icon in Linux, the command $SCHRODINGER/knime is used (Fig. 9.25).

To create a new KNIME project, click file new then select new KNIME project from the wizards list. In the next step, the project name can be entered by the user, say structural properties. A new tab by that name is created in the main window (Fig. 9.26).

Go to the node repository under Schrodinger and click to open readers/writers category; drag the smiles reader into workspace. The red light under it indicates that the node needs to be configured (Fig. 9.27).

To configure it, right-click on the node and in the dialog box that opens select the file where the molecular structures are available; here, we will choose the example molecules already loaded in directory at $SCHRODINGER/macromodel-ligprep/ samples/examples/1S_smiles.smi.

The file gets added to the properties table as shown in Fig. 9.28.

The users have a choice to import all structures or select a range. On click-ing OK, the red light turns yellow. Next, we will add the LigPrep node by typing

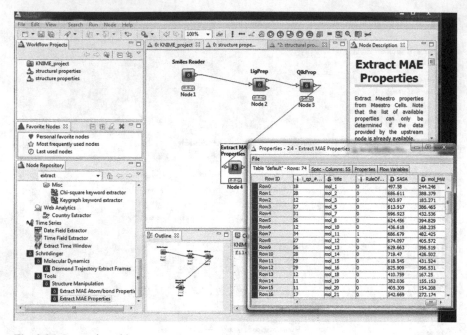

Fig. 9.36 Properties table

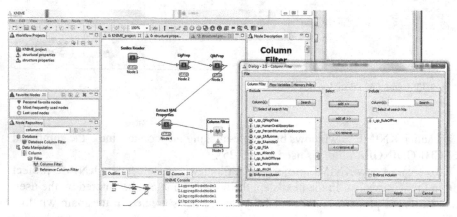

Fig. 9.37 Select compounds which obey Lipinski's rule

LigPrep into the text search box and drag it into the workspace to connect with the smiles reader node as shown in Fig. 9.29

Similarly connect the QikProp node to the LigPrep upper node to create a workflow (Fig. 9.30).

Select the QikProp node and right-click to configure it; in the configure window, select output only option (Fig. 9.31).

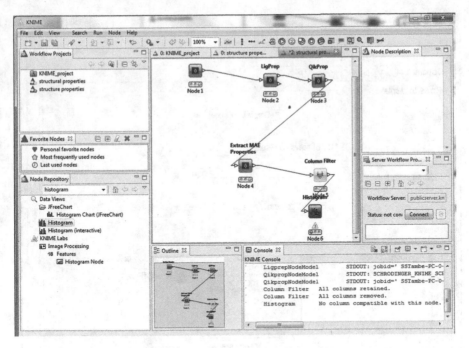

Fig. 9.38 Histogram node creation

Now the workflow is ready to be executed. Right-click on QikProp and choose execute (Fig. 9.32).

The nodes are executed in sequence beginning from the smiles reader. The green colour indicates that the task is done, while a blue bar indicates that the job is running (Fig. 9.33).

To extract the properties calculated by QikProp, drag the extract MAE properties node into the workspace and connect to QikProp node (Fig. 9.34).

The extract MAE properties node is configured. The user can select the properties to be calculated. By default, all the properties are selected for extraction. Here, we will select only four properties s_m_title, i_qp_RuleofFive, r_qp_SASA and r_qp_mol_MW (Fig. 9.35).

Select the output-only option and click OK and execute the extract MAE properties node. Then right-click on this node to choose 0 properties to display a table with extracted properties (Fig. 9.36).

Alternatively, an interactive table node can be used to display the same results. The data can be written to an excel file using xls writer node.

To visualize the obtained data, we can use column filter node to study compounds violating Lipinski's rule of five. Drag the node to the workspace and right-click to configure it. Only Lipinski's property is to be kept in the include list (Fig. 9.37).

Next, a histogram node is added to the column filter node (Fig. 9.38).

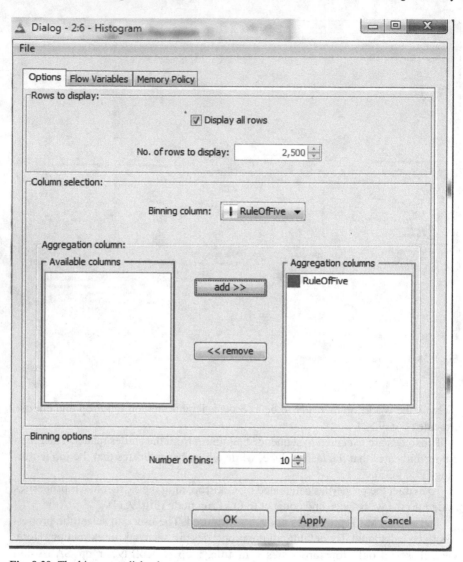

Fig. 9.39 The histogram dialog box

Right-click histogram node, choose configure in the options and select rule of five; when it is added to aggregation list, click on ok (Fig. 9.39).

Next, right-click on the histogram node and choose execute and view. The histogram is displayed. Label all the elements and put the orientation horizontal (Fig. 9.40).

We can add another column filter node to extract MAE properties node and configure it by sending remaining three properties other than SASA to the exclude list. Further add a scatter plot node to the output of the column filter node. Right-click without configuring to execute and view the scatter plot between molecular weight and SASA property (Fig. 9.41).

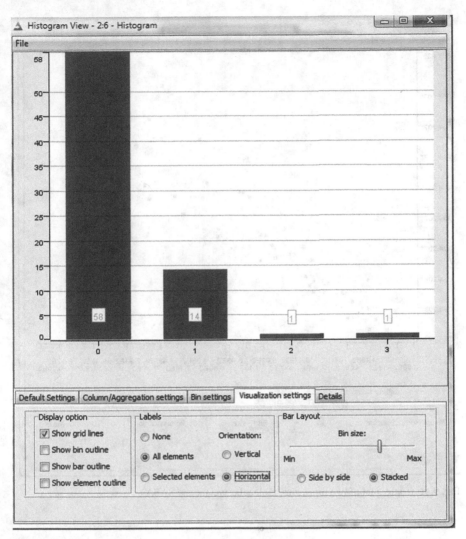

Fig. 9.40 Histogram generated for comparing properties data

9.6 Other KNIME Extensions (Fig. 9.42)

9.6.1 MOE(CCG)

Using a chemistry-aware embedded language like Scientific Vector Language (SVL), the Molecular Operating Environment (MOE) engine is not dependent on hardware and operating system [21]. More than 80 MOE nodes are included, for example, node for retrosynthetic accessibility, protonation, Murcko framework generation, Shannon entropy model creation, InChI calculation, pharmacophore

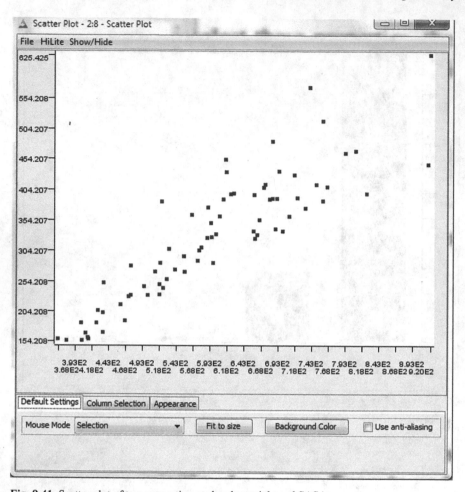

Fig. 9.41 Scatter plot of two properties, molecular weight and SASA

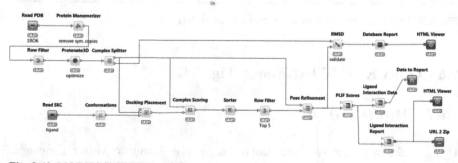

Fig. 9.42 MOE(CCG) KNIME nodes

generation etc. The MOE model ports can connect to other generic KNIME ports. Some optional ports are also supported by MOE extensions. Many chart types can be supported for the data. It also provides extensions to Chemical Computing Group (CCG) bioinformatics nodes.

9.6.2 ChemAxon

JChem Marvin KNIME extensions are also available [22]. The modules allow researchers to handle chemical structure data using ChemAxon's software tools such as Marvin, JChem and Standardizer within the open-source KNIME workflow environment. The KNIME platform provides a modular environment to visually create data flows, analyse and build predictive models. The JChem Extensions contain some nodes that are free of charge for general use. These nodes are called 'Marvin Family Nodes' which include a set of nodes for structure conversion, 'Marvin Sketch', 'Marvin View' and 'Marvin Space' which offer sophisticated rendering capabilities for chemical structures.

9.7 Protein–Ligand Analysis-Based Workflows for Drug Discovery

Target–ligand association data are growing rapidly thanks to increasing sophistication in experimental techniques like nuclear magnetic resonance (NMR) and X-ray on one hand and computational methods for homology modelling and compound library generation on the other. These enormous data are ideal for knowledge-based drug design approaches [23]. In fact, the emerging field in drug design, viz. *chemogenomics,* specifically investigates compound classes against families of functionally related proteins [24]. Proteins that have highly flexible binding sites or belong to large and diverse protein families can bind structurally dissimilar ligands [25]. Ligands that bind specifically to certain proteins can lead to enzyme inhibition or modulation of signal transduction and thus can be used as drugs [26]. By use of the properties of the ligand-binding site along with the assumption of the 'lock-and-key' and 'induced fit' principle [27], many computational techniques can be employed to identify and/or design a potential drug molecule.

The structure-based design of active compounds is based on the folding of the polypeptide backbone of the protein into the characteristic 3D structure which gives it its functional form [28]. Sequences of α helical proteins are reported to bind with ligands of similar structures which may be attributed to divergent evolution or moderate binding specificity of some proteins or experimentalists' tendency to employ only native ligand analogues for solving crystal structures [29]. On the other hand, there are examples where proteins with sequence similarity do not bind to similar ligands probably due to convergent evolution where a common fold is reinvented to perform a related function [30]. About 10 K of the biomolecular complexes in the Protein Data Bank (PDB; ~80 K entries) [31] consist of proteins with bound ligands. The diversity or similarity of ligands binding to the same protein can reflect the potential for making different interactions within the binding site. The majority of these structures provide valuable information on how the true substrates, cofactors, inhibitors or ligands bind to their cognate targets. Moreover, the structures provide some degree of comparative information, where, for example, different ligands bind to the same protein of a different species or the same ligand binds to structurally different proteins.

Analysis of protein–ligand complexes is therefore likely to reveal patterns and relationships and provide insight into the biochemical functions of proteins related to important human diseases to serve as guidelines for virtual screening. There are many instances of application of protein–ligand knowledge in fingerprint searching for ligands of corresponding targets in virtual screening as a constraint prior to docking [32]. From interacting fragments, interacting fingerprints (IF-FP) were calculated for similarity searching even for targets for which no 3D structures were available or only a few validated screening hits were known. Another important development in protein–ligand complex analysis was the use of interaction fingerprints approach—structural interaction fingerprint (SiFT) [33], profile-structural interaction fingerprint (pSIFt) [34] and weighted protein–ligand interaction fingerprint (wSIFT) [35] to translate desirable target–ligand-binding interactions into library filtering constraints [36]. There are other existing tools to analyse protein–ligand interactions but they very often involve receptor–ligand programming and the obtained interaction fingerprints are not generic for all proteins belonging to different families. G protein-coupled receptor (GPCR)-based interactions fingerprints cannot be used for drug development of kinases and vice versa.

The Protein–ligand Interaction Fingerprinting (PLIF) tool is a method for summarizing the interactions between ligands and proteins using a fingerprint scheme available in the MOE site [37]. Interactions such as hydrogen bonds, ionic interactions and surface contacts are classified according to the residue of origin and built into a fingerprint scheme which is representative of a given database of protein–ligand complexes.

The input data for PLIF can be from a variety of sources, which most commonly include X-ray crystal structures and docking results. Using fingerprints to collectively represent protein–ligand interactions for a large database is an effective way of dealing with databases which are noisy and error-prone due to the many difficulties involved in modelling ligands bound to proteins.

Fingerprints generated using PLIF are compatible with other fingerprint tools found in MOE. Standard fingerprint tools such as clustering and diverse subsets can be applied to fingerprints generated by PLIF. There is a specialized visualization interface which is designed to take into account the specific structural meaning of each fingerprint bit.

There are six types of interactions in which a residue may participate: side-chain hydrogen bonds (donor or acceptor), backbone hydrogen bonds (donor or acceptor), ionic interactions and surface interactions. The most potent of each of these interactions in each category, if any, is considered.

If no interactions of a particular category are found, or none pass the thresholds, no bits are set for that category. If the strongest interaction passes the lower interac-

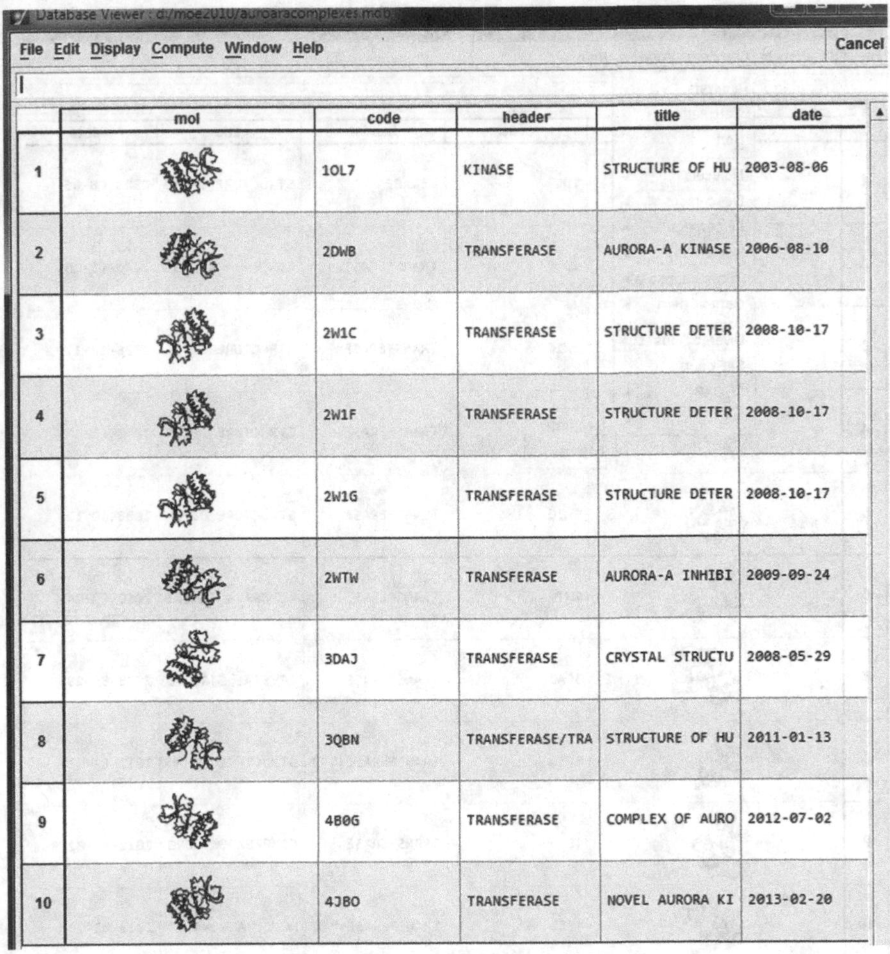

	mol	code	header	title	date
1		1OL7	KINASE	STRUCTURE OF HU	2003-08-06
2		2DWB	TRANSFERASE	AURORA-A KINASE	2006-08-10
3		2W1C	TRANSFERASE	STRUCTURE DETER	2008-10-17
4		2W1F	TRANSFERASE	STRUCTURE DETER	2008-10-17
5		2W1G	TRANSFERASE	STRUCTURE DETER	2008-10-17
6		2WTW	TRANSFERASE	AURORA-A INHIBI	2009-09-24
7		3DAJ	TRANSFERASE	CRYSTAL STRUCTU	2008-05-29
8		3QBN	TRANSFERASE/TRA	STRUCTURE OF HU	2011-01-13
9		4B0G	TRANSFERASE	COMPLEX OF AURO	2012-07-02
10		4JBO	TRANSFERASE	NOVEL AURORA KI	2013-02-20

Fig. 9.43 The protein–ligand complexes loaded in DBV in MOE

tion threshold, the low-order fingerprint bit is set. If the strongest interaction passes the higher interaction threshold, then the low-order and high-order bits are both set. Therefore, the bit patterns for each category can take on values of 00, 10 or 11, correspondingly.

Hydrogen bonds between polar atoms are calculated using a method based on protein contact statistics, whereby a pair of atoms is scored by distance and orientation. The score is expressed as a percentage probability of being a good hydrogen bond. Ionic interactions are scored by calculating the inverse square of the distance between atoms with opposite formal charge (e.g. a carboxylate oxygen atom and a protonated amine) and expressed as a percentage (100% corresponds to 1 Å distance). Surface contact interactions are determined by calculating the solvent-exposed surface area of the residue, first in the absence of the ligand, then in the presence of the ligand. The difference between the two values is the extent to which the ligand has shielded the residue from exposure to solvent, which is potentially

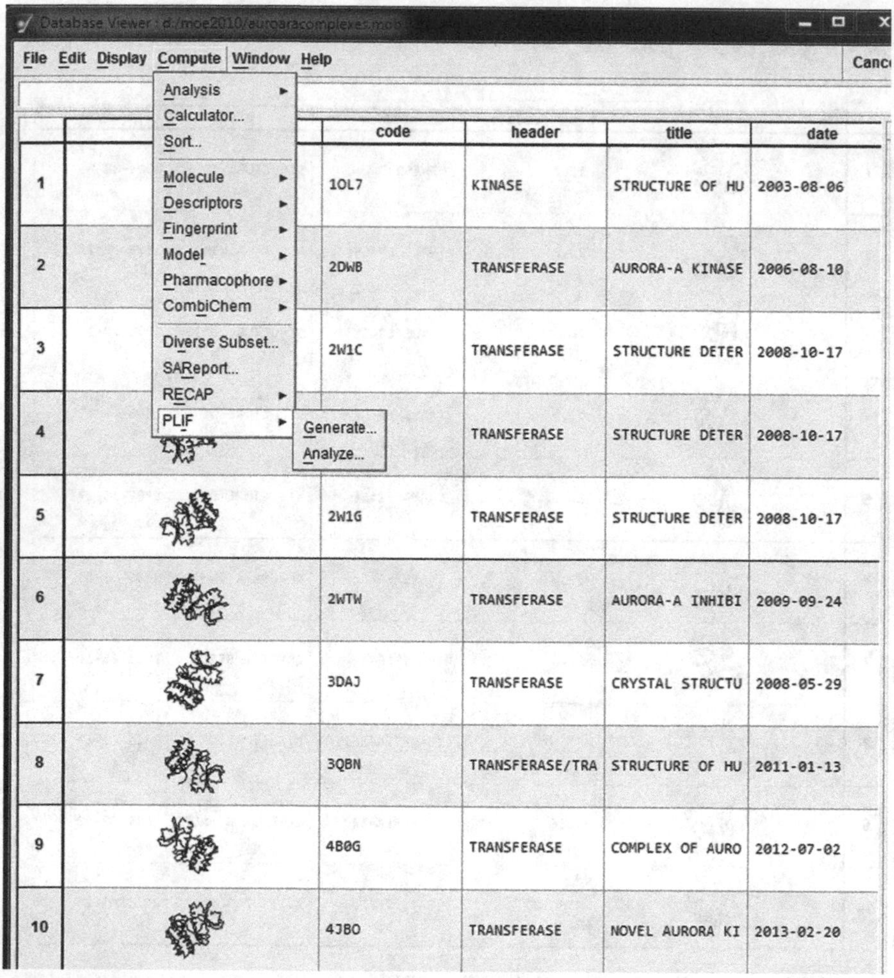

Fig. 9.44 Computing protein–ligand fingerprints of complexes using PLIF

indicative of a hydrophobic interaction. The solvent-exposed surface area is determined by adding 1.4 Å to the van der Waals radii of each heavy atom, and computing the fraction of this total surface which does not lie within the radius of any other.

9.7.1 A Practice Tutorial for Protein–Ligand Fingerprint Generation

To use PLIF, it is necessary to assemble one or more proteins to serve as the receptor species and some number of ligands with bound conformations. Here, we will take example of a data set of Aurora kinase A complexes having a bound ligand

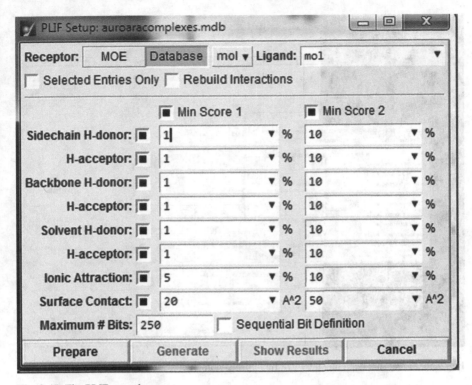

Fig. 9.45 The PLIF setup box

(PDB ids: 1OL7, 2DWB, 2WIC, 2WIF, 2WIG, 2WTW, 3DAJ, 3QBN, 4BOG, 4JBO). The data set is loaded into Database Viewer (DBV) in MOE. Both the receptor and the ligand are saved together as a single molecule field (Fig. 9.43).

In the DBV panel, go to Compute PLIF (Fig. 9.44).

In MOE there are 8 fingerprints and the maximum allocated are 250 (Figs. 9.45 and 9.46).

The computed fingerprints are written to the database field FP:PLIF. Next, they are analysed (Fig. 9.47).

The results are opened to show bar-code mode of fingerprints; the display mode can also be changed to population mode where the residues are shown in their three letter codes and can be analysed to understand key interactions. This also shows residue corresponding to fingerprint bits (Fig. 9.48).

Population display shows the frequency of occurrence of residues. The show ligand option displays all the bound ligands in a 2D depict form. Here, it shows in nine out of ten complexes, the alanine residue is interacting with the aurora kinase protein (Fig. 9.49).

The tools tab in this window has many options like bit selector, pharmacophore query generator and similarity calculator for further segregating the data (Fig. 9.50).

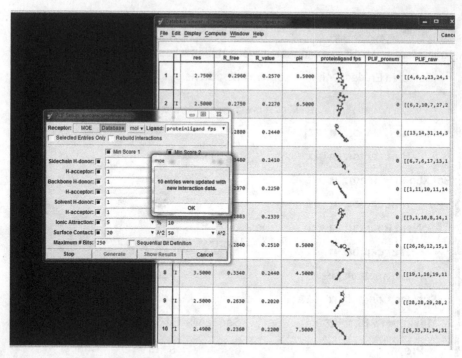

Fig. 9.46 Screenshot depicting the PLIF fingerprints computed for the complexes

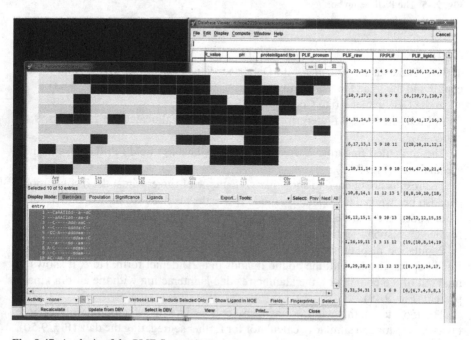

Fig. 9.47 Analysis of the PLIF fingerprints

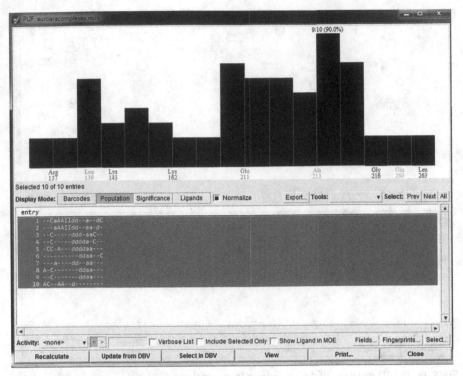

Fig. 9.48 Population option display of PLIF

The PLIF data can be used in various ways. It can be used to generate a pharmacophore query, if activity data are available for ligands and if docked complexes are being analysed.

9.8   Prolix

A tool for rapid data mining of protein–ligand interactions in large crystal databases has been developed, PROLIX [38]. It is a workflow to mine protein–ligand interactions using fingerprint representation pattern for quick searches. The front end has a query sketcher for the user to communicate with the back-end matching algorithms through xml files.

9.9   J-ProLINE: An In-house-developed Chem-Bioinformatics Workflow Application

J-ProLINE (Java-based Protein–ligand Network) is an interactive tool that detects relationships between ligand, scaffolds, protein sequence and structures which are finally validated through biomedical literature-based text mining [39]. Its func-

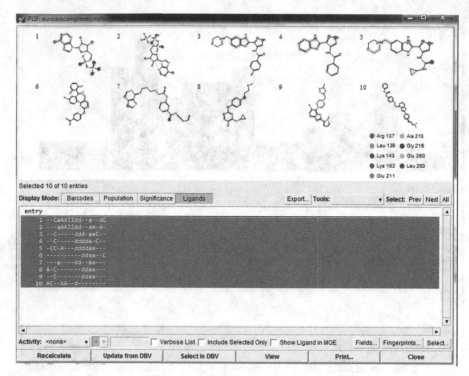

Fig. 9.49 The 2D structures of the native ligands in complexes

tion is to connect proteins, ligands and corresponding molecular scaffolds based on similarity scores among sequences and ligands. It provides the user with five major chem-bioinformatics functionalities, viz. pairwise sequence alignment, multiple sequence alignment, molecular similarity score, molecular mechanics descriptors and computing docking scores. The well-developed but simple GUI portlet enables the user to effectively communicate the queries and obtain results. It supports model building for any given set of query molecules, is capable of handling large data sets and integrates data from diverse background. The ligand similarities are identified using fingerprint-based scores. The similarity scores generated for proteins and ligands can be used for classification, network and tree building. To handle vast protein and molecular data, J-ProLINE programs were deployed on an in-house-developed Distributed Computing Environment (DCE), previously used in ChemXtreme (harvesting chemical data from Internet) [40] and ChemStar (Computing molecular properties for millions of publicly available molecules) applications [41]. The links established between the proteins and ligands were used for identification of common scaffolds and their occurrences in several databases, the results of which are presented in the subsequent sections. For this study, more than 9,000 protein complexes from Mother of All Databases (MOAD) and PDB having chain A and a co-crystallized ligand were identified. All PDB–ligand complexes from Binding MOAD (pdb id) were collected from http://www.BindingMOAD.org.

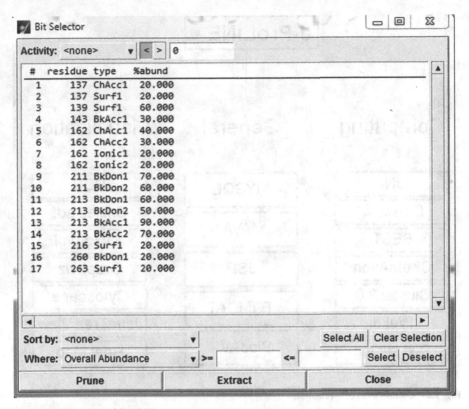

Fig. 9.50 The bit data in PLIF

The J-ProLINE architecture consists of three major components viz., General, Computing and Visualization. The program was developed using Java platform connected to RDBMS for storage of primary sequence data and other computed similarity score data. We also used several conventional similarity analysis and clustering tools in distributed computing environment (DCE) to handle the massive computational load. The GUI consists of two parts, one is a computing part and the other is a browsing/visualizing part. The home page of J-ProLINE was built using Liferay [42]. MPJ Express is an open-source Java implementation of Message Passing Interface that allows developers to write and deploy parallel applications using Java as a programming language (Fig. 9.51).

Figure 9.52 highlights the theoretical concept of J-ProLINE program (Fig. 9.53).

To understand the relationships between a class of compounds and target families, a heatmap using Tanimoto coefficient [43] for ligand similarities and sequence alignment score for protein similarities for protein–ligand complexes of six protein families was built. The heatmap was generated using R statistical package [44].

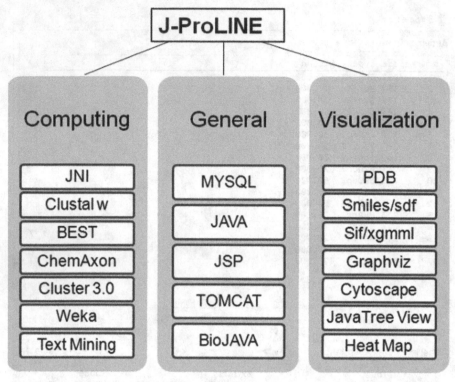

Fig. 9.51 Components of J-ProLINE

Fig. 9.52 The computational steps in J-ProLINE

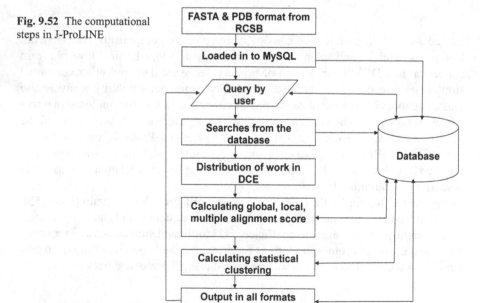

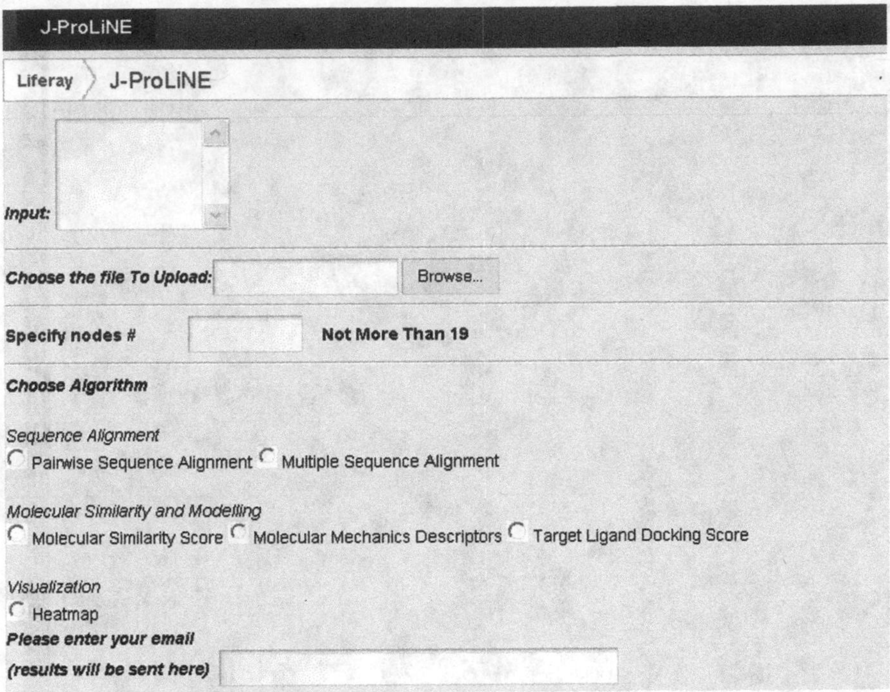

Fig. 9.53 Home page of J-ProLINE

R Input File
id1,id2,alnscore,
1CKE,1FF4,-39.0,
1CKE,1LG2,197.0,
1CKE,1QF1,207.0,
1CKE,1YST,252.0,
1CKE,2CMK,1022.0,
1FF4,1LG2,-120.0,
1FF4,1QF1,-110.0,
1FF4,1YST,-44.0,
1FF4,2CMK,-39.0,
1LG2,1QF1,348.0,
1LG2,1YST,267.0,
1LG2,2CMK,197.0,
1QF1,1YST,244.0,
1QF1,2CMK,206.0,
1YST,2CMK,247.0,

Fig. 9.54 Heatmap of 500 proteins belonging to six protein families

A heatmap is a convenient means of graphically depicting a 2D data matrix. J-Pro-LINE helps in generating heatmaps of proteins and ligands on the basis of sequence similarity and Tanimoto coefficient, respectively.

R commands file to get heatmap (Fig. 9.54).

```
prot <-read.table("rj_input.txt", sep=",");
prot <- prot[1:3];
prot_matrix <- data.matrix(prot);
png(filename = "rjava_output.png",width = 700, height = 700, units = "px",
pointsize = 12,bg = "white", res = NA, family = "", restoreConsole = TRUE,type =
c("windows", "cairo", "cairo-png"));
prot_heatmap <- heatmap(prot_matrix, scale="column", margins=c(3,3));
dev.off();
```

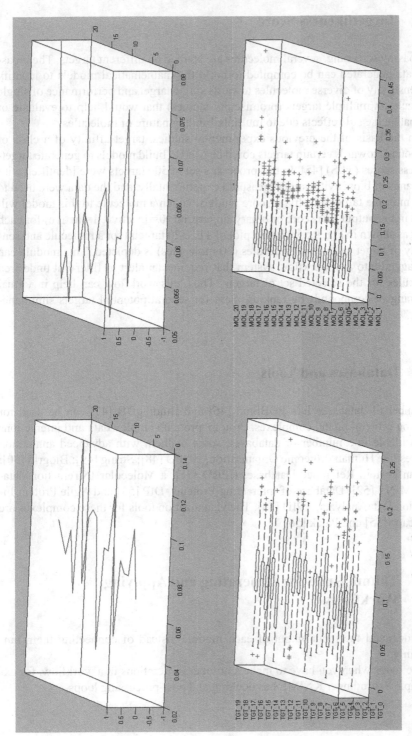

Fig. 9.55 Targetlikeness scores of molecules

9.10   Targetlikeness Score

In bioassay screening, several molecules are studied on different targets. The bioassay data generated can be compiled to build the mathematical models to identify the sensitivity of diverse molecules towards single target and performance of single molecule on multiple targets (promiscuity studies) that would help to evaluate or rationalize the side effects due to multiple-binding nature of molecules.

On the basis of the previous experimental studies, target affinity of a class of compounds towards certain targets could be used to build models to generate target-likeness scores (TLS) [45]. In this process, a set of 500 targets were identified with thousands of bioactivity values measured experimentally, and the data were used to build multiple target selectivity score models. Given a molecule to this model will result in the output as a score or binary fingerprint (0=inactive, 1=active) for each target as shown in Fig. 9.55 where a plot of TLS (20 targets) for a molecule and sensitivity of target to a set of molecules (20 molecules) is depicted. This module can be plugged into any workflow system that requires an alert to filter-out undesired molecules for the selected set of targets. Thus, this workflow can help in virtual screening of compounds by ranking them for several potential targets simultaneously.

9.11   Databases and Tools

A number of databases like PDBbind [46] and BindingDB [47] can be used for studying protein–ligand complexes. Protein–protein complex data and interactions are available in a number of databases, some of them with advanced annotation features like Human Proteome Organisation, HUPO [48]; String [49]; Biogrid [50]; Human Protein Reference Database, HPRD [51]; a Molecular INTeraction database, MINT [52]; Database of Interacting Proteins, DIP [53] and Agile Protein Interaction DataAnalyzer, APID [54]. The visualization tools for these complexes are cytoscape [55] and Pajek [56].

9.12   Thumb Rules for Generating and Applying Workflows

- Understand the input/output of each module instead of connecting them randomly.
- Use loops when you have to make a number of iterations in a workflow, for example Schrodinger KNIME has docking and post-processing loops.

9.13 Do it Yourself

1. Download KNIME
2. Understand the available modules for chemoinformatics (statistics: Weka, Molecular data: CDK, ChemAxon, Schrodinger (jaguar, glide) Database: MySQL, Oracle, postgresql)
3. Compute the chemoinformatics properties in any workflow program
4. Build a QSAR model in Knime

9.14 Questions

1. What is a workflow? Explain highlighting the need for workflow development.
2. What are the available workflows in chemistry and biology domain?
3. Explain the working of the Knime bench.
4. How can you make use of the protein–ligand analysis data for drug designing?

References

1. Wyrzykowski R, Dongarra J, Karczewski K et al (2008) Scientific workflow: a survey and research directions. Parallel processing and applied mathematics. Springer Berlin Heidelberg, pp 746–753
2. http://www.doc.ic.ac.uk/~vc100/papers/Scientific_workflow_systems.pdf. Accessed 30 Oct 2013
3. Taylor IJ, Deelman E, Gannon DB, Shields M (eds) (2007) Workflows for e-science—scientific workflows for grids. XXI, p 523
4. Zhao Y, Raicu I, Foster I (2008) Scientific workflow systems for 21st century, new bottle or new wine? 2008 IEEE Congress on Services 2008-Part I, pp 467–471
5. http://www.cs.gonzaga.edu/~bowers/papers/Bowers_et_al_SCIFLOW06.pdf. Accessed 30 Oct 2013
6. Aranguren ME, Fernandez-Breis JT, Mungall C et al (2013) OPPL-Galaxy, a Galaxy tool for enhancing ontology exploitation as part of bioinformatics workflows. J Biomed Semantics 4:2
7. Elhai J, Taton A, Massar JP et al (2009) BioBIKE: A Web-based, programmable, integrated biological knowledge base. Nucleic Acids Res 37:W28–W32
8. Kallio MA, Tuimala JT, Hupponen T et al (2011) Chipster: user-friendly analysis software for microarray and other high-throughput data. BMC Genomics 12:507
9. Ovaska K, Laakso M, Haapa-Paananen S et al (2010) Large-scale data integration framework provides a comprehensive view on glioblastoma multiforme. Genome Med 2:65
10. http://www.aosabook.org/en/vistrails.html. Accessed 30 Oct 2013
11. http://accelrys.com/products/pipeline-pilot/. Accessed 30 Oct 2013
12. http://www.idbs.com/products-and-services/inforsense-suite/chemsense/. Accessed 30 Oct 2013
13. Oinn T, Addis M, Ferris J et al (2004) Taverna: a tool for the composition and enactment of bioinformatics workflows. Bioinformatics 20:3045–3054

14. Mazanetz MP, Marmon RJ, Reisser CBT, Morao I (2012) Drug discovery applications for KNIME: an open source data mining platform. Curr Top Med Chem 12:1965–1979
15. Warr WA (2012) Scientific workflow systems: Pipeline Pilot and KNIME. J Compu Aided Mol Des 26:801–804
16. www.chemaxon.com. Accessed 30 Oct 2013
17. Kuhn T, Willighagen EL, Zielesny A, Steinbeck C (2010) CDK-Taverna: an open workflow environment for cheminformatics. BMC Bioinform 11:159
18. Thorsten M, Wiswedel B, Berthold, Michael R (2012) Workflow tools for managing biological and chemical data. In: Guha R, Bender A (eds) Computational approaches in chemoinformatics and bioinformatics, pp 179–209
19. Fourches D, Muratov E, Pu D, Tropsha, A (2011) Boosting predictive power of QSAR models Alexander Abstracts of Papers, 241st ACS National Meeting & Exposition, Anaheim, CA, United States, March 27–31
20. http://www.knime.org/files/01_Schroedinger.pdf. Accessed 30 Oct 2013
21. http://www.knime.org/files/09_CCG.pdf. Accessed 30 Oct 2013
22. http://www.chemaxon.com/library/chemaxons-jchem-nodes-on-the-knime-workbench/. Accessed 30 Oct 2013
23. Dunbar JB, Smith RD, Damm-Ganamet KL, Ahmed A, Esposito, EX, Delproposto J, Chinnaswamy K, Kang Y-N, Kubish G, Gestwicki JE (2013) CSAR data set release 2012: ligands, affinities, complexes, and docking decoys. J Chem Inf Model 53(8):1842–1852
24. Chan AWE, Overington JP (2003) Recent development in chemoinformatics and chemogenomics. Annu Rep Med Chem 38:285–294
25. Hwang KY, Chung JH, Kim SH, Han YS, Cho Y (1999) Structure-based identification of a novel NTPase from methanococcus jannaschii. Nat Struct Biol 6:691–696
26. Martin YC, Willett P, Heller SR (eds) (1995) In designing bioactive molecules. American Chemical Society, Washington DC
27. Koshland DE Jr (1994) The key-lock theory and the induced fit theory. Chem Int Ed Engl 33:2375–2378
28. Todd AE, Orengo CA, Thornton JM (1999) Evolution of protein function, from a structural perspective. Curr Opin Chem Biol 3:548–556
29. Eckers E, Petrungaro C, Gross D, Riemer J, Hell K, Deponte M (2013) Divergent molecular evolution of the mitochondrial sulfhydryl: cytochrome c oxidoreductase Erv in opisthokonts and parasitic protists. J Biol Chem 288(4):2676–2688
30. Gaston, Daniel;Roger, Andrew J (2013) Functional divergence and convergent evolution in the plastid-targeted glyceraldehyde-3-phosphate dehydrogenases of diverse eukaryotic algae. PLoS One 8(7):e70396
31. Berman HM, Westbrook J, Feng Z, Gilliland G, Bhat TN, Weissig H (2000) The protein data bank. Nucl Acids Res 28:235–242
32. Ewing T, Baber JC, Feher M (2006) Novel 2D fingerprints for ligand-based virtual screening. J Chem Inf Mod 46:2423–2431
33. Deng Z, Chuaqui C, Singh J (2003) Structural Interaction fingerprint (SIFt): a novel method for analyzing three-dimensional protein ligand binding interactions. J Med Chem 47:337–344
34. Deng Z, Chuaqui C, Singh J (2007) Generation of profile-structural interaction fingerprints for representing and analyzing three-dimensional target molecule-ligand interactions. U.S. Pat Appl Publ US 20070020642A120070125
35. Nandigam RK, Kim S, Singh J, Chuaqui C (2009) Position specific interaction dependent scoring technique for virtual screening based on weighted protein-ligand interaction fingerprint profiles. J Chem Inf Mod 49(5):1185–1192
36. Tan L, Bajorath J (2009) Utilizing target–ligand interaction information in fingerprint searching for ligands of related targets. Chem Biol Drug Des 74:25–32
37. Klepsch F, Chiba P, Ecker GF (2011) Exhaustive sampling of docking poses reveals binding hypotheses for propafenone type inhibitors of P-Glycoprotein. PLoS Comput Biol 7(5):e1002036

38. Weisel M, Bitter H-M, Diederich F (2012) PROLIX: rapid mining of protein ligand interactions in large crystal structure databases. J Chem Inf Model 52:1450–1461
39. Unpublished results
40. Karthikeyan M, Krishnan S, Pandey AK, Bender A (2006) Harvesting chemical information from the internet using a distributed approach: vhemXtreme. J Chem Inf Model 46:452–461
41. Karthikeyan, M, Krishnan S, Pandey AK, Andreas B, Alexander Tropsha A (2008) Distributed chemical computing using chemstar: an open source java remote method invocation architecture applied to large scale molecular data from pubchem. J Chem Inf Model 48(4):691–703
42. http://www.liferay.com/products/liferay-portal/overview. Accessed 30 Oct 2013
43. https://surechem.uservoice.com/knowledgebase/articles/84207-tanimoto-coefficient-and-fingerprint-generation. Accessed 30 Oct 2013
44. http://stat.ethz.ch/R-manual/R-patched/library/stats/html/heatmap.html. Accessed 30 Oct 2013
45. Unpublished work
46. http://www.pdbbind.org.cn/. Accessed 30 Oct 2013
47. http://www.bindingdb.org/bind/index.jsp. Accessed 30 Oct 2013
48. http://www.hupo.org/. Accessed 30 Oct 2013
49. http://string-db.org/. Accessed 30 Oct 2013
50. http://thebiogrid.org/. Accessed 30 Oct 2013
51. http://www.hprd.org/. Accessed 30 Oct 2013
52. Zanzoni A, Montecchi-Palazzi L, Quondam M, Ausiello G, Helmer-Citterich M, Cesareni G (2007) FEBS letters MINT: a molecular INTeraction database. Nucleic Acid Res 35:D572–574
53. Xenarios, I, Salwinski L, Duan XJ, Higney P, Kim S-M, Eisenberg D (2002) DIP, the Database of Interacting Proteins: a research tool for studying cellular networks of protein interactions. Nucleic Acids Res 30(1):303–305
54. http://bioinfow.dep.usal.es/apid/index.htm. Accessed 30 Oct 2013
55. http://www.cytoscape.org/. Accessed 30 Oct 2013
56. http://vlado.fmf.uni-lj.si/pub/networks/pajek/. Accessed 30 Oct 2013

Chapter 10
Cloud Computing Infrastructure Development for Chemoinformatics

Abstract Chemical research is progressing exponentially, thus fuelling the need to integrate data and applications and develop workflows. To support proper execution of workflows with multiple teams working on collaborative projects, we need robust portals powered by cloud computing infrastructure. A cloud computing portal provides customization configurability to users on a secured, unified and integrated platform with extensive computational power. The sheer magnitude and diversity of the chemical data require customized system-based solutions utilizing available mass storage, CPUs, GPUs and hybrid processors. Porting existing applications to a common portal to provide a single framework which can be deployed on a high-performance computing distributed computing platform for automated programmatic access to workflows. A portal enables efficient scanning, searching and annotating of the data for the users and resource monitoring for the enterprise. They also provide additional features like security, scalability, quality, data consistency and error checks. Portal development has a bright future as they can perform large-scale quantum chemical studies of molecules and become decision support tools to mine functional relationships in chemical biology. In this chapter, we first focus on the essentials of portal development with stepwise tutorials using relevant examples. Mobile computing has transformed the information technology scenario in recent times; consequently, a section is devoted to android, its open-source operating system. Few chemoinformatics-based apps are also discussed.

Keywords Portals · Mobile computing · Chemoinformatics drug design · High-performance computing · GPU computing · Cloud computing

10.1 What is a Portal?

A portal usually connotes a gateway or a door [1]. It is generally defined as a software platform for building websites and web applications [2]. Modern portals have added multiple features that make them the best choice for a wide array of web applications. Portals may be used as an integrated platform for problem solving or as a content management system.

M. Karthikeyan, R. Vyas, *Practical Chemoinformatics*,
DOI 10.1007/978-81-322-1780-0_10, © Springer India 2014

501

10.2   Need for Development of Scientific Portals

Ever-increasing publicly available chemical structure and bioactivity data have created challenges in data handling and curation [3–4]. This can be mitigated by building and using web-based portal systems for easy access, search, analysis and discovery. Portals let us integrate various data and compute applications that run together in a coordinated way. For example, ChEMBL [5] bioactivity data can be stored and bioactivity data can be exposed to users through portlets. Further, these data can be subjected to descriptor calculation or quantitative structure–activity relationship (QSAR) via portlet-to-portlet communication, thus aggregating various chemistry-specific data resources and applications that are compiled together for knowledge discovery within a unified user interface.

A portal integrates data and applications together using layout management for maintaining several applications, with drag-and-drop features which makes it more intuitive for users [6]. Portlets can communicate with each other; thus, the output of one portlet goes as an input for the other, a very important consideration for designing pipeline workflows especially in chemoinformatics. Other value-added features such as Structure Search, Chemical Data management, Research Document management, blogs or wiki for adding to the chemical knowledge space in collaboration, Community and Discussions to solve certain problems, etc. can be added [7]. This also allows researchers to focus more on the domain logic rather than the computing processes beneath. However, one should proceed with caution and not resort to deploying all applications on the portal without a proper requirement analysis as the complexity of setting up and configuration can complicate the tasks. Other considerations to be borne in mind while developing a portlet such as events and action, render phase, etc.

10.3   Components of a Portal

A software, good database management system, front-end user interface and algorithms comprise a portal [8]. Liferay Community Edition is one such Lesser General Public License (LGPL) open-source portlet container and portal server [9]. GateIn Portal, formerly known as JBoss Portal, and Drupal are other examples of open-source portal and content management systems [10–11]. Portlets are mini applications which make up a portal page [12]. They share many similarities with servlets as they are managed by specialized container and interact with web client via request and response action classes. So, a novice need not worry about other technicalities and can focus on developing logic in portlet code, which runs at the application level. In molecular informatics, portlets can be categorized into two categories, i.e. data portlets and compute portlets. Data portlets essentially deal with input, storage, distribution and display of molecular data, while compute portlets involve exact/substructure searching for hits, molecular descriptor calcula-

ChemDB Portal

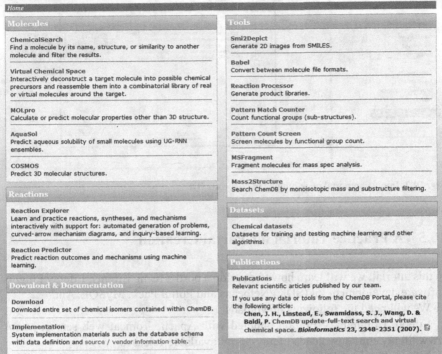

Home

Molecules

ChemicalSearch
Find a molecule by its name, structure, or similarity to another molecule and filter the results.

Virtual Chemical Space
Interactively deconstruct a target molecule into possible chemical precursors and reassemble them into a combinatorial library of real or virtual molecules around the target.

MOLpro
Calculate or predict molecular properties other than 3D structure.

AquaSol
Predict aqueous solubility of small molecules using UG-RNN ensembles.

COSMOS
Predict 3D molecular structures.

Reactions

Reaction Explorer
Learn and practice reactions, syntheses, and mechanisms interactively with support for: automated generation of problems, curved-arrow mechanism diagrams, and inquiry-based learning.

Reaction Predictor
Predict reaction outcomes and mechanisms using machine learning.

Download & Documentation

Download
Download entire set of chemical isomers contained within ChemDB.

Implementation
System implementation materials such as the database schema with data definition and source / vendor information table.

Tools

Smi2Depict
Generate 2D images from SMILES.

Babel
Convert between molecule file formats.

Reaction Processor
Generate product libraries.

Pattern Match Counter
Count functional groups (sub-structures).

Pattern Count Screen
Screen molecules by functional group count.

MSFragment
Fragment molecules for mass spec analysis.

Mass2Structure
Search ChemDB by monoisotopic mass and substructure filtering.

Datasets

Chemical datasets
Datasets for training and testing machine learning and other algorithms.

Publications

Publications
Relevant scientific articles published by our team.

If you use any data or tools from the ChemDB Portal, please cite the following article:
Chen, J. H., Linstead, E., Swamidass, S. J., Wang, D. & Baldi, P. ChemDB update-full-text search and virtual chemical space. *Bioinformatics* 23, 2348-2351 (2007).

Fig. 10.1 Homepage of chemDB portal

tion, statistical model building, data mining, target–ligand docking, fingerprinting to name a few.

10.4   Examples of Portal Systems

Recently many portals have been created like the enzyme portal, which performs data mining related to enzymes, biological pathways, small molecules and diseases [13]. A protein and structure analysis workbench Expasy is the most well-known portal for proteomics with several software tools and databases [14]. Wolf2Pack portal has been deployed with force-field optimization package to enable users to integrate force fields from different research areas [15]. MolClass portal helps users to develop computational models from given data sets based on structural feature identification [16]. The drug discovery portal [17] enables virtual screening in a collaborative manner. ChemDB portal [18] has integrated several OpenEye [19] and ChemAxon tools [20] to provide chemoinformatics functionalities like searching chemical, virtual library generation, three-dimensional (3D) molecular structure generation, predicting properties, reactions etc (Fig. 10.1).

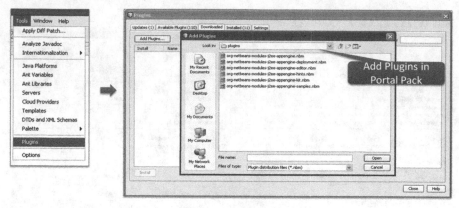

Fig. 10.2 Importing portal pack plugins in Netbeans IDE

10.5   A Practice Tutorial for Portal Creation

In this tutorial, we will learn how to develop a portlet using Liferay, Liferay Plugins software development kit (SDK)/Netbeans and Portal Pack, MySQL, Ant.

We will need the following downloads:

Liferay Community Edition bundled with Apache Tomcat web server: http://www.liferay.com/downloads/liferay-portal/available-releases

Liferay Plugins SDK for development: http://www.liferay.com/downloads/liferay-portal/additional-files

Mysql Community Server: http://dev.mysql.com/downloads/mysql/

Apache Ant: http://ant.apache.org/bindownload.cgi

Download and unzip Liferay Tomcat Installation zip to <path to liferay>; go to bin\startup.sh to test it at http://localhost:8080. This will open Liferay Portal with default Liferay page. To login as admin, click on 'Login as Bruno Admin' link. For user-defined portlets, various development tools like Liferay Plugins SDK and integrated development environment (IDE) such as Eclipse or Netbeans are available. Using Netbeans for portlet development, download and install Netbeans IDE 7.2.1 (with Java EE, Tomcat support) roughly around 204 MB in size and Netbeans Portal Pack 3.0.5 Beta available at http://netbeans.org/downloads and https://contrib.netbeans.org/portalpack/pp30/download305.html, respectively. After IDE installation, follow the following screenshots to add Liferay Portal Plugins, configure and add server, create a web application with Portlet Support and start server from within IDE (Figs. 10.2, 10.3, 10.4, 10.5, 10.6, 10.7, 10.8 and 10.9).

We will use Liferay Plugins SDK [19] for portlet development. Unzip Liferay Plugins SDK in a similar way as liferay portal to <path to plugins>. To point it to correct installation folder, i.e. Liferay Portal, we need to change uncommented property '*app.server.dir*' at <path to plugins\build.properties> to <path to liferay/tomcat>. While doing this, include forward slashes (/) to define path in Unix style instead of Windows-specific back slash (\). To get started, navigate to <path to plugins\portlets> and type

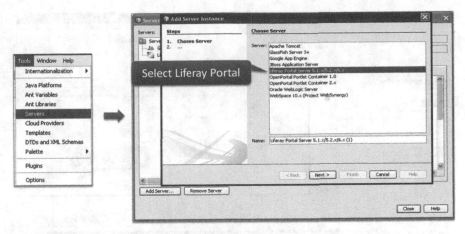

Fig. 10.3 Adding liferay portal server in Netbeans IDE

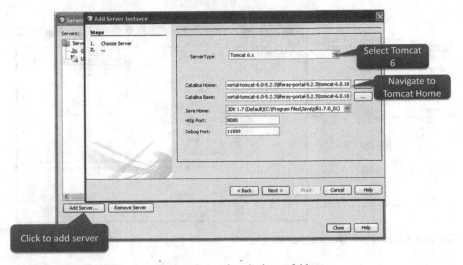

Fig. 10.4 Configuring Tomcat 6 and setting path to its home folder

```
<path to plugins>\portlet # ant –Dportlet.name=firstPortlet –Dportlet.display.name="First Portlet"
create
```

This will create a portlet folder named 'firstPortlet-portlet'. It should have the following files

docroot: root of portlet and web application

docroot/WEB-INF: standard folder with configuration files

docroot/WEB-INF/portlet.xml; liferay-portlet.xml: description of portlet properties

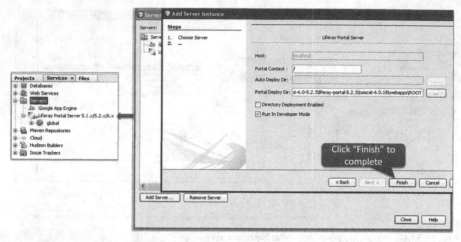

Fig. 10.5 Server addition completed

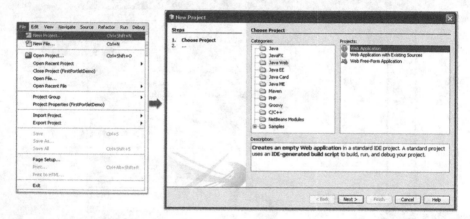

Fig. 10.6 Creating new web application

docroot/WEB-INF/liferay-display.xml: display of portlet in applications menu of portal.

docroot/WEB-INF/liferay-plugin-package.properties: file containing packaging options for the project

docroot/WEB-INF/src: java source files

docroot/WEB-INF/view.jsp: defines the user interface and interacts with the underlying java code

Sample view.jsp

```
<%@ taglib uri="http://java.sun.com/portlet_2_0" prefix="portlet" %>
<portlet:defineObjects />
This is <b>My First Portlet</b>
```

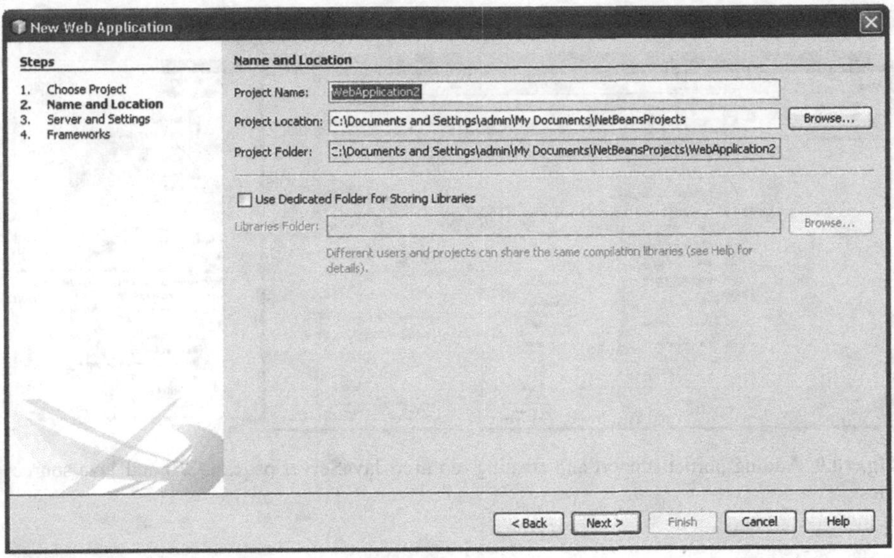

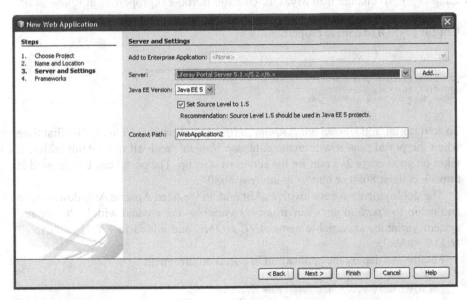

Fig. 10.7 Specifying project name as 'WebApplication2'

Fig. 10.8 Server and settings set to Liferay portal server

Sample JSPPortlet.java processAction method

```
public void processAction(ActionRequest actionRequest, ActionResponse
actionResponse) throws IOException, PortletException {
//User defined code goes here
}
```

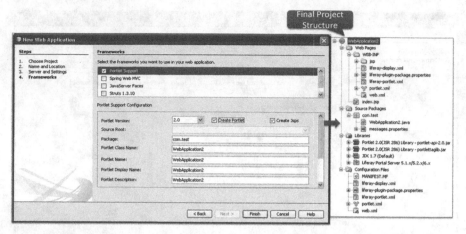

Fig. 10.9 Adding portlet support and creating required JavaServer pages (*JSP*) and Java source files

The default database connection will be to HSQL (Hypersonic Structured Query Language). To change it to MySQL, edit file portal-ext.properties at *<path to liferay\tomcat-6.0.18\webapps\ROOT\WEB-INF\classes>* and add the following content for MySQL database properties.

```
jdbc.default.driverClassName=com.mysql.jdbc.Driver
jdbc.default.url=jdbc:mysql://localhost/lportal?useUnicode=true&characterEncoding=U
TF-8&useFastDateParsing=false
jdbc.default.username=<db_username>
jdbc.default.password=<db_password>
```

To start portal, run *startup.bat* at *<path to liferay\tomcat-6.0.18\bin>*. The first time when the portal runs it will create database '*lportal*' with all the default tables. It takes on an average 2–3 min for the server to start up. The portal can be accessed at http://localhost:8080 or http://<ip-address:8080>.

The deployment process involves Ant and so we need Apache Ant; download it and unzip to *<path to ant>*. Environment variables are created with both user and system variables as variable name *ANT_HOME* and added in PATH as *%ANT_HOME%\bin*.

Finally, we deploy the application in portal environment

<path to plugins>\portlet\firstPortlet-portlet# ant deploy

The firstPortlet we created will be deployed in a few seconds and to add it to the portal, we can create a page 'First Portlet' by clicking on *Add Page* in the right top corner as shown in Fig. 10.10. Now, we can add the first application to the page we just created (Fig. 10.11).

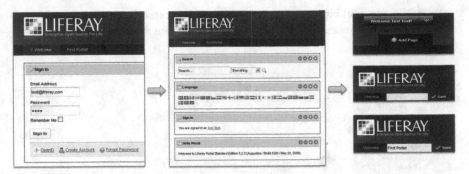

Fig. 10.10 Single sign In and adding page to portal

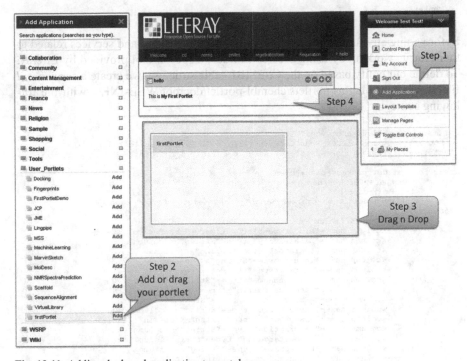

Fig. 10.11 Adding deployed application to portal page

10.5.1 Custom Database connection and Display Table with Paginator via portlet in Liferay Portal

For example, ChEMBL table in lportal database is created with bioactivity data and needs to be accessed via portlet.

Primarily for working connection, simply we can start by setting *context.xml* at *<path to liferay\tomcat-6.0.18\conf>* with appropriate resource properties mentioned as below.

```
<Context>
<Resource name="jdbc/lportal"
      auth="Container"
      type="javax.sql.DataSource"
      maxActive="100"
      maxIdle="30"
      maxWait="10000"
      username="<db_username>"
      password="<db_password>"

      driverClassName="com.mysql.jdbc.Driver"
      url="jdbc:mysql://localhost:3306/lportal?autoReconnect=true"/>

</Context>
```

Before going any further, we need to build database entities and services related to them for its working by using *Service Builder*, a special tool provided by Liferay. To define our entity based on the ChEMBL table structure, we create a *service.xml* file at <path to plugins/portlets/chembl-portlet/docroot/WEB-INF/> with the following content.

```
<?xml version="1.0" encoding="UTF-8"?>
<!DOCTYPE service-builder PUBLIC "-//Liferay//DTD Service Builder 5.1.0//EN"
"http://www.liferay.com/dtd/liferay-service-builder_5_1_0.dtd">

<service-builder package-path="chembl">
      <namespace>CHEMBL</namespace>
      <!-- Project -->
      <entity name="Item" table="chembl" local-service="true" remote-
service="false">
            <!-- PK fields -->
            <column name="bid" type="int" primary="true"></column>
            <column name="bioactivity" type="String"></column>
            <column name="operator" type="String"></column>
            <column name="value" type="String"></column>
            <column name="units" type="String"></column>
            <column name="compoundname" type="String"></column>
            <column name="canonicalsmiles" type="String"></column>
            <column name="assaychemblid" type="String"></column>
            <column name="assaysource" type="String"></column>
            <column name="assaytype" type="String"></column>
            <column name="description" type="String"></column>
            <column name="chembltargetid" type="String"></column>
            <column name="targetname" type="String"></column>
            <column name="organism" type="String"></column>
            <column name="reference" type="String"></column>
      </entity>
</service-builder>
```

Edit view.jsp as
    // Declarations and Imports

```
<%@ taglib uri="http://java.sun.com/portlet_2_0" prefix="portlet" %>
<%@ taglib uri="http://liferay.com/tld/ui" prefix="liferay-ui" %>

<%@ page import="java.util.ArrayList" %>
<%@ page import="java.util.List" %>

<%@ page import="javax.portlet.PortletURL" %>
<%@ page import="javax.portlet.PortletPreferences" %>
<%@ page import="javax.portlet.WindowState" %>

<%@ page import="com.liferay.portal.kernel.dao.search.ResultRow" %>
<%@ page import="com.liferay.portal.kernel.dao.search.SearchContainer" %>
<%@ page import="com.liferay.portal.kernel.dao.search.SearchEntry" %>

<%@ page import="chembl_14.model.Item" %>

<%@page import="chembl_14.service.ItemLocalServiceUtil"%><portlet:defineObjects />

<%
    PortletURL portletURL = renderResponse.createRenderURL();
```

// Define list of ChEMBL table headers

```
    List<String> headerNames = new ArrayList<String>();

    headerNames.add("bioactivity");
    headerNames.add("operator");
    headerNames.add("value");
    headerNames.add("units");
    headerNames.add("compoundname");
    headerNames.add("canonicalsmiles");
    headerNames.add("assaychemblid");
    headerNames.add("assaysource");
    headerNames.add("assaytype");
    headerNames.add("description");
    headerNames.add("chembltargetid");
    headerNames.add("targetname");
    headerNames.add("organism");
    headerNames.add("reference");
```

// Creating search container, used to display table

```
    SearchContainer searchContainer = new SearchContainer(renderRequest,
                                                          null, null,
SearchContainer.DEFAULT_CUR_PARAM,

SearchContainer.DEFAULT_DELTA,
                                                          portletURL,
                                                          headerNames,
                                                          "There No Records To
Display");
    portletURL.setParameter(searchContainer.getCurParam(),
String.valueOf(searchContainer.getCurValue()));
```

// Get count of total records and list of records to display on current page

```
    int totalChemblRecordCount = ItemLocalServiceUtil.getItemsCount();
    List<Item> chemblRecordList =
ItemLocalServiceUtil.getItems(searchContainer.getStart(),
                                                searchContainer.getEnd());
```

// Set count into search container per page

searchContainer.setTotal(totalChemblRecordCount);

// Fill table

```
List<ResultRow> resultRows = searchContainer.getResultRows();
    for (int i=0; i < chemblRecordList.size(); i++) {
        Item chemblRecord= chemblRecordList.get(i);
        ResultRow row = new ResultRow(chemblRecord, chemblRecord.getBid(), i);

        row.addText(chemblRecord.getBioactivity(), "");
        row.addText(chemblRecord.getOperator(), "");
    row.addText(chemblRecord.getValue(), "");
        row.addText(chemblRecord.getUnits(), "");
        row.addText(chemblRecord.getCompoundname(), "");
        row.addText(chemblRecord.getCanonicalsmiles(), "");
        row.addText(chemblRecord.getAssaychemblid(), "");
        row.addText(chemblRecord.getAssaysource(), "");
        row.addText(chemblRecord.getAssaytype(), "");
        row.addText(chemblRecord.getDescription(), "");
    row.addText(chemblRecord.getChembltargetid(), "");
    row.addText(chemblRecord.getTargetname(), "");
    row.addText(chemblRecord.getOrganism(), "");
    row.addText(chemblRecord.getReference(), "");
        resultRows.add(row);
    }
```

// and finally display it

```
%>
<liferay-ui:search-iterator searchContainer="<%= searchContainer %>" />
```

The above code is adapted from Pet Catalog tutorial available at the following link. Refer for a detailed understanding:

http://www.emforge.net/web/liferay-petstore-portlet/wiki/-/wiki/Main/Step1%3 A+From+DB+to+simple+UI;jsessionid=AEE788CF2575EFA63F452A081BAA3 8B6

10.6   A Practice Tutorial for Development of Portlets for Chemoinformatics

10.6.1   *Marvin Sketch Portlet*

Marvin Sketch is an advanced chemical editor used for drawing structures, queries and reactions [20]. This tool can be integrated into the portlet by using javascript as follows:

Before using the following javascript code, we should download Marvin for JavaScript available at chemaxon.com and point src attribute to the desired location.

```
<script languge="JavaScript1.1" src="http://localhost:8080/clouddesc-
portlet/marvin/marvin.js">
</script>
<script languge="JavaScript1.1">
msketch_name = "MSketch";
msketch_begin("http://localhost:8080/clouddesc-portlet/marvin/", 600, 480);
if(window.opener.document.all.smiles.value!=''){
     msketch_param("molFormat", "smiles");
     msketch_param("mol", window.opener.document.all.smiles.value);
     }
else{
     msketch_param("mol", "");
     }
msketch_param("preload", "MolExport");
msketch_end();
</script>
```

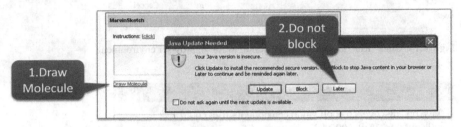

Fig. 10.12 Click 'Later' to skip Java update

Fig. 10.13 Enter proxy details

Steps for using Marvin Sketch Portlet available at http://moltable.ncl.res.in

Step 1: Click 'Draw Molecule' to start MSketch Molecule Editor (Fig. 10.12)

Step 2: If you are working behind a proxy server, use login details to load web application (Fig. 10.13)

Step 3: Click 'Run' to run MSketch application in your browser (Fig. 10.14)

Step 4: Click 'No' to avoid blocking the application from running (Fig. 10.15)

Step 4: Finally, draw molecule of interest and click 'Submit' to get smiles in the text input box of MSketch portlet (Fig. 10.16)

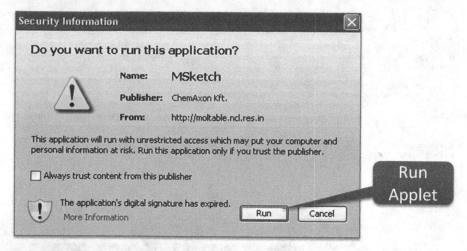

Fig. 10.14 Click 'Run' to authorize application to run in browser environment

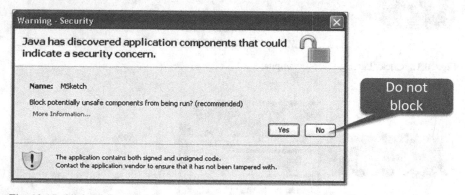

Fig. 10.15 Click 'No' to continue

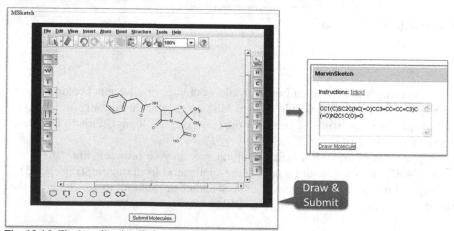

Fig. 10.16 Final application ID loaded and ready for use

Fig. 10.17 JME molecular editor loading with smiles window

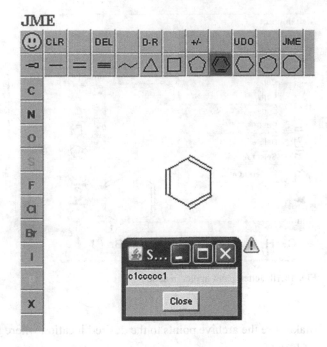

10.6.2 JME Portlet

JME Molecule Editor is a Java applet to draw/edit structures and reactions [21]. It also displays molecules on screen in display panel and generates output formats like Simplified Molecular-Input Line-Entry System (SMILES) and MDL molfile. To use JME in your portlet, use the following applet code. Include the JME distribution containing JME.jar for referencing.

```
<applet code="JME.class" name="JME" archive="jme/JME.jar" width="360" height="335">
<param name="options" value="list of keywords">
Enable Java in your browser !
</applet>
<font face="arial,helvetica,sans-serif"><small><a
href="http://www.molinspiration.com/jme/index.html">JME Editor</a> courtesy of
Peter Ertl, Novartis</small></font>
```

Steps for using JME Portlet available at http://moltable.ncl.res.in

Step 1: Click 'Draw Molecule in JME' to run JME applet in your web browser (Fig. 10.17)

Step 2: Draw molecule and click Smiley in top left corner to generate SMILES automatically

10.6.3 Jchempaint Portlet

Jchempaint is a free, open-source and platform independent chemical editor written in Java [22]. Following is the applet code to embed Jchempaint into portlet.

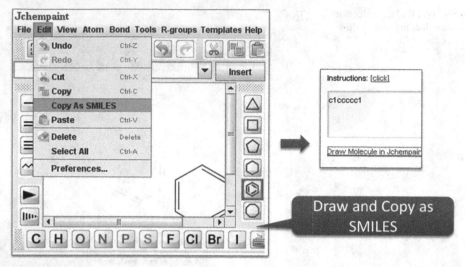

Fig. 10.18 Jchempaint applet in action

Make sure the archive points to the desired location where jchempaint-applet-core. jar resides.

```
<applet
code="org.openscience.jchempaint.applet.JChemPaintEditorApplet"
name="JME" archive="jchempaint/EditorApplet_files/jchempaint-applet-core.jar"
width="360"
height="335">
<param name="options" value="list of keywords">
Enable Java  in your browser !
</applet>
```

Steps for using Jchempaint Portlet available at http://moltable.ncl.res.in

Step 1: Click 'Draw Molecule in Jchempaint' to run Jchempaint applet in your web browser (Fig. 10.18)

Step 2: Edit→Copy As Smiles to get smiles format of the molecule drawn

10.7   Mobile Computing

Mobile computing has been defined as 'the ability to use computing capability without a predefined location and/or connection to a network to publish and/or subscribe to information' [23]. It is a technique which has revolutionized the world of hand-held devices like personal digital assistants (PDAs), tablet PCs and smartphones [24]. The standard mobile phone application environment is supplied by Android [25]. The Android operating system released by Google in 2007 is an open catalogue of applications which users can download over the air or directly load via

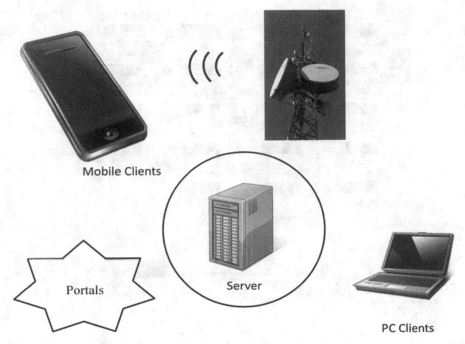

Fig. 10.19 Components of a mobile computing platform

a USB connection to their phone [26]. The users can create their own creative applications from the existing ones; the operating system takes care of which application to use for a specific task (Figs. 10.19 and 10.20).

There are certain limitations of mobile computing, for instance the computing resources are restrained by the battery size and can handle only few GB of data [27]. However, these limitations are likely to be overcome in the near future. Another consideration is security as personal data are generally stored on smartphones and are susceptible to attack. Internet speed is slower on a mobile compared to a direct Internet connection. Of course there are other general concerns as usual associated with the effect of radiations in human vicinity.

10.7.1 Android Applications for Chemoinformatics

Any android application in general requires the installation of four components, Java Development Kit (JDK), Eclipse (Integrated development environment for JAVA), android SDK and Android development tool (ADT) [28]. An emulator is required for testing and debugging the software. The executable code for android is termed as Activity which corresponds to display screens.

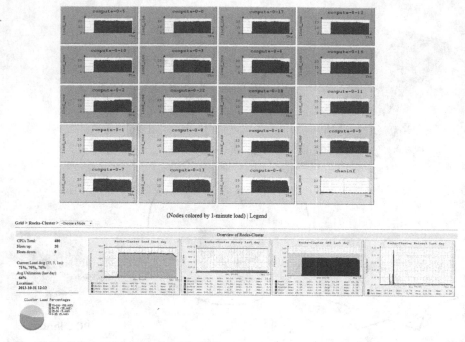

(Nodes colored by 1-minute load) | Legend

Fig. 10.20   Computer clusters with 480 CPUs

10.7.1.1   iMolview a Mobile App for iPhone/iPad and Android

iMolview is an app for browsing protein, DNA and drug molecules in 3D via direct links to Drug Bank and Protein Data Bank (PDB) database [29]. One can toggle the molecules for better visualization using a touch screen rather than the conventional keyboard–mouse combination. The app can be downloaded from Apple App store or into any android device. It is still in developmental stages with new features being added like surface representation, colour selection, 2D labels, electron density maps, etc.

10.7.1.2   In-house-developed ChemInfo App

An app has been developed using android for computing properties of biologically important molecules using a mobile [30] (Figs. 10.21 and 10.22).

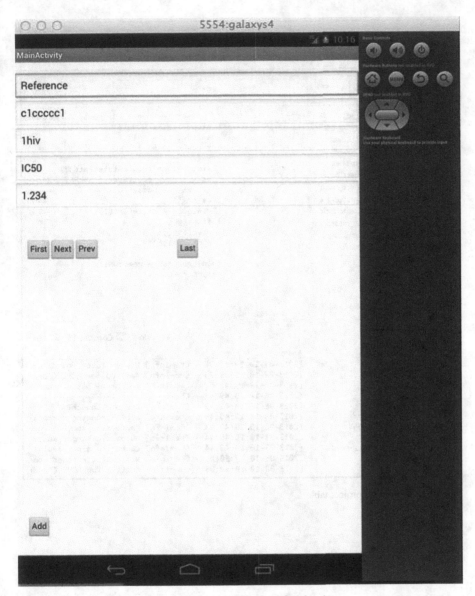

Fig. 10.21 Mobile app interface for property prediction of molecules

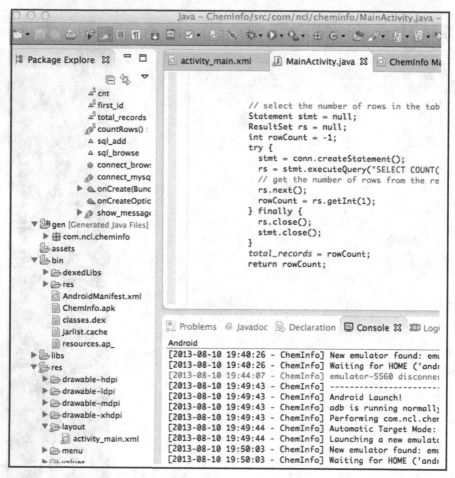

Fig. 10.22 Cheminfo project table

10.7.1.3   Code for Android Application Development

```
package com.ncl.cheminfo;

import java.security.interfaces.RSAKey;
import java.sql.Connection;
import java.sql.DriverManager;
import java.sql.ResultSet;
import java.sql.SQLException;
import java.sql.Statement;
import java.io.*;
import java.util.*;

import com.mysql.jdbc.PreparedStatement;

import android.os.Bundle;
import android.app.Activity;
import android.app.AlertDialog;
import android.content.DialogInterface;
import android.view.Menu;
import android.view.View;
import android.view.View.OnClickListener;
import android.widget.Button;
import android.widget.EditText;

public class MainActivity extends Activity {

        String sql_add="";
        String sql_browse="";
        static int cnt=1, total_records=0,first_id=0;


        public static int countRows()   {


        Connection conn = null;
          Statement st = null;
          String driver ="com.mysql.jdbc.Driver";
          String ip=""; //get dynamically
          if(ip.length()==0){
                ip="localhost";
          }


          String url = "jdbc:mysql://"+ip+":3306/test";
          String user = "root";
          String password = "*****";


          try {


                Class.forName(driver).newInstance();


              conn = DriverManager.getConnection(url, user, password);


              // select the number of rows in the table
              Statement stmt = null;
              ResultSet rs = null;
              int rowCount = -1;
              try {
                stmt = conn.createStatement();
                rs = stmt.executeQuery("SELECT COUNT(*) FROM cheminfo.bioactivity" );
                // get the number of rows from the result set
                rs.next();
                rowCount = rs.getInt(1);
              } finally {
                rs.close();
```

```
                stmt.close();
            }
        total_records = rowCount;
        return rowCount;


    } catch (Exception e) {



    }

        return -1;


    }


@Override
public void onCreate(Bundle savedInstanceState) {
    super.onCreate(savedInstanceState);
    setContentView(R.layout.activity_main);
    cnt=1;
    countRows();

    Button ff = (Button) findViewById(R.id.button1);
    ff.setOnClickListener(new OnClickListener()
    {
        public void onClick(View v)
        {


                cnt=1;
                connect_browse(cnt);
        }
    });
    Button fn = (Button) findViewById(R.id.Button01);
    fn.setOnClickListener(new OnClickListener()
    {
        public void onClick(View v)
        {
         if(cnt < total_records)
        cnt++;
         else
                cnt=total_records;
        connect_browse(cnt);
        }
    });
    Button fp = (Button) findViewById(R.id.Button02);
    fp.setOnClickListener(new OnClickListener()
    {
        public void onClick(View v)
        {
                if(cnt > 1)
                        cnt--;
                else
                        cnt=1;
        connect_browse(cnt);
        }
    });
    Button fl = (Button) findViewById(R.id.Button03);
    fl.setOnClickListener(new OnClickListener()
    {
        public void onClick(View v)
        {
                cnt=total_records;
                connect_browse(cnt);
        }
    });
    Button fa = (Button) findViewById(R.id.button2); //add
    fa.setOnClickListener(new OnClickListener()
    {
```

```java
            public void onClick(View v)
            {
                    connect_mysql();
                    countRows();
            }
    });
    Button fb = (Button) findViewById(R.id.Button04); //browse
    //fb.setVisibility(1);
    fb.setOnClickListener(new OnClickListener()
    {
            public void onClick(View v)
            {

            }
    });

}

@Override
public boolean onCreateOptionsMenu(Menu menu) {
    getMenuInflater().inflate(R.menu.activity_main, menu);
    return true;
}

public void show_message_box(String title,String msg)
{
   AlertDialog alertDialog;
   alertDialog = new AlertDialog.Builder(this).create();
   alertDialog.setTitle(title);
   alertDialog.setMessage(msg);
   alertDialog.setButton("OK", new DialogInterface.OnClickListener() {
            public void onClick(DialogInterface dialog, int id) {
                dialog.cancel();
          }
          });
   alertDialog.show();

}

public void connect_browse(int id)
{
   Connection con = null;
    Statement st = null;
    String driver ="com.mysql.jdbc.Driver";
    ResultSet rs = null;

    String url = "jdbc:mysql://localhost:3306/test";
    String user = "root";
    String password = "*****";
    int id_cnt= 1      ;

    try {

        Class.forName(driver).newInstance();


        con = DriverManager.getConnection(url, user, password);


        st = con.createStatement();


        //show_message_box("Connect","Connected=" + count);

        String select_query = "Select * from cheminfo.bioactivity";


        rs = st.executeQuery(select_query);

        rs.first();
        first_id =  rs.getInt("id");
```

```
            select_query = "Select * from cheminfo.bioactivity" ;

            rs = st.executeQuery(select_query);

            while (rs.next())
            {
              if(id == id_cnt)
              {
                  //show_message_box("Connect","Connected=" + select_query + " \n"
+ rs.getString(1));
                  EditText citation = (EditText)findViewById(R.id.editText1);
                  EditText smiles = (EditText)findViewById(R.id.EditText01);
                  EditText protein = (EditText)findViewById(R.id.EditText02);
                  EditText ActivityType =
(EditText)findViewById(R.id.EditText03);
                  EditText ActivityValue =
(EditText)findViewById(R.id.EditText04);
                  citation.setText(rs.getString(1));
                  smiles.setText(rs.getString(2));
                  protein.setText(rs.getString(3));
                  ActivityType.setText(rs.getString(4));
                  ActivityValue.setText(rs.getString(5));

                  break;
              }
                  id_cnt++;
            }

            //rs.

            /*

            String insert_qry = "insert into cheminfo.bioactivity
(citation,SMILES,Protein,ActivityType,ActivityValue) values ('" +
citation.getText().toString() + "','" + smiles.getText().toString()  + "','" +
protein.getText().toString()  + "','" + ActivityType.getText().toString()  + "','"
+ ActivityValue.getText().toString()  + "')";

            st.executeUpdate(insert_qry);
            show_message_box("Recor d","Inserted Record");
            //if (rs.next()) {
            //     System.out.println(rs.getString(1));
            // }
            *
            */

      } catch (Exception e) {
            show_message_box("Connect Error","" + e);

      }
   }

   public void connect_mysql()
   {
      Connection con = null;
      Statement st = null;
      String driver ="com.mysql.jdbc.Driver";
      ResultSet rs = null;
      String ip="";
      if(ip.length()==0){
            ip="localhost";
      }
      String url = "jdbc:mysql://"+ip+":3306/test";
      String user = "root";
      String password = "******";

      try {

            Class.forName(driver).newInstance();

            con = DriverManager.getConnection(url, user, password);
```

```
          st = con.createStatement();

          //show_message_box("Connect","Connected");

          EditText citation = (EditText)findViewById(R.id.editText1);
          EditText smiles = (EditText)findViewById(R.id.EditText01);
          EditText protein = (EditText)findViewById(R.id.EditText02);
          EditText ActivityType = (EditText)findViewById(R.id.EditText03);
          EditText ActivityValue = (EditText)findViewById(R.id.EditText04);

          String insert_qry = "insert into cheminfo.bioactivity
(citation,SMILES,Protein,ActivityType,ActivityValue) values ('" +
citation.getText().toString() + "','" + smiles.getText().toString()  + "','" +
protein.getText().toString()  + "','" + ActivityType.getText().toString()  + "','"
+ ActivityValue.getText().toString()  + "')";

          st.executeUpdate(insert_qry);
          show_message_box("Record","Inserted Record");
          //if (rs.next()) {
          //    System.out.println(rs.getString(1));
          // }

      } catch (Exception e) {
          show_message_box("Connect Error","" + e);

      }


    }
}

====

<manifest xmlns:android="http://schemas.android.com/apk/res/android"
    package="com.ncl.cheminfo"
    android:versionCode="1"
    android:versionName="1.0" >

    <uses-sdk
        android:minSdkVersion="8"
        android:targetSdkVersion="15" />
    <uses-permission android:name="android.permission.INTERNET"/>
<uses-permission android:name="android.permission.ACCESS_NETWORK_STATE" />
    <application
        android:icon="@drawable/ic_launcher"
        android:label="@string/app_name"
        android:theme="@style/AppTheme" >
        <activity
            android:name=".MainActivity"
            android:label="@string/title_activity_main" >
            <intent-filter>
                <action android:name="android.intent.action.MAIN" />

                <category android:name="android.intent.category.LAUNCHER" />
            </intent-filter>
        </activity>
    </application>

</manifest>
```

10.8   Need of High-Performance Computing in Chemoinformatics

Harnessing high-end technology for solving problems in biology and chemistry is one of the recent emerging trends in modelling. Building efficient platforms to perform large-scale data modelling of the large data being produced by high-performance computing (HPC) assumes high importance in view of the tremendous applications, some of which are mentioned below.

- Evaluation of Virtual Library
- Prediction of spectral data
- Text mining medical literature
- Harvesting chemical data from Internet
- Structure–activity relationship studies
- Lead identification and optimization
- Linking species (AYURVEDA) to modern medicine
- Image analysis
- Statistical machine learning
- Quantum mechanics/quantum chemistry (QM/QC) methods (reaction modelling)

A multicomponent platform ChemInfoCloud for enabling rapid virtual screening by integrating new and existing molecular informatics applications has been built [31]. It is provided with many bioinformatics and chemoinformatics functionalities and computational flexibility for automated workflows (Fig. 10.23).

10.9   Thumb Rules for Developing and Using Scientific Portals and Mobile Devices for Computing

- Build basic infrastructure compatible for open-source tools and computing resources
- Get access to publicly available molecular data and preprocess them for reusability
- Always think of the utility and developmental efforts required for building a portal before just pressing on anything technology has to offer. A portal need not be built for each and every computational task
- Follow good software engineering practices (security, version control)

10.10   Do it Yourself Exercises

- Build a portlet for computing molecular properties using Liferay
- Get access to cloud computing infrastructure (free or paid services)
- Build open-source tools for evaluation of virtual libraries

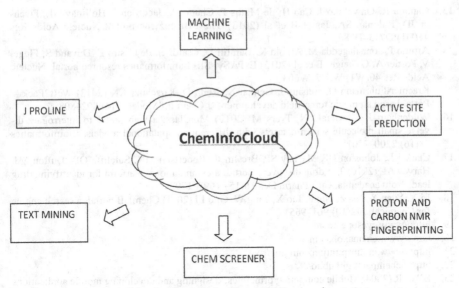

Fig. 10.23 The functionalities deployed on the ChemInfoCloud engine

10.11   Questions

- What is a portal? Give some examples of portals in chemoinformatics domain.
- What do you understand by the term mobile computing?
- Enumerate the steps required for building an android application.

References

1. http://www.infoworld.com/d/developer-world/new-enterprise-portal-131
2. http://www.liferay.com/products/what-is-a-portal/web-platform
3. http://portals.apache.org/
4. http://www.javaworld.com/javaworld/jw-10-2002/jw-1011-maven.html
5. Willighagen EL, Waagmeester A, Spjuth O, Ansell P, Williams AJ, Tkachenko V, Hastings J, Chen B, Wild DJ (2013) The ChEMBL database as linked open data. J Cheminform 5:23
6. Wong AK, Park CY, Greene CS, Bongo LA, Guan Y, Troyanskaya OG (2012) IMP: a multi-species functional genomics portal for integration, visualization and prediction of protein functions and networks. Nucleic Acids Res 40(W1):W484–W490
7. https://msmedicaid.acs-inc.com/help/envision_web_portal.html
8. http://www.ibm.com/developerworks/library/us-portal/
9. http://www.liferay.com/
10. http://www.jboss.org/jbossportal/
11. https://drupal.org/project/portal
12. http://www.javaworld.com/javaworld/jw-08-2003/jw-0801-portlet.html

13. Cantara R, Onwubiko J, Cao H, de Matos P, Cham JA, Jacobsen J, Holliday GL, Fischer JD, Rahman SA, Jassal B et al (2013) The EBI enzyme portal. Nucleic Acids Res 41(D1):D773–D780
14. Artimo P, Jonnalagedda M, Arnold K, Baratin D, Csardi G, de Castro E, Duvaud S, Flegel V, Fortier A, Gasteiger E et al (2012) ExPASy: SIB bioinformatics resource portal. Nucleic Acids Res 40(W1):W597–W603
15. Kraemer-Fuhrmann O, Neisius J, Gehlen N, Reith D, Kirschner KN (2013) Wolf2Pack—Portal based atomistic force-field development. J Chem Inf Model 53(4):802–808
16. Wildenhain J, FitzGerald N, Tyers M (2012) MolClass: a web portal to interrogate diverse small molecule screen datasets with different computational models. Bioinformatics 28(16):2200–2201
17. Clark RL, Johnston BF, Mackay SP, Breslin C, Robertson MN, Sutcliffe OB, Dufton MJ, Harvey AL (2010) The drug discovery portal: a computational platform for identifying drug leads from academia. Curr Pharm Des 16(15):1697–1702
18. Li X, Yuan X, Xia Z, Nie F, Tao X, Tang W, Guo Li (2011) ChemDB portal, a search engine for Chemicals. 74(10):961–965
19. http://www.eyesopen.com/
20. http://www.chemaxon.com/
21. http://www.molinspiration.com/jme/
22. http://jchempaint.github.io/
23. B'Far R (2004) Mobile computing principles: designing and developing mobile applications with UML and XML
24. Poslad S (2009) Ubiquitous computing: smart devices, environments and interactions. Wiley
25. http://www.android.com/
26. http://www.openhandsetalliance.com/android_overview.html
27. http://www.aisec.fraunhofer.de/content/dam/aisec/Dokumente/Publikationen/Studien_TechReports/deutsch/AISEC-TR-2012-001-Android-OS-Security.pdf
28. Rogers R, Lombardo J, Mednieks Z, Meike B (2009) Android application development. O Reilly Media, USA
29. http://www.molsoft.com/iMolview.html
30. Unpublished results
31. Karthikeyan M, Pandit D, Bhavsa A, Vyas R (2013) Design and development of ChemInfoCloud: an integrated cloud enabled platform for virtual screening. Chem Comb High T Scr xx:xx

Index

A

Active site, 83, 221, 222, 229–231, 311, 369
  in blind docking, 226
  chemical features of, 297
  of drug targets, 121
  identification of, 219
  molecular modelling approaches for, 32
  prediction of, 272–275
    online tools for, 279
    using MOE, 276
    using sitemap, 272
  in protein–ligand docking, 224
  of proteins, 202
  role in database screening, 235
  role in protein function, 297
  structural features of, 298
  studies on, 309
  use in chemoinformatics, 300
ADME
  calculation of, 104
  modelling of, 185
  properties of, 102
  screening applications of, 122
  use in QikProp module, 102
Android, 2
  application development code for, 521
  applications for chemoinformatics, 517
  ChemInfo app for, 518
  operating system, 516
  iMolview mobile app for, 518
Ant, 504, 508
AquaSol, 109
Artificial intelligence, 136, 152, 179, 351, 453
  branches of, 136
  in biocomputing, 453
  in ontology, 351
  workflow for, 451
Artificial neural network (ANN), 137, 178,
      189, 272

Autodock, 219, 221, 222
  as a docking-based screening tool, 126, 211
  open-source software, 124, 211
  steps involved in, 212
Autodock VINA, 220, 222
  docking using, 124, 211, 220
  use of, 220

B

Barcode, 57
Beer Lambert law, 376, 377
Beilstein, 15, 57, 76, 363, *see also* Beilstein
      Information System
  chemical information service providers, 15
  database, 76
  registry number, 57, 76
Beilstein Information System, 363
Biomedical, 84, 417, 425, 439
  sciences, 80
  domains, 421, 435
  entities, 435
  text mining, 434, 435, 444, 489
BLAST, 287

C

C
  main directory of, 139
  in MATLAB, 178
  programs, 5
C++, 5, 7
  advantage of, 5
  in annotation tool development, 98
  main directory of, 139
  in MATLAB, 178
Chem Robot, 57
ChEMBL, 126, 405, 444, 502
ChemDB, 66–68
  portal, 109, 503

Chemdraw, 12, 16, 235, 384
    software, 17
    use in chemistry, 16
    use in pharmacophore modelling, 235
    using, 32
Chemical entity recognition (CER), 436
Chemical markup language (CML), 9, 351,
    454
Chemical shift, 377, 378, 395
    measurement of, 390
    in NMR, 377, 392
    prediction of, 384
    in quantum chemistry programs, 409
    role in binary fingerprints, 405
    values, 377
Chemical structure, 80, 357, 435, 483
    analysis of, 59
    ChemAxon tool, 17
    definition of, 21
    drawing tools for, 10, 12, 454
    encoding of, 57
    fingerprints-based approach for, 354
    formats, 21
    in PubChem, 77
    IUPAC nomenclature, 16
    prediction programs for, 391
    representation of, 8, 33
    in sub-structure searching, 21, 47, 79
    of unknown metabolites, 401
ChemicalTagger, 437, 438
Cheminfo app, 518
Chemistry development kit (CDK), 9, 93, 94,
    453
ChemScreener, 124
    program, 126, 444
    virtual screening platform, 126, 127
ChemStar, 83, 490
    chemical computing, 438
    chemical properties computation by, 438,
        105
ChemXtreme, 56
    in molecular properties computing, 490
    program, 83
CLiDE, 58, 59
Cloud computing, 59, 502–526
Clustering, 60, 137, 141, 424, 425, 440, 484,
    493
    in biomedical domains, 421
    role in molecular diversity, 55
    of text, 417
    tools used for, 153
    using Scaffold hunter, 112
Corina, 30–32, 88
Correlation coefficient, 174, 187, 188

Cross docking, 226, 229
Cytoscape, 441, 496

D
Data mining workflows, 152, 417
Database, 42, 60–79, 496
    bibliographic, 75, 76
    chemical, 45, 46, 58, 74, 77, 79
    creation of, 67, 68
    hosting of, 71
    management of, 68
    management systems, 502
    reaction, 363–365
    screening of, 242
    structures, 49, 51
    query, 49, 62
Decision learning, 134
Diels–Alder reaction, 318, 319, 324, 333, 361,
    366, 368
Disease, 421, 426, 435, 439
    role of protein–protein interaction in, 231
    role of protein–ligand complexes in, 484
Distributed computing, 56, 438
    of chemical properties, 83, 105
    text mining in, 438
Diversity image to structure tools, 58
Docking pharmacophore, 426
Dragon, 104
Drug bank, 518
Drugs
    database of, 79
    development of, 2
    discovery of, 137
    indexing of, 78
    side effects of, 93
    text mining for, 424
    use of ligands as, 483

E
Eclipse, 6, 457, 504, 517
eMolecules, 80, 81
Empirical methods, 319, 384
Environmental Protection Agency, 79, 110,
    126
EPA, see Environmental Protection Agency

F
Fasta, 286, 313
Formulize, 185, 186
Fragments, 33, 41, 113, 224
    detection of, 377
    in hashed fingerprints, 43
    molecular, 9, 308
    sub-structural, 39

G

GAMESS, 385
Gaussian, 186, 328, 333, 384, 390
    program, 331, 385
    software, 325
Genetic programming, 137, 179, 182, 184
Geometry optimization, 324, 328, 384, 410
GLIDE, 196, 197, 202, 211, 225, 229
GPU computing, 7, 501

H

Heatmap, 491, 494
High performance computing, 2, 7, 302, 405
    in chemoinformatics, 526
Homology modelling, 165, 225, 282–285, 483
    practice tutorial for, 285–293
    thumb rules for, 312
Hybrid computing, 7

I

IBM SPSS, 176, 177
IDBS, 453–455
iMolview, 518
InChI, 21, 57, 58, 77, 401, 437, 454
International Chemical Identifier, *see* InChI
Induced fit docking, 224, 225
    advantages of, 225
Intrinsic reaction coordinate (IRC), 317, 323,
        326
IUPAC name, 15, 16, 19, 58, 78, 436, 437

J

Jaguar, 338, 340, 385, 390
Java, 6, 141, 418, 441, 515
    as a programming language, 493
    based project, 160
    based tool, 112, 124, 425
    in general text-mining tools, 424
    in MATLAB, 178
JChem, 33, 35, 46, 73, 364, 483
JChemPaint, 8, 9, 515, 516
JME, 10, 11, 79
    distribution, 515
    molecular editor, 10, 59, 79, 515
    structure-generating programs, 56
JOELib, 3, 97
J-ProLINE, 489–496
Java server pages, 74, 508
JSP, *see* Java server pages

K

Kernel, 140, 152, 156, 190
KNIME, *see* Konstanz Information Miner

Konstanz Information Miner, 452, 453, 455,
        456, 462, 470, 477, 480, 483

L

LigPrep, 102, 200, 209, 477, 478
LingPipe, 425, 426, 438, 440
Linux, 2–4, 7, 365, 477

M

Machine learning (ML), 60, 91, 93, 131, 134,
        150, 270, 297, 298
    free tools for, 152, 153
    methods, 133, 134, 136
    models, 134
    predictive studies on, 132
    thumb rules for, 189
Marvin
    sketch, 11, 99, 364, 483, 512
    space, 483
    view, 102, 154, 253
Mass spectrum , 382, 401, 402
MATLAB, 142, 149, 178, 453
Matrix, 21, 142, 178, 288, 323, 330, 358, 390
MegaMiner, 438–441, 444
Mobile computing, 516, 517–525
Molconvert, 35, 36
Molecular
    dynamics, 196, 231, 321
    mechanics, 30, 104, 320–322, 384, 490
    networks, 32
Molecular operating environment (MOE),
        104, 164, 272, 481
    using CombiGen, 122
Molinspiration, 11, 108
MOLTABLE, 82, 83, 154, 161, 357
MySQL, 62, 68, 122, 438, 439
    cluster, 440
    code for connecting to, 63
    database server, 62

N

Named entity recognition (NER), 417
Natural language processing (NLP), 80, 416
Netbeans, 6, 123, 429, 504
    for portlet development, 504
Nuclear magnetic resonance (NMR), 16, 232,
        376, 483
Nvidia, 7, 302

O

Open Babel, 38, 39, 126
OpenEye, 100, 113, 503
Organic synthesis, 317, 363, 366, 368, 375
OSCAR, 426, 436, 437

P

PaDEL, 98
Pairwise alignment, 288
Patents, 16, 33, 82, 369, 416
PDB, 77, 78, 197, 220, 282, 490
 coordinates, 396
 file format, 24
 database entries, 66, 518
 structure, 220, 221, 288
 search options, 287
 source of protein coordinates, 303
pdbqt, 220–222
Perl, 6, 60, 453
Pipeline pilot, 364, 452, 453
 Accelrys, 453
 programs, 452
POS tagging, 419, 421, 425
Practical Extraction Report Language, *see*
 PERL
preADMET, 109
Preprocessing, 190, 419
 of data, 141, 186
 of text, 420
Prodrugs, 369
Programming languages, 3, 6, 7
PROLIX, 489
Protein, 30, 78, 195, 202, 211, 225, 226,
 229–233, 272, 286, 298, 300, 308,
 439, 483, 484
Protein-ligand complexes portals life ray, 484,
 493, 496
Protein–protein interaction, 231, 233, 424
PyRx, 125
Python, 6, 60, 102, 124, 146, 453
 codes, 39

Q

Q-site, 279, 280, 282
Quantitative structure activity relationship
 (QSAR), 31, 135, 235, 451, 469,
 502
Quantum chemical methods, 322, 323, 384

R

r[2], *see* Correlation coefficient
Ramchandran plot, 294, 295, 302, 308, 310,
 312
Random forest, 133, 136, 149, 300
Rapid miner, 160
 classification, 160
 graphic user interface, 161
 machine learning models in, 161
 text mining in, 434
Reaction modelling, 2, 321, 322, 326, 327,
 367

computational methods in, 318
 Diels–Alder, 338
 steps in, 328
Reaction ontology, 353
Reactor, 364, 365
RTECS, *see* Registry of Toxic Effects of
 Chemical Substances
Registry of Toxic Effects of Chemical
 Substances, 79, 126

S

Scaffold, 59, 60, 111, 120, 121, 125
 hopping, 112, 117, 128
 hunter, 112
 replacement, 117
Spectral Database for Organic Compounds,
 402, 403, 409
SDBS, *see* Spectral Database for Organic
 Compounds
Semi-empirical methods, 319, 368, 384
Sequence, 4, 6
 alignment tools, 8
 in databases, 79
 of proteins, 198, 283, 286
 in homology modelling, 287
 in drug discovery, 483
Similarity searching fingerprints
 hashed fingerprints, 42–44
Sitemap, 272–275
Smarts, 21, 38, 48, 77, 347, 364
SMILES, 9, 10, 15–21, 56–59, 77, 93, 120,
 347, 515
Software development kit (SDK), 504
Spartan, 344, 396, 399
Spectroscopy, 78, 375–377, 399
 IR, 376, 377
 NMR, 78, 377, 392
 organic, 376
 UV, 376
Stemming, 419, 420
String, 20, 38, 39, 75, 309
Structure activity relationship (SAR), 55, 108,
 168, 171
Structure
 drawing tools, 10, 454
 formats, 20
 searching, 21, 47, 48, 361, 400, 502
Supervised learning, 421
 algorithms, 136, 137
Support vector machine (SVM), 133–137,
 272, 425
 in LibSVM, 141
 library for, 139
SYLVIA, 369
Synthetic accessibility, 369, 481

T

Tanimoto, 355, 359
  coefficient, 52–54, 491, 494
  metric, 357, 358
Target, 494
  proteins, 30
  for lead identification, 121
  in protein–protein docking, 231
  -ligand association, 483, 484
Taverna, 437, 453, 455
Term weighting, 421
Text mining, 416, 417, 424, 439, 489
  applications of, 441
  biomedical, 434
  chemically intelligent tools for, 435
  in chemoinformatics, 438
  free tools for, 434
  methods, 49
  R program for, 430
Tomcat apache server, 71–73, 504
Toxicity, 79, 80, 83, 94, 109, 110, 135, 469
  prediction, 104

Transition state modelling, 324
  of reactions, 322
  practice tutorial for, 326

U

Unsupervised learning, 60, 136, 137, 421

V

Virtual library, 57, 82, 121–123, 126
  enumeration, 59, 111
  screening platforms, 123
  synthesis, 119, 443

W

Waikato Environment for Knowledge
      Analysis, *see* WeKa
WeKa, 140, 141, 144, 146, 149, 425
Wolf2Pack, 503
Workbench, 456–458, 473
  Expasy, 503
  KNIME, 457, 458